Cognitive Work Analysis

Toward Safe, Productive,
and Healthy Computer-Based Work

Cognitive Work Analysis

Toward Safe, Productive,
and Healthy Computer-Based Work

Kim J. Vicente
University of Toronto

CRC Press
Taylor & Francis Group
Boca Raton London New York

CRC Press is an imprint of the
Taylor & Francis Group, an **informa** business

First Published by
Lawrence Erlbaum Associates, Inc., Publishers
10 Industrial Avenue
Mahwah, NJ 07430

Transferred to Digital Printing 2009 by CRC Press
6000 Broken Sound Parkway, NW Suite 300, Boca Raton, FL 33487
270 Madison Avenue New York, NY 10016
2 Park Square, Milton Park Abingdon, Oxon OX14 4RN, UK

Cover design by Kathryn Houghtaling Lacey

Cover photo: "Motions necessary to move and file sixteen boxes full
of glass" taken by Frank Gilbreth, from the Special Collections of the
Purdue University Library. Copyright © Purdue Research Foundation,
West Lafayette, IN 47907.

Library of Congress Cataloging-in-Publication Data

Vicente, Kim J.
Cognitive work analysis : toward safe, productive, and healthy computer-based work / by Kim J. Vicente.
      p. cm.
Includes bibliographical references and index.
ISBN 0-8058-2396-4 (cloth : alk. paper) — ISBN 0-8058-2397-2
(pbk. : alk. paper).
   1. Human-computer interaction. 2. Work environment. 3. Industrial
safety. I. Title.
QA76.9.H85V515   1999
004' .01'9—dc21   98-31670
                                              CIP

Publisher's Note
The publisher has gone to great lengths to ensure the quality of this reprint
but points out that some imperfections in the original may be apparent

*To Neville, who gave me my start in human factors
and taught me that research should be fun.*

Contents

Foreword

During recent decades, we have witnessed a dramatic change in the conditions of work. The very fast pace of technological change has influenced work conditions in all application domains, such as transportation, shipping, manufacturing, and process industries. At the same time, the rapid development of transportation systems, information technology, and just-in-time schemes have led to a high degree of integration and coupling of systems. Under these challenging conditions, a single decision can have dramatic effects that propagate rapidly and widely through global society. Furthermore, most work organizations today live in a very aggressive and competitive environment and success is granted only to those that exploit the benefits of operating at the fringes of usual, accepted practices. Closing in on and exploring the boundaries of established practice under pressure necessarily runs the risk of crossing the limits of safe practices.

This evolution brings with it a concurrent change in the approaches to be taken to the design of work support systems, and consequently, to the analysis of work and related academic research paradigms. There has been a move from formulation of "the one best way" associated with scientific management theories toward "user-centered" tool design based on analyses of actual performance during work.

Recently, however, the pace of change has become so fast, and the need for high reliability of new systems so high, that the design of new work tools and systems cannot be based only on empirical, incremental evolution based on studies of existing work conditions. Predictive models of work behavior become necessary, models that can serve to predict work behavior by explicitly identifying the behavior-shaping features of a new work environment. In Vicente's terms, we need to replace the normative and descriptive models of work analysis by formative models.

This is the aim of Kim Vicente's book on *cognitive work analysis* (CWA). It is an important book that advances cognitive engineering by providing a pedagogic introduction to the formative analysis of work requirements. It is evident that Vicente now has developed a teaching curriculum that has been refined by years of interaction with students.

In this effort, Vicente opens up several important questions related to current approaches to academic research and design. First, he advocates research based on a fundamental respect for people engaged in practical work. Through the years, academic studies of work performance have often led to the conclusion that technicians behave in unsystematic and irrational ways and that designers are clumsy, or even stupid, resulting in proposals of "more rational" ways to work. However, over and over again, careful studies of work in the field have shown that, when the behavior of experts was found to be irrational or unsystematic, the frame of reference applied for judgment was wrong. Researchers had failed to identify the criteria that actually shaped workers' behavior.

People at work are generally very clever and rational from their local, contextualized perspective. A deep respect for this competence shines through Vicente's text. The communication of this respect for people at work to students within the various sciences involved in the development of information technology is probably the most important objective served by this book and by Vicente's teaching efforts.

A second important and interesting aspect of Vicente's text emerges when it is compared with our 1994 book (Rasmussen, Pejtersen, & Goodstein, 1994). The framework that we described there was developed by the Risø group mentioned in the historical review in Vicente's Appendix. This comparison across books provides a demonstration of the influence of environmental constraints on research methods and publication practices. The two books illustrate the difference between the behavior-shaping features of an academic teaching environment and those of a cross-disciplinary design environment.

Vicente presents the concepts and methods underlying CWA in a pedagogical fashion. He places these ideas in the context of a careful review of current academic research paradigms and illustrates those ideas with examples from domains that are familiar to university students. This presentation follows from the constraints of a university environment: Typically, academic research is based on the work of PhD students who should primarily be well versed within the central paradigms of their profession. It is necessary, therefore, to focus on problems within a discipline that can be formulated with reference to established concepts and paradigms, that are manageable within a limited time horizon, and that will add up to a recognizable professional competence and status. In academic research, comparing and evaluating paradigms are therefore important teaching topics that characterize differences in the paradigms of different groups and lines of thinking. Among the problems found in "real-world" organizations, academic research will normally select those that are best suited to illustrate the current paradigms and to demonstrate the relevance of the research to external funding bodies. Consequently, the basic aim and organization of academic research necessarily lead to a focus on issues that are captured by the central professional paradigms of a particular discipline. Thus, advice concerning the design of work systems will have a very paradigm-dependent bias. Furthermore, in academic research, success is determined by the status of concepts and models in the professional peer competition and success, including tenure track performance, depends on the resulting volume of refereed publications.

In contrast, our 1994 book clearly reflects the constraints of a cross-disciplinary, problem-oriented research setting that is reviewed in the Appendix of Vicente's book. From the beginning, our group had to trace a solution of a safety problem through several disciplines, and for that, we had the opportunity to create a multidisciplinary research team. We had no requirement for refereed publication volume to get tenure or funding. Instead, we could establish a fast-paced publication service with technical reports to provoke a quick, selective response from experts whose research paradigms matched our needs. We did not face the academic need of systematic paradigm reviews and comparison. Neither did we face the need to communicate our findings in a pedagogical way to students that were not familiar with complex work domains. We only needed to communicate with experts from the various domains that could help us prepare for a foreseen risk analysis task. As a result, the publication practice was aimed at the creation of contacts with experts in a way that was not very well suited for communication to students without work domain expertise.

Early during our field studies at Risø, a similar research program to study information search behavior in libraries was initiated by Pejtersen at the Royal Danish School of Information and Library Science. A cooperation was established to explore the potential to generalize a shared theoretical framework that would span both technical and humanistic application domains. Pejtersen's comprehensive field results also turned out to be a very promising basis for evaluating this framework. A full-scope, prototype retrieval system was designed to be tested under actual work conditions. For this design effort, the overly restrictive constraints of an academic teaching institute materialized very clearly. Thus, to realize this plan, Pejtersen had to join our cross-disciplinary, problem-oriented research team at Risø.

Given this background, I find Vicente's pedagogical presentation of the CWA framework placed in the context of current research paradigms to be very important. The relationship between our problem-oriented approach and Vicente's very successful teaching approach probably demonstrates one successful solution to the problem of how to overcome the mismatch between the requirements of cross-disciplinary, design-oriented research and the constraints of an academic university environment.

A third important issue is emphasized by Vicente's book — the CWA framework is not a prescriptive method. Rather, it is a point of view, a state of mind, and a demonstration of the various dimensions of the problem of analyzing work performance in a dynamic society. The framework is intended to inform the design of support systems that match workers' performance criteria and leave them space to learn and develop their expertise. CWA is not the one and only approach to work analysis, but is instead a framework to be modified and adapted to the particular needs and horizon of a research or design team.

—Jens Rasmussen

Preface

A TRUE STORY: PROCEDURAL COMPLIANCE VERSUS FLEXIBLE ADAPTATION

Some nuclear power plant operators have to periodically go to a simulator and pass a certification test. The purpose of the test is to make sure that operators have the expertise that is required to operate the plant safely and effectively. The focus during testing is on evaluating whether operators can deal effectively with severe, abnormal events. These fault situations rarely occur on the job, but they can seriously threaten the integrity of the plant and the welfare of the public. During these events, operators are supposed to follow a set of written procedures that tell them what actions to take to cope with different kinds of abnormalities.

At one plant, operators would not always follow the written procedures when they went to the simulator for recertification. They deviated from them for one of two reasons. In some cases, operators achieved the same goal using a different, but equally safe and efficient, set of actions. In complex sociotechnical systems, like nuclear power plants, there is rarely one best way of achieving a particular goal. In other cases, the operators would deviate from the procedures because the desired goal would not be achieved if the procedures were followed. It is very difficult to write a procedure to encompass all possible situations. A small change in context might require different actions to achieve the very same goal. In either case, the operators' actions seemed justifiable, particularly in the latter set of circumstances. The people who were evaluating the operators in the simulator did not agree, however. They criticized the operators for "lack of procedural compliance." Despite this admonishment, the operators got their licenses renewed.

This happened several times. Eventually, the operators became frustrated with the evaluators' repeated criticism because they felt it was unwarranted. The operators decided that, the next time they had to go into the simulator for recertification, they would do exactly what the procedure said—no matter what. One team of operators followed this "work-to-rule" approach in the simulator and became stuck in an infinite loop. At one point, an emergency procedure told operators to switch to another procedure, but then that procedure eventually sent operators back to the first one. The operators dutifully followed the procedures, and thus wound up in a cycle, repeating the same set of actions several times. The evaluators were not amused. They eventually turned off the simulator, ending that particular test.

Later, the evaluators wrote a letter to the utility that employed this group of operators. In that letter, the evaluators criticized the operators yet again, this time for "malicious procedural compliance."

PURPOSE

> Jens Rasmussen's contribution to the study of complex human-machine
> interaction is, I think, nothing less than a paradigm shift.

–Moray (1988, p. 12)

> The conceptual analyses carried out by Rasmussen ... represent an important
> converging link between cognitive models of thinking and problem solving on the
> one hand and human computer interactions on the other. It is clear, however, that
> research in this area has not gone far beyond the stage of conceptual modeling.
> Given the tremendous complexities of both human cognition in problem solving
> and of the system itself, there would appear to be a great potential for gains to be
> made by developments in this area.

–Wickens (1984, p. 484)

This book describes, for the first time in pedagogical form, an approach to
computer-based work in complex sociotechnical systems developed over the last 30
years by Jens Rasmussen and his colleagues at Risø National Laboratory in Roskilde,
Denmark. This approach is represented by a framework called *cognitive work analysis*
(CWA). Its goal is to help designers of complex sociotechnical systems create
computer-based information support that helps workers adapt to the unexpected and
changing demands of their jobs. In short, CWA is all about *designing for adaptation*.
Rather than replace existing knowledge, however, CWA instead tries to provide an
overarching framework within which the contributions of various "technical" and
"humanistic" disciplines can all be integrated in a way that makes their implications for
systems design more clear. The need for this type of framework is vividly illustrated by
the true story just described.

The set of ideas to emerge from Risø is widely known and frequently cited in several
circles in both academia and industry. Nevertheless, it still has not had the influence it
deserves, primarily because the foundations and concepts that make up CWA have not
been described in a pedagogical form. As a result, it is not difficult to find distortions and
misrepresentations that indicate a lack of understanding of these concepts (e.g.,
Bainbridge, 1984, 1997; Cacciabue, 1997; Hollnagel, 1993; Jones & Mitchell, 1987;
Krosner, Mitchell, & Govindaraj, 1989; Mitchell, 1996; Venda & Venda, 1995). The
primary purpose of this book is to present this framework in a systematic manner and
thereby make these ideas more accessible to a much broader audience than they have
been in the past. By making the nature of this view more transparent, I in turn hope to
facilitate and encourage its application by a broader range of people to a wider spectrum
of problems. I believe that the dissemination and application of these ideas will be
instrumental in designing safer, more productive, and healthier computer-based work.
Given the rapidly increasing prevalence of computer technology in contemporary life
and the trends toward a global knowledge-based economy, the value of these potential
contributions will only increase in the future.

WHY SHOULD YOU READ THIS BOOK?

If you identify with one or more of the following characterizations, then you will
definitely find something of interest and value here:

•You are dissatisfied with traditional approaches to analyzing human work to inform the design of computer-based information systems, and you feel that a new approach is necessary to meet the challenges of today's and tomorrow's demanding work environments.

•You have collected a great deal of data about workers' current practices using a myriad of measurement techniques (e.g., field observations, interviews, usability tests, questionnaires, simulator studies, etc.) but you have found that it is very difficult, if not impossible, to organize this mass of data into a coherent and systematic form that readily suggests implications for the design of improved computer-based information systems.

•You have faced the challenging problems associated with designing information systems for complex work environments whose demands are primarily cognitive and social in nature, and you are searching for a more appropriate hammer with which to tackle this rather unique nail.

•You think that the research and design of computer-based work are frequently conducted in an impressionistic and ad hoc fashion, and that a more coherent and well-defined set of concepts is needed to achieve cumulative progress in this area.

In addition to these four general motivations, there are a number of more specific reasons why you would want to read this book:

•As mentioned earlier, the book describes, for the first time in pedagogical form, the paradigm shift in computer-based work developed over the last 30 years by Rasmussen and colleagues at Risø National Laboratory in Roskilde, Denmark.

•The book contains a detailed case study that illustrates each of the phases of CWA by example. The case is from the "technical" domain of process control, thereby complementing the more "humanistic" case study of library information retrieval presented in Rasmussen et al. (1994). By choosing a different type of case, I hope to illustrate the breadth of the concepts described in the book.

•The book places, also for the first time, the Risø framework for CWA in the context of existing research from several related approaches, including: activity theory, contextual design, distributed cognition, Francophone ergonomics, naturalistic decision making, participatory design, scenario-based design, situated action, and task analysis. This grounding should help you better understand the relationship between the Risø framework and ideas with which you may already be more familiar.

•The concepts in the book are driven by three practical problems of great social significance: safety, productivity, and worker health. Corporations dealing with complex, sociotechnical systems are currently having difficulty dealing with each of these perspectives. Improvements along these dimensions can lead to enormous benefits, whether they be measured in terms of reduced threat to human life and to the natural environment, greater financial returns, or better quality of working life. Thus, each of these perspectives provides a criterion for evaluating the effectiveness of computer-based work. The framework for CWA described in this book has the potential to produce significant improvements on all three fronts simultaneously.

WHO IS THE BOOK INTENDED FOR?

This book is based on a graduate industrial and systems engineering course I have taught for the past 6 years. Thus, one of its intended roles is as a textbook for graduate-level courses in disciplines concerned with applying information technology to complex work environments. But because the view I am advocating is not well understood, the book is also intended to be a resource for researchers (and perhaps inquisitive practitioners) from those same disciplines. I realize that this choice represents a trade-off that will probably not completely satisfy either audience. On the one hand, students may find that the text has too many citations and too many passing allusions to ideas with which they may not be familiar. Hopefully, researchers will find those citations to be helpful in guiding further reading, and those allusions to be useful in placing the ideas in the book in a broader intellectual context. On the other hand, researchers may find that the book has too many examples and that it sometimes states the obvious. Hopefully, students will appreciate the pedagogical value of those examples as well as my effort at making all presuppositions clear and explicit.

Because the book is written for a mixed audience, the writing alternates between two styles: (a) the more formal style we usually expect in a research monograph, and (b) the less formal style that is more typical of a textbook that tries to guide students' learning processes. Pedagogical pointers and advice are presented in the latter style. Researchers may find these comments distracting, so I have explicitly identified such pedagogical text by placing it in a box. To the student, a box will indicate text that should prompt some critical self-reflection. To the seasoned researcher, a box may indicate text that can be skipped over.

I have tried to write the book for as broad an interdisciplinary audience as I know how. The intent is to make these ideas accessible to interested parties who are involved in research on computer-based work and identify themselves with any one of the following communities: anthropology, cognitive engineering, cognitive ergonomics, cognitive psychology, cognitive science, communications, computer-supported cooperative work, computer science, control engineering, ecological psychology, human-computer interaction (HCI), human factors, information science, information systems, sociology, and systems engineering. You will find references to each of these areas of research, some more than others. Given the length of this list, I do not expect to reach all of you with equal success. Nevertheless, I do hope to make all of you aware that there is a useful, coherent, and well thought out framework for addressing the problems with which you are concerned.

WHAT WILL YOU NOT FIND IN THE BOOK?

Just as it is important to make explicit what I have tried to do, it is at least as important to make explicit what I have not tried to do. Specifically, there are a number of topics that I have deliberately chosen not to tackle, such as: (a) procedural guidance for how to use the concepts described in the book, (b) practical advice on how to implement this framework in organizations in industry, given the usual constraints that govern corporate design efforts, (c) formalisms that can be used to codify the concepts described in the book, and (d) computer-based tools that can be used to reduce the effort associated with applying these concepts to industry-scale problems. All of these topics are very important, and in fact, they all represent significant obstacles to the widespread adoption of the view described in this book.

Why, then, are these issues not addressed? The reason is simple and can be found in my primary motivation described earlier. Because this framework is not well understood, all of these other bottlenecks are currently of secondary importance. This approach needs be to explained to a wider audience to show its potential value in a convincing fashion. As progress is made along this front, then more attention can be devoted to these other, more pragmatic issues (see Hurst & Skilton, 1997, and Potter, Roth, Woods, & Elm, 1998 for a start on the issue of computer-based tools for conducting CWA).

It should be clear from this caveat that this is not a "how to" book. Progress on how best to introduce information technology into the modern workplace must take place on several time scales. In the near term, we need approaches that improve the design of existing products by taking into account the notorious "harsh realities" of design in industry (e.g., Burns & Vicente, 1995). If you are primarily interested in this short-term problem, you should consult sources such as Rouse (1991) and Beyer and Holtzblatt (1998), both of which describe approaches that have already been adopted in numerous organizations and that appear to have changed, for the better, the way in which companies design computer-based information systems to support human work. In the long term, however, we need approaches that are less bound to existing design practices in industry and that provide a vision for how information systems could be designed in the future so as to make fuller use of the power of information technology. The framework described in this book is, I believe, a very promising candidate for achieving this longer term objective.

HOW CAN YOU READ THE BOOK?

The book is divided into four parts. Part I (Glossary and chaps. 1 and 2) provides a motivation by introducing three themes that tie the book together: safety, productivity, and worker health. The ecological approach that serves as the conceptual basis behind the book is also described. In addition, a glossary of terms is provided. Part II (chaps. 3–5) situates the ideas in the book in a broader intellectual context by reviewing alternative approaches to work analysis. The limitations of normative and descriptive approaches are pointed out, and the rationale behind the formative approach advocated in this book is explored. Part III (chaps. 6–12) describes the concepts that comprise the cognitive work analysis framework in detail. Each concept is illustrated by a case study, and the implications of the framework for design and research are illustrated by example. Part IV (Chapter 13 and Appendix) tries to pull everything together by revisiting the themes of safety, productivity, and health, and by showing why the need for the concepts in this book will only increase in the future. In addition, a historical addendum briefly describes the origins of the ideas described in the book.

The book was written to be read in a linear fashion, but given this structure, it can also be read in several other ways. For example, if you just want an overview of the ideas, you can read chapters 1, 5, 12, and 13. If you want to use this book as a text in a graduate-level course, you can use the case study described in Part III as a challenging project to give students an opportunity to apply CWA practice (but tell the students to work out their own answers before reading the solutions provided here!). Conversely, if you want to focus just on the concepts, and are not interested in a detailed case study, then you can skip all of chapters 6 and 12 and the latter parts of chapters 7, 8, 9, and 11. If you are most interested in the implications that CWA has for systems design or research, then pay particular attention to chapter 12. If you are not interested in alternative approaches to work analysis, then you can skip chapters 3 and 4, although by doing so you will probably lose some of the rationale behind the approach described in the remainder of the book. Finally, from a graduate student's perspective, you can use this book, especially Parts II and III, to identify a topic for a master's thesis or doctoral dissertation. Although they are not always explicitly labeled as such, if you look closely you will find plenty of interesting and important research topics that have yet to be addressed.

ACKNOWLEDGMENTS

Although I did not know it at the time, the sequence of events that led me to write this book began when I spent 1 year as a Visiting Scientist in the Section for Informatics and Cognitive Science of Risø National Laboratory during 1987–1988. It was during that time that I became a serious student of the ideas presented in this book. This immersion was followed by several extended stays at Risø, during which my understanding was reinforced, expanded, and deepened. I am forever grateful to Jens Rasmussen for making this invaluable learning opportunity possible, and to João, Lisbeth, Sandra, Tine, Jan, and Gunnar for introducing me to the Roskilde Festival and for turning my time in Denmark into an experience of a lifetime.

I would also like to thank the following people for very helpful and constructive comments or discussions concerning this book: Amy Bisantz, Suzanne Bødker, Sara Chen-Wing, Donald Cox, Cindy Dominguez, John Flach, Simon Goss, Saul Greenberg, Stephanie Guerlain, Karen Holtzblatt, Alex Kirlik, Tim Lethbridge, Gavan Lintern, Marshall McClintock, Chris Miller, Neville Moray, Bonnie Nardi, Annelise Mark Pejtersen, Suzanne Rochford, Penny Sanderson (who saved me from countless misunderstandings by suggesting the term *formative*), Janice Singer, Tom Stoffregen, Lucy Suchman, Fumiya Tanabe, David Woods, and Yan Xiao.

I am also indebted to the students who took the class on which this book is based. Without their probing questions and insights, this book would not be what it is. In addition, I owe an unpayable debt to all of my former and current colleagues in the Cognitive Engineering Laboratory at the University of Toronto. Their insightful contributions in exploring the ideas described in this book have taught me a great deal. Particular thanks go to the following for commenting on earlier drafts: Peter Benda, Cathy Burns, Sandra Chery, Renée Chow, Klaus Christoffersen, John Hajdukiewicz, Greg Jamieson, Elfreda Lau, Gerard Torenvliet, and Xinyao Yu.

There are a few people who deserve special mention. First, many thanks to Erling Johannsen and Annelise Mark Pejtersen of Risø National Laboratory, and John Hajdukiewicz of the University of Toronto, for either creating, or providing me with, many of the figures in the book. Second, special thanks to Marshall McClintock of Microsoft for providing very thorough feedback and for pointing me to relevant and valuable literature of which I was not aware. His pointers considerably broadened the scope of the book. Third, many thanks to Chris Miller of Honeywell Technology Center, who spent a 1-year sabbatical in our lab while this book was being written. His critical, comprehensive, and constructive feedback was truly invaluable. Fourth, many thanks to Gerard Torenvliet of the University of Toronto for creating the author and subject indices. Fifth, thanks to Anne Duffy, Sara Scudder, and the staff at Erlbaum for all their help. Sixth, I am deeply grateful to my parents, André and Marlene, for their unending love, support, and encouragement throughout the 2 years of writing the book.

Finally, a very special heartfelt thank you to Annelise Mark Pejtersen and Jens Rasmussen. I have been very lucky to count them not only as invaluable colleagues, but as warm friends as well. Anyone familiar with their work will recognize their fingerprints on virtually every page of this book. Only their busy schedules and modesty have kept them from being co-authors of this book.

–Kim J. Vicente

In the progress of the division of labour, the employment of the ... great body of the people, comes to be confined to a very few simple operations; frequently to one or two. But the understandings of the greater part of men are necessarily formed by their ordinary employments. The man whose whole life is spent in performing a few simple operations, of which the effects, too are, perhaps, always the same, or very nearly the same, has no occasion to exert his understanding, or to exercise his invention in finding out expedients for removing difficulties which never occur. He naturally loses, therefore, the habit of such exertion, and generally becomes as stupid and ignorant as it is possible for a human creature to become. The torpor of his mind renders him, not only incapable of relishing or bearing a part in any rational conversation, but of conceiving any generous, noble, or tender sentiment, and consequently of forming any judgment concerning many even of the ordinary duties of private life.

—Adam Smith, *The Wealth of Nations*
(cited in Reed, 1996, p. 83)

INTRODUCTION I

What's in a Word?
(Glossary)

Clearly task analysis is a large, important and confusing area. . . . Acronyms proliferate and are easily confused. The papers describing the techniques are scattered over a number of journals, books and technical reports. Often the techniques have only been used by their authors. Terms are used by different writers to mean different things. A small sample of the terms used includes: plans, goals, methods, operations, actions, objects, procedures, tasks, subtasks and projects. Central, of course, to task analysis is the notion of a "task". However, some authors argue that a task is device-independent (Bösser, 1987) and others that it is device-dependent and it is the goal which is device independent (e.g., Shepherd, 1989).
 —Benyon (1992a, p. 105)

Theories provide patterns within which data appear intelligible. They constitute a conceptual Gestalt. A theory is not pieced together from observed phenomena; it is rather what makes it possible to observe phenomena as being of a certain sort, and as related to other phenomena.
 —Hanson (1958, p. 90)

PURPOSE

Because this book is describing a relatively new paradigm for work analysis, it is important that we define the key terms that we use. In doing so, we hope to provide some conceptual coherence, consistency, and clarity to the sometimes-impressionistic fields of human factors, human-computer interaction, and cognitive engineering. Ideally, all of our terms should be defined formally. Unfortunately, our thinking is not mature enough yet to achieve this ambitious goal, so the terms we use have instead been defined in a less formal, although hopefully still clear, manner. Despite this less-than-ideal level of rigor, we have tried to be systematic in our use of terms throughout the book. Eventually, we hope to work toward a more formal set of definitions (see R. A. Miller, 1982, for an exemplary initial effort). Consequently, this Glossary should be considered more as a snapshot of work in progress than as a definitive, stable product.

Our intent is well captured, and in fact inspired, by the landmark effort of Dewey and Bentley (1949) to develop an efficient basis for scientific inquiry (see also Bentley, 1954). The premises of their work were as follows:

Science uses its technical names efficiently. Such names serve to mark off certain portions of the scientific subjectmatter [sic] as provisionally acceptable, thereby freeing the [scientist's] attention for closer consideration of other portions that remain problematic. The efficiency lies in the ability given the [scientist] to hold such names steady—

to know what he properly names with them—first at different stages of his own procedure and then in interchange with his associates. (Dewey & Bentley, 1949, p. 47)

The fields of human factors, human-computer interaction, and cognitive engineering have not reached this level of efficiency, as evidenced by the "linguistic chaos" (Dewey & Bentley, 1949, p. 37) that currently characterizes them. Like all other researchers in these areas, we too "undertake development definitely and deliberately within an atmosphere (one might perhaps better call it a swamp) of vague language" (Dewey & Bentley, 1949, p. 74). However, our goals are quite modest in that we do not aim to resolve this terminological problem once and for all. Instead, "what we advocate is . . . a passage from loose to firm namings" (Dewey & Bentley, 1949, p. 47). Moreover, clarity and consistency should not be confused with correctness: "We are not attempting to legislate concerning the proper use of a word, but are stating the procedure we are adopting" (Dewey & Bentley, 1949, p. 76).

Although it is customary to present a glossary at the end of a book rather than at the beginning, we present our glossary here to emphasize the systematic and coherent nature of the concepts comprising the framework described in this book. The deep meaning of many of these terms will only become apparent in subsequent chapters. Nevertheless, we encourage you to browse through the Glossary now. You will find some familiar terms with familiar definitions, some familiar terms with unfamiliar definitions, and some unfamiliar terms too. If we are successful, then the significance of each of these terms should become apparent by the end of the book. Perhaps even more important, the relationship between these terms should become obvious as well, because it is precisely this set of relationships that defines the framework for CWA that we are advocating.

As you progress through the book, you should periodically revisit the Glossary to make sure that these definitions make sense to you, in the light of what you have been reading. In particular, when you disagree with a claim being made in the book, have a look at the Glossary. It might be that the disagreement results from the fact that we have defined some terms in a way that is different from what you are accustomed to. Thus, you can use the Glossary as a resource to self-check your comprehension of the concepts we introduce throughout the book.

DEFINITIONS

All terms that have been defined are capitalized.

Abstraction Hierarchy—A stratified hierarchy (see chap. 6 for a description) defined by a means–ends relation between adjacent levels. A framework for describing Work Domains. Same as Means–Ends Hierarchy. Compare with Part–Whole Hierarchy.

Action—The Goal-Directed Behavior of an Actor. The object of Actions can be another Actor or the Work Domain itself.

Actor—A Worker or Automation.

Adaptive Behavior—Behavior that is tailored to, or exploits, a goal-relevant constraint governing the System being controlled.

Affordance (J. J. Gibson, 1979)—A goal-relevant description of the world (e.g., Work Domain) that describes an opportunity for Action defined with respect to the capabilities of a particular Actor.

Analyst—A person who is responsible for performing a work analysis.

Application Domain—A sector of industry, business, or government that can serve as the object of a work analysis (e.g., libraries, medicine, aviation, design, nuclear).

Automation—A mechanical, electrical, chemical, or computerized Actor. Automation acts on other Automation or on the Work Domain.

Behavior—The dynamic State of a System, whether it be the System being acted on (Work Domain) or the System doing the acting (Worker or Automation). Compare with Action.

Behavior-Shaping Constraints—Intrinsic Work Constraints that rule out, but do not usually uniquely specify, the Actions of Workers or Automation. See Intrinsic Work Constraints.

Category—A set of elements that share one or more attributes or relations.

Closed System—A System that does not interact with its Environment (i.e., a System without Disturbances). Compare with Open System.

Cognitive Constraints—Work Demands associated with Worker cognitive characteristics. Compare with Environment Constraints.

Cognitive Engineering—The multidisciplinary area of research that is concerned with the analysis, design, and evaluation of Complex Sociotechnical Systems.

Cognitive Work Analysis—The framework for Work Analysis described in this book. It is based on the concept of Behavior-Shaping Constraints and contains models of the Work Domain, control tasks, Strategies, social-organizational factors, and Worker competencies in a single, integrated framework.

Cognitivist Approach—An approach to Work Analysis that starts with, and gives priority to, Cognitive Constraints. Information-Processing Approach is an example. Compare with Ecological Approach.

Coherence-Driven Work Domain—A Work Domain that does not impose dynamic, Environmental Constraints on the Action of Actors. Compare with Correspondence-Driven Work Domain.

Complex Sociotechnical Systems—Sociotechnical Systems that rate highly on several of the following dimensions: large problem space, social, heterogeneous perspectives, distributed, dynamic, potentially high hazards, many coupled subsystems, automated, uncertain data, mediated interaction via computers, disturbance management (see chap. 1 for a description of each of these dimensions).

Conceptual Perspective—The types of concepts to be used for representing work and the types of Representations that should be developed by the time the Work Analysis is complete. Compare with Methodological Perspective.

Constraint-Based Task Analysis—A Category of Task Analysis techniques that specifies what Actions should not be performed if the Goal is to be achieved. An Exemplar is Input–Output Task Analysis techniques.

Constraints—Relationships between, or limits on, Behavior. Constraints remove Degrees of Freedom.

Context-Conditioned Variability—Adaptive variability in Action that is caused by the fact that different Actions are required to achieve the same Goal on different occasions because of the presence of context-specific Disturbances.

Controls—The part of an Interface that is used by Workers to Act on Automation or on the Work Domain.

Correspondence-Driven Work Domain—A Work Domain that imposes dynamic, Environmental Constraints on the Goal-Directed Behavior of Actors. Compare with Coherence-Driven Work Domain.

Current Practice—The Actions, Tasks, and Strategies that Workers currently adopt to perform their jobs with the existing Device.

Decision Ladder—A template that describes categories of Actor information-processing activity. A framework for conducting an Input–Output Task Analysis (see chap. 8).

Degrees of Freedom—Possibilities for Behavior. Compare with Constraints.

Descriptive Model—A Model that describes how a System actually behaves. Compare with Normative and Formative Models.

Device—The computer-based information system available to Workers, including both the Interface and Automation.

Device-Dependent—An adjective describing Work Analysis techniques whose results are contingent on a particular Device, or class of Devices.

Device-Independent—An adjective describing Work Analysis techniques whose results are not contingent on a particular Device, or class of Devices.

Discretion—The number of Degrees of Freedom Workers have to do their job.

Displays—The part of an Interface that is used by Workers to obtain information about a Sociotechnical System.

Disturbances—Factors that have not been be anticipated by Analysts but that can affect the State of a System.

Ecological Approach—An approach to Work Analysis that starts with, and gives priority to, Environment Constraints. Compare with Cognitivist Approach.

Environment—That which is outside of the boundaries of the System drawn by the Analyst.

Environment Constraints—Work Demands associated with factors that are external to the Worker (e.g., physical or social reality). Compare with Cognitive Constraints.

Equivalence Class—A Category defined by an Equivalence Relation. Elements belong to a common Equivalence Class if they are identical along the criterion specified by the Equivalence Relation (e.g., the class of objects that are red). Equivalence Classes are defined by Homomorphic Mappings because they retain some of the properties of the elements being categorized (the properties defined by the Equivalence Relation), and discard all other properties. Equivalence Classes are used to create Models (i.e., Representations) that abstract, and represent, a finite subset of the attributes of a Natural System being modeled. See Rosen (1978) for a formal definition.

Equivalence Relations—Criteria that can be used to categorize elements in a set into Equivalence Classes. See Rosen (1978) for a formal definition.

Event—A happening or occurrence.

Event-Dependent—An adjective describing Work Analysis techniques that are contingent on a finite class of initiating Events.

Event-Independent—An adjective describing Work Analysis techniques that are not contingent on a finite class of initiating Events.

Exemplar—A particular member of a Category.

Formalism—A mathematical language without semantics or context-dependence that can be used to construct a Model.

Formative Model—A Model that describes requirements that must be satisfied so that a System could behave in a new, desired way. Compare with Normative and Descriptive Models.

Function—A goal-relevant structural property of a Work Domain. An Affordance that is relevant to the Purposes for which the Work Domain was designed. Affordances that are relevant for other Purposes are not considered Functions.

Goal—A State to be achieved, or maintained, by an Actor at a particular time. Note that goals are attributes of Actors, not Work Domains, and that they are dynamic (unlike the Purposes for which a Work Domain is built, which are relatively permanent). Compare with Purposes.

Goal-Directed Behavior—Actor Behavior that is intended to achieve a Goal.

Homomorphic Mapping—A faithful many-to-one mapping.

Information-Processing Approach—An Exemplar of the Cognitivist Approach. Compare with Ecological Approach.

Input–Output Task Analysis—A form of Task Analysis that identifies the inputs that are required to perform a Task, the outputs that are achieved after it is completed, and the Constraints that must be taken into account in selecting the Actions that are required to achieve the Goal, but usually not the specific Actions that should be adopted. An exemplar of Constraint-Based Task Analysis.

Instruction-Based Task Analysis—A category of Task Analysis techniques that specify what Actions should be performed. Exemplars include Sequential Flow and Timeline Task Analyses.

Interface—The computer-based means by which Workers obtain information about, and control the State of, a Sociotechnical System. Composed of Displays and Controls.

Intrinsic Work Constraints—Work Constraints that are Device-Independent. The set of all Behavior-Shaping Constraints.

Isomorphic Mapping—A faithful one-to-one mapping.

Label (Rosen, 1985a)—A description denoting a particular Equivalence Class.

Means–Ends Hierarchy—Same as Abstraction Hierarchy. Compare with Part–Whole Hierarchy.

Means–Ends Relation—The relationship between adjacent levels in a Means–Ends Hierarchy. The level below a given level describes the structural means that are available for achieving the level above. The level above a given level describes the structural ends (or Functions) that can be achieved by the level below.

Methodological Perspective—The process and the data collection methods an Analyst can or should adopt in actually performing a Work Analysis. Compare with Conceptual Perspective.

Model—A description of a Natural System using some type of Formalism consisting of variables and the Constraints between those variables. Same as Representation. A Model is low-dimensional (i.e., it is an abstraction) because it is based on Equivalence Classes. Only those variables defined by the Equivalence Relations that are used to build the Model, and the Constraints between those same variables, are included. An infinite number of Models of the same Natural System can be constructed by using different Equivalence Relations to define each Model. Therefore, the Model should never be confused with the Natural System being modeled (i.e., "don't eat the menu!"; cf. Golomb, 1968). Compare with Natural System.

Natural System (Rosen, 1985a, 1985b)—A System (e.g., Worker, Automation, or Work Domain) as it exists in Nature, as opposed to a Model of such a System. Natural Systems are high-dimensional because they possess an unbounded number of properties. Compare with Model or Representation.

Normative Model—A Model that prescribes how a System should behave. Compare with Descriptive and Formative Models.

Open System—A System that interacts with its Environment (i.e., a System with Disturbances). Compare with Closed System.

Part–Whole Hierarchy—A hierarchy defined by a Part–Whole Relation between adjacent levels. By moving down the hierarchy, larger elements are decomposed into smaller elements. By moving up the hierarchy, smaller elements are aggregated into larger groups or "chunks." A framework for defining Work Domains. Compare with Abstraction Hierarchy or Means–Ends Hierarchy.

Part–Whole Relation—The relationship between adjacent levels in a Part–Whole Hierarchy. The level below a given level describes the subelements that make up elements at the level above. The level above a given level describes the superelements that are made up of the elements at the level below.

Path—A bounded collection of adjacent Trajectories in a State Space.

Process Model—A generative Model describing the mechanism by which the Behavior of a System is produced (i.e., "how," not just "what"). A Model of System Structure rather than Behavior. Compare with Product Model.

Product Model—A black-box Model describing the Behavior of a System but not the process or mechanism by which that Behavior is generated (i.e., "what," but not "how"). A Model of System Behavior rather than System Structure. Compare with Process Model.

Purposes—The overarching intentions that a Work Domain was designed to achieve. Note that Purposes are properties of Work Domains, not Actors, and that they are relatively permanent (unlike the Goals of Actors, which change over time). Compare with Goal.

Representation—Same as Model.

Self-Organization (Haken, 1988)—A change in the functional, spatial, or temporal Structure of a System without specific interference from its environment. By specific it is meant that the change in structure is not impressed onto the System.

Sequential Flow Task Analysis—A form of Task Analysis that identifies a temporally ordered sequence of Actions that are required to Achieve the Goal of interest. An exemplar of Instruction-Based Task Analysis.

Signals—A Worker's interpretation of the environment in terms of time-space properties.

Signs—A Worker's interpretation of the environment in terms of familiar perceptual cues.

Situated Action—Same as Context-Conditioned Variability.

Skills, Rules, Knowledge Taxonomy—A taxonomy of human performance that identifies three qualitatively different ways in which Workers can interact with the environment: skill-based behavior, rule-based behavior, and knowledge-based behavior (see chap. 11).

Sociotechnical System—A System composed of technical, psychological, and social elements.

State—A point in a State Space describing the Behavior of a System at a particular time.

State Space—The multidimensional space defined by the set of State Variables describing a System. Each State Variable defines one dimension of the State Space. The set of all possible Trajectories.

State Variable—A variable that describes the State of a System.

Strategy—A category of cognitive task procedures that transforms an initial state of knowledge into a final state of knowledge. An example of a Process Model for an Actor.

Structure—A relatively permanent relational property of a System.

Symbols—A Worker's interpretation of the environment in terms of meaningful relational Structures.

System—A set of interrelated elements that share a common Goal or Purpose. The boundary between the System and its Environment is specified by the Analyst, and is not inherent in the System. Several equally meaningful boundaries can be drawn, although some boundaries may be more useful than others for certain problems. The System can be an Actor (i.e., Worker or Automation), the thing being acted on (i.e., a Work Domain), or both (i.e., a Sociotechnical System).

Task—Actions that can or should be performed by one or more Actors to achieve a particular Goal.

Task Analysis—A form of Work Analysis that specifies what Goals should be achieved, or also how they should be achieved. Can be classified into three types: Input–Output, Sequential Flow, and Timeline. Compare with Work Domain Analysis.

Timeline Task Analysis—A form of Task Analysis that identifies a temporally ordered sequence of Actions that are required to achieve a Goal, with duration estimates for each Action. An exemplar of Instruction-Based Task Analysis.

Trajectory—A line in the State Space describing the Behavior of a System over time.

Unanticipated Variability—Same as Context-Conditioned Variability.

Work—The professional responsibilities and activities of Workers.

Work Analysis—Any technique for analyzing human Work. Exemplars include Task Analysis and Work Domain Analysis.

Work Demands—Constraints that govern Work. Examples include Cognitive Constraints and Environment Constraints.

Work Domain—The System being controlled, independent of any particular Worker, Automation, Event, Task, Goal, or Interface.

Work Domain Analysis—A form of Work Analysis that identifies the Functional Structure of the Work Domain. Compare with Task Analysis.

Workarounds—Worker Behaviors that circumvent the limitations in a computer-based information system and that allow Workers to achieve Goals in a required, or preferred, manner.

Workers—The people who participate in and act in a Sociotechnical System. Synonymous with operators and users (see chap. 1, Footnote 1).

What's the Problem? Scope and Criteria for Success 1

> There's no solution that applies to every problem, and what may be the best approach in one circumstance may be precisely the worst in another.
> —Weinberg (1982, p. 22)

PURPOSE

This chapter identifies the scope of the problems that we are concerned with in this book, and introduces the criteria that we adopt in searching for a solution to these problems. The book describes a framework for CWA consisting of a set of concepts that have been developed in the area of research known as *cognitive engineering*. This framework is explicitly tailored to the relatively unique demands imposed by complex sociotechnical systems. Our intention is to target a broad range of application areas, including those that are characterized primarily by cooperation between people mediated by machines, as well as those that are characterized primarily by collaborative human and machine control of some technical system. We argue that complex sociotechnical systems pose new challenges that are not effectively addressed by traditional approaches to human factors or HCI. In the second half of the chapter, we propose that the primary criteria for effectiveness in complex sociotechnical systems are safety, productivity, and health. A wealth of data is presented to demonstrate that these criteria are not currently being dealt with effectively by industry. Finally, we summarize the criteria that an analysis framework will have to satisfy if it is to lead to safe, productive, and healthy computer-based work.

WHAT IS COGNITIVE ENGINEERING?

Definition

The concepts described in this book are part of cognitive engineering, an emerging field that is concerned with the analysis, design, and evaluation of complex sociotechnical systems (Andriole & Adelman, 1995; Dowell & Long, 1998; Hollnagel & Woods, 1983; Norman, 1981, 1986; Rasmussen, 1986c; Rasmussen et al., 1994; Woods & Roth, 1988b). Figure 1.1 provides a generic illustration of the properties of a *sociotechnical system*. There are many examples of sociotechnical systems, some that accentuate the social (e.g., the stock market, a bank, an engineering design firm, military command and control), and others that accentuate the technical (e.g., a flexible manufacturing system, a commercial aircraft, a hospital, a nuclear power plant). As Fig. 1.1 illustrates, such systems are composed of several layers. Traditionally, many disciplines have viewed their technical core as comprising the entire

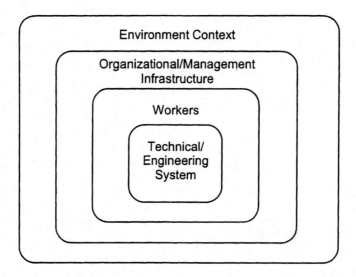

FIG. 1.1. The various layers of a complex sociotechnical system (adapted from Moray & Huey, 1988).

system. For example, the medical profession has traditionally been of the opinion that knowledge of anatomy, physiology, anesthesiology, and so on, provides the necessary and sufficient conditions for safe, competent medical practice (Bogner, 1994; Leape, 1994). Similarly, the premise in the nuclear industry before the Three Mile Island accident was that reactor physics, thermodynamics, and other such technical factors, provided the necessary and sufficient knowledge to design a safe and productive plant (Moray & Huey, 1988).

Operating experience in these and other application domains has unequivocally shown that the assumption that the technical core comprises the entire system is fundamentally misguided (see the section later on safety). As Fig. 1.1 indicates, there are several other issues that need to be understood and taken into account to achieve an acceptable level of performance. For example, the capabilities and limitations of the individual workers[1] interacting with the technical system must also be taken into account. This class of design issues has been the traditional purview of human factors engineering (e.g., Sanders & McCormick, 1993; Wickens, 1992). But as Fig. 1.1 indicates, technical and psychological factors are not the whole story. Social-organizational factors also play a crucial role in system performance. For example, Ontario Hydro—one of the largest electrical utilities in North America—recently decided to shut down 7 of its 20 nuclear power plants at an estimated cost of $8 billion Canadian.

---

[1]Unfortunately, there is no single widely accepted term that is used to refer to the people who operate, or work in, a sociotechnical system. In the HCI literature, it is common to refer to such people as *users*. In the human factors literature, the term *operators* is sometimes used instead. Because of the application domains that each of these disciplines has generally been concerned with, the term users is generally associated with the service sector (e.g., office information systems) whereas the term operators is generally associated with the industrial sector (e.g., process control). As the examples in subsequent chapters make clear, the framework presented in this book is intended to be broad enough to apply to complex sociotechnical systems in both sectors. To avoid inadvertently limiting our scope, we use the more neutral term *workers* throughout this book. This label should be interpreted as subsuming the connotations typically associated with the terms users and operators in HCI and human factors, respectively.

The motivation for this unprecedented step was not technological problems, but rather inadequate management that led to a minimally acceptable level of safety (Andognini, 1997). There is now a great deal of evidence to show that organizational factors play a very important role in system performance and safety (e.g., Anderson, 1997; Brynjolfsson & Hitt, 1998; Leveson, 1995; Pool, 1997; Reason, 1990).

A sociotechnical system is also affected by its environment. Depending on the industry in question, this environment can consist of international trade laws (as in the case of manufacturing), regulatory requirements (as in the case of the nuclear power industry), other companies competing for a significant share of the same market (as in the case of commercial aviation), or the physical environment (e.g., climate). Collectively, these factors constitute the various layers of a sociotechnical system.

Because it is based on a sociotechnical system viewpoint, cognitive engineering is characterized by a broad systems approach. The key is to understand the various layers in Fig. 1.1, and even more important, the interactions among them. Although these layers are shown independently in Fig. 1.1, in practice they are all intimately intertwined. All of the layers come together to shape the performance of the system as a whole. As a result, a holistic approach to systems analysis is absolutely essential.

Cognitive Work Analysis

Although cognitive engineering is concerned with analysis, design, and evaluation, this book focuses primarily on analysis.[2] Any computer-based system for human work is based on a certain set of assumptions about the work that it is intended to support. Rather than making those assumptions implicitly, the design of information systems should be based on an explicit analysis of work. This point has been understood for many years in the human factors community, as indicated by the emphasis on various work analysis methods, such as task analysis (e.g., Kirwan & Ainsworth, 1992; Meister, 1985; R. B. Miller, 1953).[3] The value of work analysis has also been recognized more recently by the HCI community (e.g., Diaper, 1989; Nardi, 1996b; Suchman, 1987, 1995). However, work analysis is not an end in itself but rather a means to derive implications for design. Although we do not discuss design principles per se in this book, we try to show the strong connection between work analysis and design through the use of examples. It is critical to illustrate the value of this linkage because, without it, work analysis can easily be perceived to be a wasteful investment. Work analysis is valuable only to the extent to which it gives designers insights into how to create tools that effectively support human work. Finally, evaluation per se is not discussed at all in this book, despite its crucial and unique role in the systems design life cycle. We refer you to Rouse (1991) and Rasmussen et al. (1994) for treatments of evaluation from the perspective of cognitive engineering.

---

[2]This distinction between analysis, design, and evaluation is an abstraction and does not capture the actual practice of designers. If systems are to be built in an integrated fashion, then all three activities must all be intimately intertwined and mutually informing each other. Nevertheless, the distinction is useful for pedagogical purposes because it reveals important conceptual differences.

[3]The generic term *work analysis* is used in this book to refer to any technique for analyzing human work. There are different types of work analysis techniques, including task analysis and work domain analysis, both of which are introduced in chapter 3.

WORK ANALYSIS FOR COMPLEX SOCIOTECHNICAL SYSTEMS

Before we discuss the demands associated with work analysis in complex sociotechnical systems, it is worthwhile describing the complexity in such systems in more detail. After all, the methods that we chose for analyzing a class of systems should be based on an understanding of the characteristics of that class.

What Makes Systems Complex?

Complexity is, in general, a multidimensional concept and sociotechnical systems are no exception. The nature and origin of the complexity in such systems can vary substantially. To take one example mentioned earlier, some complex sociotechnical systems are characterized primarily by cooperation between people mediated by machines, whereas others are characterized primarily by collaborative human and machine control of some technical system. The following list of interrelated characteristics is intended to be broad enough to subsume the different types of complexity that we can find in sociotechnical systems (cf. Perrow, 1984; Woods, 1988): (a) large problem spaces, (b) social, (c) heterogeneous perspectives, (d) distributed, (e) dynamic, (f) potentially high hazards, (g) many coupled subsystems, (h) automated, (i) uncertain data, (j) mediated interaction via computers, (k) disturbances. The qualitative ways in which each of these factors contributes to increasing the demands on workers and system designers is described next.

Large Problem Spaces. First, complex sociotechnical systems tend to be composed of many different elements and forces. As a result, the number of potentially relevant factors that designers and workers need to take into account can be enormous. For example, in medicine, the number of identified illnesses—and hence the number of potential diagnoses—is on the order of 500,000. Similarly, in the domain of finance, portfolio managers must deal with the already large and ever-increasing number of investment objects in world markets. Large problem spaces also create difficulties for designers who must ensure that the information systems they develop allow workers to deal with the entire space of possibilities without exceeding their resource limitations.

Social. Second, complex systems are usually composed of many people who must work together to make the overall system function properly. This creates a strong need for clear communication to effectively coordinate the actions of the various parties involved. For example, a hospital cannot provide the most effective care for its patients unless there is good communication between its doctors, interns, nurses, administrators, and other staff. This is no easy feat when the number of people involved can literally reach into the thousands, as it does in large engineering design projects.

Heterogeneous Perspectives. Third, the workers in a complex sociotechnical system frequently come from different backgrounds and thus represent the potentially conflicting values of a diverse set of disciplines. Consequently, the social negotiation process is made more difficult by the fact that different value structures have to be resolved during the decision-making process. For example, an engineering design organization cannot design truly effective products unless the various engineers, architects, accountants, advertisers, and managers are able to resolve their differences in priorities to achieve a common goal (Bucciarelli, 1994). This is not an easy task when the set of values involved is large and heterogeneous.

Distributed. Fourth, the demands associated with social coordination can be complicated by the fact that the people involved may be located in different places, sometimes even different countries. Not only does geographical separation make communication more difficult, but it can introduce cultural factors as well. An extreme case of the distributed nature of complex sociotechnical systems is the work domain of engineering design. For example, the design of the Boeing 777 aircraft involved 238 design teams and almost 60 companies spread over 17 time zones across the globe (Norris, 1995; Petroski, 1995). It is difficult enough to get hundreds or thousands of people to communicate clearly and coordinate their actions and decisions when they are geographically distributed. The fact that these people come from different cultures with diverse expectations and values can introduce additional complications (Burns & Vicente, 1995).

Dynamic. Fifth, complex systems are usually dynamic and can have long time constants; it can take minutes or even hours for the work domain to completely respond to an action from its workers. Because the effects of their actions are delayed, workers have to anticipate the future state of the work domain and act well before the time when a response is desired. For example, large oil tankers have very slow and sluggish dynamics. Thus, if the tanker is to be maneuvered accurately, workers must anticipate these lags by initiating a control action well before the point where they want the tanker to change direction. This creates a challenging situation for workers, especially in tight corners where an error in anticipation can cause a tanker to run aground. It is precisely for this very reason that international ports have pilots who are very familiar with the navigational demands of a particular harbor and ship dynamics. These workers come aboard large ships to take over the controls from the captain to safely maneuver the ship into the harbor and to its stopping point.

Hazard. Sixth, there is also a high degree of potential hazard in operating complex sociotechnical systems, because inappropriate human beliefs or actions can have catastrophic economic consequences (e.g., a large disturbance in world money markets), jeopardize public safety (e.g., a release of radioactivity), or challenge the integrity of the natural ecology (e.g., a leak from an oil tanker). This suggests that workers may have to evaluate carefully the consequences of their actions based on a conceptual understanding of work domain functioning before actually implementing those actions. This will be particularly critical in abnormal or unusual situations. Moreover, designers cannot afford to rely on trial-and-error approaches. Because of the potential hazard involved, there is a very strong requirement to try to "get it right the first time."

Coupling. Seventh, complex sociotechnical systems also tend to be composed of many subsystems that are highly coupled (i.e., interacting). This makes it very difficult to predict all of the effects of an action, or to trace all of the implications of a disturbance because there are many, perhaps diverging, propagation paths

(Dörner, 1989/1996). For example, if a particular action is selected to affect goal X, workers must also consider whether that same action will also affect goals Y and Z. These other effects may not be desirable, so workers must consider them before acting. Reasoning in a highly coupled work domain puts a great burden on workers because of all of the factors that need to be considered at the same time.

Automation. Eighth, sociotechnical systems also tend to be highly automated (e.g., automated trading in the stock market, distributed control systems in petro-chemical plants). Much of the time, computer algorithms control the work domain, and the workers' responsibility is to monitor the state of the automation and the work domain itself. During the abnormal situations that the automation cannot handle effectively, workers play the role of problem solvers who must compensate for the lack of robustness in the automation. During these infrequent but challenging situations, the worker's demands are primarily cognitive in nature, rather than psycho-motor. Because of the potential hazard involved, there is a great deal of pressure to deal with the problem effectively. And because workers do not have to intervene very frequently, they are not accustomed to performing these compensatory or reconfiguration activities. Consequently, although automation can make some parts of the job easier, it can also create new challenges and place new demands on workers (Hirschhorn, 1984; Wiener, 1988).

Uncertainty. Ninth, there tends to be uncertainty in the data that are available to workers (e.g., imperfect physiological sensors in an operating room, uncertain economic indicators to an investment banker). Because of this impoverished input, the true state of the work domain is never known with perfect certainty. Furthermore, workers must distinguish changes that are caused by events in the work domain from those that are caused by random drift or failure of the sensors. Sometimes, the data may seem to indicate that an abnormality has occurred but, in fact, the root cause may just be a misleading indication from a sensor. Thus, there will frequently be a need for problem solving and inference (i.e., to "go beyond the information given").

Mediated Interaction. Tenth, it is often the case that the goal-relevant properties of a complex sociotechnical system cannot be directly observed by human perceptual systems unaided (e.g., the intentions of another company competing for the same market, the current work activities of a remotely located collaborator, and the blood pressure of a patient in an operating room). In these cases, it is usually not possible for people to go out and directly gather information using the powerful perceptual systems that serve them so well in the natural environment (Vicente & Rasmussen, 1990). As a result, the computer interface must serve as a "window" into the sociotechnical system, providing workers with a mediating representation of the work domain. Thus, the everyday skills that people use to routinely explore the natural environment are not sufficient to deal with the demands associated with mediated interaction. More complex cognitive resources are sometimes needed to get the job done.

Disturbances. Finally, if this were not enough, workers are also responsible for dealing with unanticipated events (e.g., a rush order in a flexible manufacturing system, a large and unexpected change in foreign exchange for a financial analyst, a fault in a process control plant that was not anticipated by designers). They must improvise and adapt to the contingencies of an unanticipated event quickly to maintain system safety or productivity. Because their normal work procedures no longer apply in these cases, workers must generate an appropriate response based on a conceptual understanding of the work domain. As a result, information system design cannot be based solely on expected or frequently encountered situations. Instead, complex sociotechnical systems must also operate effectively even—or especially—under idiosyncratic rare events that are not anticipated by workers or designers.

Conclusions. There are a number of different dimensions that characterize complex sociotechnical systems. It is important to emphasize, however, that not all complex systems rate highly on all of these dimensions. There are important differences across application domains (e.g., nuclear power vs. cooperative office work), so some dimensions may not even be particularly relevant for some systems. Nevertheless, all complex sociotechnical systems will rate highly on at least some of these dimensions, and will also usually exhibit several other dimensions of complexity albeit to a lesser extent.

Because of this complexity, there is an enormous set of demands placed on the people who work in, and the designers who create, these systems. The job of the workers is especially demanding because they must deal with this complexity online, in real time. This makes the role of computer-based support systems all the more important. The job of system designers is particularly challenging because they must try to "get it right the first time around" and design a system that is capable of functioning effectively even—or better yet, especially—under rare, unanticipated events. To paraphrase Alexander (1964), what is required is revolutionary rather than evolutionary design: "[The designer] has to make clearly conceived forms without the possibility of trial and error over time" (p. 4). As we see in subsequent chapters, these requirements are relatively unique and different from those associated with other systems with which most human factors and HCI professionals have traditionally been concerned. Thus, it should not be surprising to find that approaches that work well for other systems may not be effective or appropriate for systems with the characteristics listed previously.

New Problems Demand New Approaches

Evolution of Work Demands. It is only relatively recently that systems with many of the characteristics just described have appeared in our society. This is an important observation because it suggests that the methods and practices that constitute traditional systems design were developed during a time when work had a very different set of characteristics.

This conclusion can be supported by examining very briefly the evolution of human work (Vicente, 1995c). As Rasmussen (1988b) observed, industrial development has undergone several phases, each posing a unique set of challenges for work analysis methods. The first phase, the Industrial Revolution, began with the *mechanization* of work and the division of labor. For this phase, the work methods analyses pioneered by Taylor (1911) at the end of the 19th century (see Kanigel, 1997, for an incisive biography, and Mandel, 1989, for a photographical essay) seemed appropriate because they were well suited to the demands associated with physically demanding manual labor (although they ignored other important considerations; see Karasek & Theorell, 1990; Reed, 1996). The second phase corresponded with the introduction of *automation* and *centralization*. With this phase, the nature of work evolved so that workers' job responsibilities became more complex in nature, requiring more than just physical force. Thus, new methods were required to analyze the new types of demands being imposed on workers. This change led R. B. Miller (1953) to develop task analysis, a broader method for analyzing work that describes task demands as a function of system inputs and outputs. This class of work analysis methods has proven itself over time to lead to valuable insights that have direct implications for design, training, and selection. In fact, task analysis methods are one of the methodological backbones of human factors engineering, being discussed in many textbooks (e.g., Meister, 1985). But again, the nature of work has changed, this third phase corresponding to a trend toward *computerization*. In this phase, the role of people began to evolve from one of manual laborer, requiring primarily perceptual-motor skills, to intellectual worker, thereby requiring more conceptual knowledge and cognitive skills. Thus, the evolution of work has led to a greater demand for communication, collaboration, and problem solving, thereby increasing the discretion, and therefore the variability, in worker action (Hirschhorn, 1984). It should not be surprising to find, therefore, that the task analysis methods that had worked so well in the past are limited in the extent to which they can usefully capture the demands imposed by computer-based work in complex sociotechnical systems (see chap. 3).

Computer Technology in the Workplace. Given this pattern of evolution in human work, information technology usually plays a very prominent role in complex sociotechnical systems. Computer-based information systems are found in virtually every application domain. In part, this may be motivated by an attempt to adopt information technology to control the complexity inherent in these sociotechnical systems. In other cases, information technology is an intrinsic part of the system in the sense that the functions or services provided by the sociotechnical system could not exist without that technology.

The prevalence of computer-based information systems in the workplace is worth documenting briefly to emphasize its scale. In 1996, U.S. companies spent an estimated $213 billion—43% of their capital investment—on computer hardware alone (Gibbs, 1997). If the costs associated with software, networks, computer support, and training are included, this figure rises to about $500 billion for the United States and about $1 trillion globally (Gibbs, 1997). It should not be surprising, therefore, that 50 million American office workers spent a significant portion of their work day interacting with a computer terminal of some sort, or that General Motors expected that 90% of all of their new capital investments would be in computerized numerical control machines (Zuboff, 1988).

More recently, many companies have begun to take advantage of information technology originally designed for the World Wide Web (WWW) to create *intra*nets, computer-based tools to support work within companies (e.g., sharing of information, communicating between workers, creating application front-ends, and connecting to customers and suppliers). The introduction of intranets into the workplace is very fast-paced, as evidenced by the following data obtained from interviews with the chief information officers from 50 Fortune 1,000 companies: In February 1996, 16%

of these companies had intranets; 5 months later, by July 1996, 64% of these companies had intranets and an additional 32% were in the process of building them (Deutsch, Callahan, & Edwards, 1996). To give an idea of the level of investment involved, the creation of an intranet for a company with 10,000 employees requires an investment of over $15 million over an 18-month period (Deutsch et al., 1996).

Computer technology is even making substantial inroads into systems that have traditionally been mechanical in nature, part of society's transition "from atoms to bits" (Negroponte, 1995). For example, the Boeing 777 has over 2.6 million lines of computer code (Norris, 1995). Not only is computer technology used in the operation of many different types of complex systems, but it is also being increasingly used in the design of these systems. Again, the Boeing 777 serves as a prime example—it was designed completely on computer, with an infrastructure of some 7,000 work-stations and 3.5 terabytes of memory, at an estimated cost of $4 billion (Petroski, 1995).

Many industries are investing heavily in information technology with the hope of achieving fiscal restraint. For example, the health care industry has found that cost containment goals cannot be satisfied by traditional administrative systems (Anderson, 1997). This problem has become increasingly important in the United States, as managed care organizations put more and more pressure on health care providers to cut costs (Anderson, 1997). Consequently, the industry is making a strong commitment to computer-based information systems, spending between $12 billion and $16 billion on them in 1996 alone (Raghupathi, 1997). It seems that, no matter where we look, computer technology is making rapid advances into the workplace.

Large investments have also been made in researching how best to utilize information technology. Examples of such multiyear, multimillion-dollar research programs include: the ESPRIT projects funded by the European Community, the FRIEND '21 program funded by the Japanese government (FRIEND '21, 1995), and the series of workshops on risk management in sociotechnical systems sponsored by the World Bank (Rasmussen & Batstone, 1991). Trends such as these are sure to increase in number and size in the future.

Do We Know How to Use It? There is no doubt that information technology is here to stay. What is in doubt is our capacity to effectively utilize such technology. An excellent example is the impact of computerized trading on the stock market crash of Black Monday, October 19, 1987 (Waldrop, 1987). Although the general public may not have been aware of it at the time, computer programs were (and still are) routinely used to trade very large numbers of stocks in large portfolios. With computerized trading, "decisions" to buy or sell stocks are made in a split second, without reflection, because they are triggered by automated rules. In the aftermath of Black Monday, computer experts and financial analysts indicated that computerized trading actually amplified the magnitude of the market's crash. This behavior was completely unexpected. As one government official observed: "On Monday, we all discovered that program trading was like an *incompletely designed new car*" (Waldrop, 1987, p. 603, emphasis added). Waldrop went on to point out that "computers can be thrown into a world where their assumptions are false, and where they can end up blindly following strategies that border on the lunatic" (pp. 603–604). There is little in the arsenal of traditional human factors engineers or HCI specialists that would have allowed them to identify such a potential problem, let alone safely design against it.

This example illustrates one of the basic theses of this book, namely that new concepts and methods are required to complement (not replace) traditional human factors and HCI concepts and methods. Changes in the nature of human work have led to new design problems and questions, for which traditional methods are not very well equipped (Meister, 1995b; Schraagen et al., 1997). As Norman (1993) observed: "In the past, technology had to worry about fitting people's bodies; today it must fit people's minds. This means that the old approaches will no longer work" (p. 9). We could even argue that this is precisely what has led to the development of cognitive engineering, which can be viewed conceptually as an expansion of human factors engineering or HCI to suit the challenging needs of our time. We need to look for new ways of analyzing human work, ways that are well suited, or better yet, explicitly tailored, to the unique set of challenges posed by complex sociotechnical systems. Only by developing such novel concepts and methods can we take full advantage of the vast potential of information technology.

CRITERIA FOR EFFECTIVENESS: SAFETY, PRODUCTIVITY, AND HEALTH

If we are to design effective computer-based information systems to facilitate work in complex sociotechnical systems, we need an understanding of what *effective* means. That is, we need to develop an explicit statement of the performance criteria that we must strive to satisfy. Three criteria in particular stand out: safety, productivity, and health. Although these criteria are correlated in practice, each provides a different perspective for thinking about a complex sociotechnical system, and therefore, a different frame of reference for evaluating effectiveness. Ideally, we would like to consider all three because a truly effective system should not pose a safety threat, should be economically viable, and should enhance the quality of life of its workers. As we see in the following discussion, existing complex sociotechnical systems have rarely satisfied all of these criteria. In fact, there are important deficiencies in the ways in which each of these perspectives is currently being dealt with in industry. It is important to understand the nature of these deficiencies if we are to develop a way of analyzing human work that will lead to substantial improvements on all three fronts. We begin with the safety perspective, because this has been the traditional concern in cognitive engineering (see Appendix).

Safety: Reducing Risk

We might think that the safety perspective is of interest only in industries where there is a potential for large-scale catastrophic accidents. This perception has been reinforced by the fact that nuclear power plants have traditionally been used as the prototypical example when discussing safety in complex sociotechnical systems. But with the trend toward computerization in many diverse application domains (see earlier discussion), safety can be meaningfully defined in quite different ways. An example that has already been mentioned is the stock market. There, the risk is not ecological or life threatening, but economic. Thus, a safe system is one where the risk of economic loss is somehow ameliorated by systems design.

Halbach (1994) provided some insight into what it would take to achieve this goal. His analysis suggests that financial institutions, such as the stock market, are at risk in situations that have not explicitly been anticipated by system designers. The software programs that are frequently used to automate buy and sell orders are based on only a few parameters and on assumptions of linearity. But of course, the financial world in which these programs operate is actually governed by a very large number of variables that are interconnected by dynamic, nonlinear relationships. Most of the time, the assumptions on which the programs are based are good enough to lead to good, meaningful decisions. However, the limitations of these assumptions are brought to the fore when the state of the financial world changes drastically and abruptly (as it did on Black Monday). In these unanticipated situations, the algorithms governing financial transactions are no longer appropriate and can lead to devastating results (Waldrop, 1987). Halbach's analysis suggests that unanticipated events pose a threat to the safety of financial institutions, such as the stock market. But is this conclusion equally valid for industrial systems (e.g., nuclear power, commercial aviation, space flight, maritime transportation, and chemical process plants), where a threat to safety also means a threat to human life or the natural ecology, sometimes on a tragically large scale?

Before we answer this question, it is important to briefly address the magnitude of the potential safety hazards associated with industrial accidents. Although such consequences are frequently difficult to quantify, threats to safety also have enormous economic consequences that are more amenable to objective documentation. To take but one example, it has been estimated that abnormal situations in the petrochemical industry alone have an annual preventable impact of $20 billion on the United States (U.S.) economy (Bullemer & Nimmo, 1994). If we combine such numbers with the tragic outcomes that have been associated with accidents such as Chernobyl, Bhopal, Challenger, Seveso, Exxon Valdez, and others (cf. Casey, 1993; Leveson, 1995; Perrow, 1984; Pool, 1997; Reason, 1990), then the case for safety can hardly be overstated. Every design should be based on a work analysis framework that tries to reduce the risk of catastrophic failure.

To achieve this goal, it would be useful to understand some of the factors that contribute to industrial accidents. There is a relatively large body of operating experience over the last few decades that sheds some light on this issue. Rasmussen (1969) was perhaps the first to observe the feature shared by large-scale accidents, namely that they were prompted by situations that were both unfamiliar to workers and that had not been anticipated by system designers. In these situations, workers cannot rely on a procedure or on automation to deal with the abnormality because designers did not anticipate the particular event. Instead, workers have to creatively improvise an innovative solution to the problem with which they are faced. Rasmussen came to this conclusion after examining 29 cases with major consequences to either plant or personnel in the nuclear domain and 100 accidents in air transportation. In the 30 years since his technical report was published, additional evidence has pointed to the importance of unanticipated events to system safety. For example, the various accidents reviewed by Perrow (1984), Reason (1990), and Leveson (1995) all fall into this category (see also Hirschhorn, 1984; Pool, 1997). Similarly, data from the petrochemical industry obtained by Bullemer and Nimmo (1994) indicate that 27% of incidents involving people and work context are caused by situations for which there are inadequate or no procedures. Also, one investigation from the nuclear industry found that 100% of the abnormal events surveyed that were not handled successfully by workers consisted of situations for which there was no procedure or for which the existing procedure had to be modified (Kauffman, Lanik, Spence, & Trager, 1992). This datum is particularly significant because the "event" category represents situations that pose a more severe threat to safety than the "incident" category investigated by Bullemer and Nimmo.

In summary, data from both the service and industrial sectors show that events that are unfamiliar to workers and that have not been anticipated by designers pose the greatest threat to system safety. Therefore, any framework for analyzing human work must provide a way of identifying what type of information and support people need in order to be able to deal successfully with unanticipated events. As the "malicious procedural compliance" story in the Preface illustrates, the current practice of expecting procedures to deal with all situations will, by definition, not work (see chap. 3).

Productivity: Making the Business Case

Although any organization has a responsibility with respect to safety, it will not stay in business for very long unless it also makes a profit. Thus, productivity is a perspective to which all executives and managers are very sensitive. Money talks. Given the large capital investments that are required to introduce information technology into the workplace, we would expect and would like to see a corresponding improvement in productivity. But as Landauer (1995) showed, this has not been the case (see also Gibbs, 1997). To take but one of the many examples he cited, the Internal Revenue Service in the United States spent $50 million on personal computers for its agents in an effort to improve productivity. The result? The number of cases processed per agent per week decreased by 40%! Thus, just as there is a problem with meeting the safety criterion, there is also a problem, or a paradox as Landauer put it, in meeting the productivity criterion as well. This also poses a challenge for analyzing human work because, if a work analysis framework does not generate designs that lead to productivity improvements, those designs will not be deemed to be successful.

Landauer's (1995) argument showing that information technology has generally not led to improvements in productivity growth is very important, so we provide an extended summary of it here, borrowing heavily from the economic evidence that he gathered to support his thesis. The starting point for his argument is represented in Fig. 1.2, which shows the growth of productivity in the United States over the last 130 years or so. As the figure indicates, productivity growth decreased from the early 1970s onward. This period happens to coincide with the large investment in computer technology by businesses in various sectors of the economy. And as Fig. 1.3 shows, this trend is not unique to the United States, but in fact has been experienced by other heavily industrialized countries as well. Coincidence? Perhaps. The only way to tell is to understand the factors that contributed to these changes in productivity.

Interestingly, before the 1970s, economists had a relatively good understanding of the measurable influences on productivity. However, as Fig. 1.4 shows, something changed after about 1973. The factors that had traditionally accounted for virtually all of the annual changes in productivity account for only 40% of the slowdown exhibited in Figs. 1.2 and 1.3. Apparently, there is a new factor that accounts for

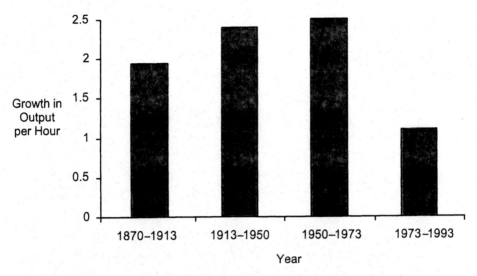

FIG. 1.2. Labor productivity growth in the U.S., 1870–1993, measured as GNP per hour worked. Reprinted from T. K. Landauer (1995), *The trouble with computers: Usefulness, usability, and productivity*, published by MIT Press. Copyright © 1995 Massachusetts Institute of Technology. Reprinted by permission.

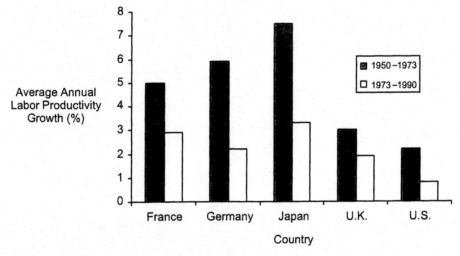

FIG. 1.3. Productivity growth before and after 1973 for five major industrial nations. Reprinted from T. K. Landauer (1995), *The trouble with computers: Usefulness, usability, and productivity*, published by MIT Press. Copyright © 1995 Massachusetts Institute of Technology. Reprinted by permission.

60% of the changes in productivity observed after 1973. Landauer (1995) claimed that this new factor is the widespread introduction of information technology into the workplace. That is, the reason why productivity growth has declined since 1973 is *because* of computerization. Could this be true?

Figure 1.5 provides another relevant piece of evidence. This graph shows the annual change in productivity and the annual change in information technology capital in the service sector of the economy. The conclusion is straightforward.

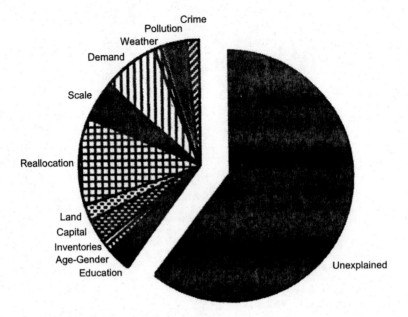

Sources of Post-1973 Productivity Growth Decline

FIG. 1.4. Estimates of contributions to the post-1973 slowdown in productivity growth in the U.S. Reprinted from T. K. Landauer (1995), *The trouble with computers: Usefulness, usability, and productivity*, published by MIT Press. Copyright © 1995 Massachusetts Institute of Technology. Reprinted by permission.

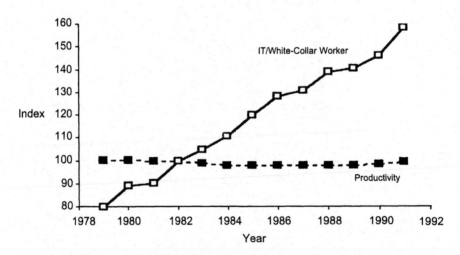

FIG. 1.5. Information technology and productivity in the service sector (non-goods-producing industries). While information technology investment went up rapidly, productivity growth slowed. Reprinted from T. K. Landauer (1995), *The trouble with computers: Usefulness, usability, and productivity*, published by MIT Press. Copyright © 1995 Massachusetts Institute of Technology. Reprinted by permission.

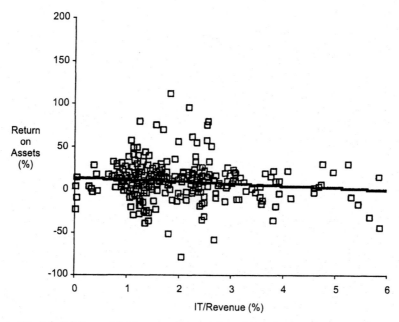

FIG. 1.6. Relations of business success to information technology investment, here measured as return on assets (net income over total worth of company) as a function of the proportion of gross income spent on computers and other information technology. Reprinted from T. K. Landauer (1995), *The trouble with computers: Usefulness, usability, and productivity*, published by MIT Press. Copyright © 1995 Massachusetts Institute of Technology. Reprinted by permission.

Although investment in information technology has increased at a roughly constant rate, productivity has not kept up. On the contrary, it has actually declined slightly over the same period. A skeptic might argue that this result may be an artifact of the particular measure of productivity that has been adopted. Figure 1.6 shows the relationship between the proportion of a company's gross income spent on information technology and a different metric of productivity that is a meaningful measure of business success, namely return on assets. The general trend is the same. There is a negative correlation between information technology investment and return on assets. Figure 1.7 provides data on a more pragmatic measure of business success, namely stockholder return. The graph also shows a negative correlation, this time between investment in information technology per employee per year and economic return to stockholders.

Landauer's (1995) conclusion, based on these and many other data, is that the missing piece of the productivity slowdown puzzle in Fig. 1.4 is information technology. Although computers have allowed us to do things that were never before possible, they have not generally led to an increase in productivity growth. On the contrary, they are responsible for the decline in productivity growth observed in industrialized nations since the early 1970s.[4] Why is this? Landauer devoted much

---

[4]There are some who would disagree with Landauer's conclusions. For example, there is a minority of studies that shows a positive relationship between investment in information technology and productivity (e.g., Kelley, 1994; Siegel, 1997). Also, some have argued that this area of research is plagued by a lack of good measures of productivity, but it seems that measurement problems do not provide a complete explanation for the productivity paradox (Willcocks & Lester, 1996).

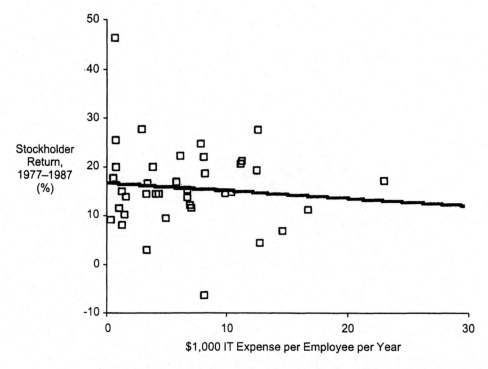

FIG. 1.7. Return to stockholders compared to money spent by individual firms for information systems on a per employee annual basis. Reprinted from T. K. Landauer (1995), *The trouble with computers: Usefulness, usability, and productivity*, published by MIT Press. Copyright © 1995 Massachusetts Institute of Technology. Reprinted by permission.

of his book to answering this question, so only a very brief synopsis can be provided here.

Part of the problem is that activities that can be completely reduced to a set of algorithms or rules have already been automated. In other words, the low-hanging fruit has already been picked. What remains is a much more challenging problem— how to support human intellectual activities that we are not capable of automating fully. Such activities usually require workers to engage in open-ended and creative, discretionary decision making (which is largely why these activities cannot be fully automated). However, designers have not been able to design effective information support for these intellectual tasks (see also Clegg et al., 1996). Note that Landauer (1995) was not just referring to the usability concerns that have dominated the field of HCI, but more important to the issues associated with usefulness (or functionality). That is, the trouble with computers is not just that they are not very easy to use, but also that they generally do not provide people with the functionality that is required to get their jobs done efficiently. Thus, if we are going to make large strides in productivity growth, we need to design computer systems that provide workers with new types of functionality that are required to perform discretionary intellectual activities effectively and reliably.

There are more recent data that are also consistent with Landauer's (1995) claim that information technology is responsible for the productivity paradox (Gibbs, 1997). For example, it is estimated that companies lose $5,590 per year per computer because of the time that employees waste by "futzing" with their computers rather than performing productive work with them. Similarly, a survey of 6,000 office workers found that an average of 5.1 hours per week is spent on such unproductive computer-centered activities. A different set of surveys has shown that nontechnical employees spend about 4% to 10% of their working hours helping colleagues solve computer-related problems. These figures, although only estimates, reveal that there are hidden costs associated with the introduction of information technology into the workplace.

In summary, the well-documented argument put forth by Landauer (1995) is of great importance because it adds another constraint on a viable framework for analyzing human work. A work analysis method must address the issues of usefulness and usability associated with discretionary decision-making tasks if it is to overcome the productivity paradox that has plagued the introduction of information technology into the workplace. The data collected by Landauer indicate that the consequences of not extracting the full capabilities of computer technology are great and can be understood in terms that are readily appreciated and valued by business executives.

Health: Improving the Quality of Life

The final perspective that we discuss, worker health (e.g., longevity, absence of stress and disease), is one that has been comparatively ignored in the cognitive engineering literature. However, its importance cannot be denied. People spend a substantial portion of their lives working (cf. Terkel, 1972), and the extent to which their workplace is designed to induce health has an impact on the quality of life as a whole, not just on the quality of working life (Reed, 1996). Although it is impossible to put a price on the health of a work force, employee health affects other factors that can be measured economically. One of these factors is the direct health care costs paid by employers. To give an idea of how substantial these costs are, consider the following facts:

- In the United States, 12% of company profits are spent on health care (Karasek & Theorell, 1990).
- In 1987, $360 billion were spent on third-party health premiums in the United States alone (U.S. Department of Health and Human Services, 1988).
- The largest supplier (in economic terms) for General Motors is not a steel supplier, as we might expect, but rather Blue Cross (Sloan, Gruman, & Allegrante, 1987)!

If we can reduce even a small portion of these costs by designing healthier computer-based work, then the savings are clearly enormous. Based on the data they had available, Karasek and Theorell estimated that up to $80 billion per year in direct health care costs may be preventable in the United States alone by designing healthier work.

Direct health care costs are not the only way in which worker health can impact corporate productivity, however. Two other important factors are absenteeism due to either illness or lack of motivation, and job turnover due to dissatisfaction. Finally, even when workers are on the job, they may not be as effective as they otherwise might be if work were designed in a healthier manner (see Fig. 1.12, later). This represents an untapped potential for improving productivity. Matteson and Ivancevich (1987) estimated that the preventable portion of the costs arising from these factors that can be attributed to job stress are $300 billion per year in the United States alone, not including the preventable direct health care costs mentioned earlier.

Although these estimates are subject to the typical limitations associated with quantifying the effects of such large-scale social factors (see Nickerson, 1992), there can be no doubt that the perspective of healthy work is inextricably tied to that of productivity. Investing in healthy work is good business sense, as some companies are beginning to find out. Healthier workers are generally more productive (Karasek & Theorell, 1990). Thus, ideally, every computer-based support system should be based on a work analysis framework that is explicitly geared toward improving worker health.

How can we accomplish this goal? To answer this question, we rely on the ground-breaking book by Karasek and Theorell (1990), which summarizes the findings of existing research on healthy work. Their book is important because it is an interdisciplinary effort, based on a systems perspective, to integrate a wide variety of findings on the epidemiology of work into a succinct theory that can be used to reorganize work, and therefore ameliorate some of the problems that currently threaten workers' health. Because this research is not widely known in the cognitive engineering literature, we again provide a detailed summary of the most salient and relevant findings.

Intuitively, we would think that health, or conversely, stress, is primarily a function of psychological demands. That is, workers whose jobs put greater demands on them should experience a greater level of stress that negatively impacts their health over the long run. Karasek and Theorell (1990) convincingly argued that this intuitive theory is not only incomplete, but, more strongly, does not represent the most important factor influencing health. In its place, Karasek and Theorell developed a *demand-control model,* illustrated in Fig. 1.8, that consists of two dimensions. The first is the level of *psychological demands* experienced by workers on the job. The

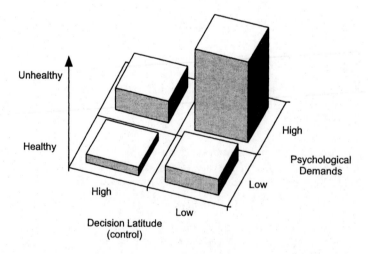

FIG. 1.8. Expected interaction between psychological demands and decision latitude on worker health, as predicted by Karasek and Theorell's (1990) demand-control model.

second is the degree of control or *decision latitude* that individual workers have over the way in which they can deal with their job demands. This second dimension is actually composed of two subfactors, autonomy and skill discretion. *Autonomy* refers to the extent to which workers are allowed to improvise or adapt in doing their job, whereas *skill discretion* refers to the extent to which workers are allowed to exploit their skills on the job. Skill allows workers to effectively cope with task demands in an autonomous manner, and providing autonomy allows workers to develop their skills. Conversely, unskilled workers will likely have difficulty effectively exercising autonomy, yet it is difficult to see how workers who lack autonomy can develop their skills (cf. Reed, 1996). As a result, autonomy and skill discretion are mutually reinforcing and essentially inseparable in practice. Thus, they are treated as a single dimension, decision latitude.

Karasek and Theorell's (1990) model predicts an interaction between psychological demands and decision latitude such that health is poorest when demands are high and latitude is low. Figure 1.9 provides data from one representative study evaluating the model. The dependent variable in this case was prevalence of heart disease. The results clearly support the model's predictions. Prevalence of heart disease is highest when workers experience a great deal of psychological demands, but have little control over how to do their job. The figure also shows another pattern that contradicts the intuitive model described earlier. Although psychological demand is a significant source of risk, it is not the primary source. Instead, lack of control over how workers do their job and how they can exploit their skills are the primary risk factors. This pattern can be seen by comparing the impact of increasing psychological demands for the high- and low-decision latitude groups. For workers who have high control over how to do their job, increasing psychological demands leads to only a marginal decrement in health. In contrast, for workers who have low control over how they do their job, increasing psychological demands leads to a very substantial

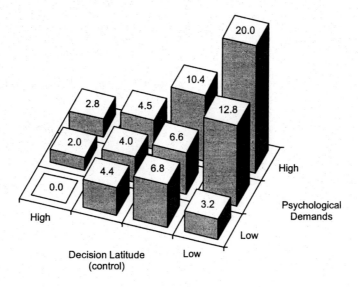

FIG. 1.9. Job characteristics and heart disease prevalence (Swedish males, 1974, N = 1,621). Number on vertical bar is percentage in each category with symptoms (Karasek & Theorell, 1990).

decrement in health. The primacy of decision latitude has been consistently observed in a number of other studies evaluating the demand-control model. Put succinctly, "it is not the bosses but the bossed who suffer most from job stress" (Karasek & Theorell, 1990, p. 16).

This pattern of findings has been replicated for a variety of dependent variables measuring worker health. For example, Fig. 1.10 shows the impact of psychological

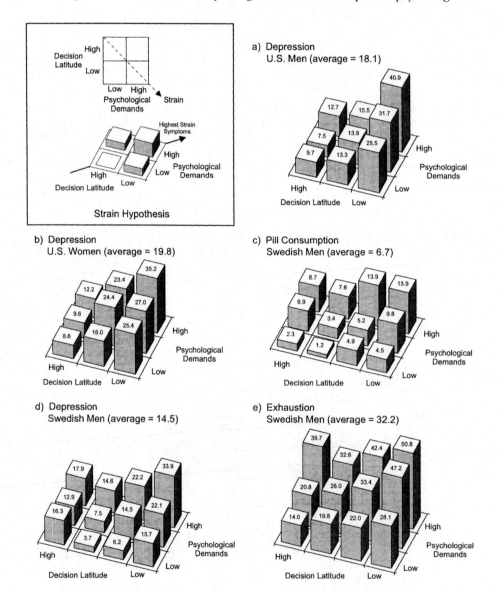

FIG. 1.10. Job characteristics and psychological strain: multiple symptom indicators. Number on vertical bar is percentage in each job category with symptoms (Karasek & Theorell, 1990).

demands and decision latitude on three dependent variables: depression, pill consumption, and exhaustion. Figure 1.11 shows the impact of demand and control on systolic blood pressure. In either figure, we see that it is the critical combination of low control and high demands that leads to poorest health. Additional empirical studies have confirmed this pattern for incidence of myocardial infarction (heart attacks) and cardiovascular deaths as well (Karasek & Theorell, 1990). It is difficult to conceive of a more significant set of dependent variables.

Figure 1.12 provides some insight into the nature of the relationship between demand and control and worker health. The dependent variable here is the level of skill underutilization (the number of years of formal education that workers have, subtracted by the number of years of education required by the job). A positive number represents a job that does not require workers to fully exercise their skills, whereas a negative number indicates a job that challenges workers to go beyond their formal education. The evidence in Fig. 1.12 illustrates that jobs with high control actually push people beyond the competencies they acquired through formal education. These jobs are making full use of workers' skills, and may even be encouraging learning as well. In contrast, jobs with low control, and especially those that have high demands as well, lead to the highest observed levels of skill underutilization. These jobs are therefore wasting the skills that workers already possess. These data are particularly distressing when we consider that the workers whose skills are being underutilized actually have fewer years of formal education than those workers whose skills are being fully utilized and challenged (usually managers and professionals). Despite the fact that their level of education is lower, the capabilities of the former group of workers are being wasted.

Research on the demand-control model has shown that the job characteristics have an impact outside of work as well. Figure 1.13 provides data on the effects of demands and control on patterns of involvement in leisure and political activity. The results show that workers who have low-control jobs have the lowest rates of

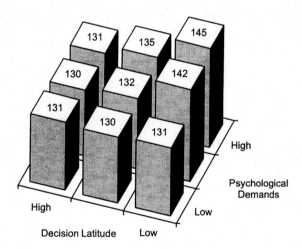

FIG. 1.11. Job characteristics and systolic blood pressure at work (Males in diverse occupations in New York City, N = 206). Number on vertical bar is blood pressure measured by ambulatory monitor every 15 minutes (Karasek & Theorell, 1990).

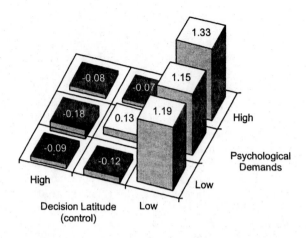

FIG. 1.12. Job characteristics and skill underutilization (U.S. males, 1972, 1977, N = 1,749). Number on vertical bar is underutilized skills in terms of years of education

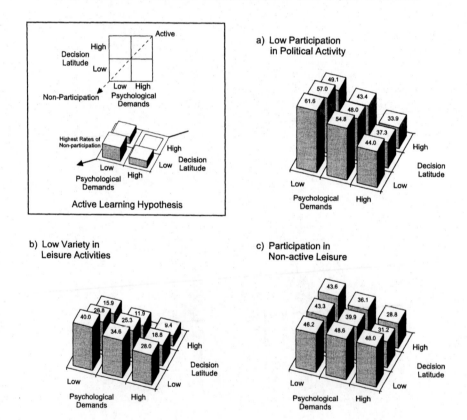

FIG. 1.13. Job characteristics, leisure, and political participation (Swedish males, 1968, N = 1,466). Number on vertical bar is percentage in each job category with low active participation (Karasek & Theorell, 1990).

participation, whereas those workers who have high control-high demand jobs have the highest. This result suggests that job characteristics have an important impact on the extent to which we are capable or willing to enjoy life and to contribute to society, thereby illustrating the pervasive repercussions of healthy—or better, unhealthy—work on society as a whole (cf. Reed, 1996).

The findings just presented take on an even more impressive status when we consider the following methodological points. First, the studies that support the demand-control model are large-scale efforts, with sample sizes usually over 1,000 and in one case comprising one fourth of the working population of Sweden (a sample size of 958,096). These large samples lend further credence to the reliability and generalizability of the findings obtained. Second, much of this research made an explicit attempt to control for other conventional risk factors, such as age, smoking patterns, weight, education, and cholesterol level. These other factors played a smaller role in predicting health than did the dimensions specified by the demand-control model (Karasek & Theorell, 1990). Thus, psychological demands, and especially decision latitude, play a unique and significant role in determining worker health.

This body of research is of great importance for computer-based work, the focus of our book, because research has shown that information technology acts as an amplifier of the effects represented by the demand-control model (Karasek & Theorell, 1990). That is, computerization combined with high psychological demands and low decision latitude leads to even more negative effects on worker health. Therefore, the body of research reviewed by Karasek and Theorell is of great value because it adds another constraint on a framework for analyzing human work. From the viewpoint of health, a work analysis methodology must strive to improve decision latitude by providing workers with the autonomy to make decisions and the opportunity to exercise and develop skill. The quantitative data that we presented earlier in this subsection indicate that enormous economic gains may be realized by adopting such an approach. Even more important, however, explicitly designing for healthy work will lead to benefits that cannot be quantified, such as humane jobs that allow people to participate fully in life and society.

Summary

What implications can we draw from this section? If a framework for work analysis is to lead to effective computer-based work, it must strive to: (a) support workers in adapting to, and coping with, events that are unfamiliar to them and that have not been anticipated by system designers, (b) identify the functionality that is required to accomplish intellectual tasks requiring discretionary decision making (i.e., usefulness issues), (c) be based on an understanding of human capabilities and limitations (i.e., usability issues), and (d) improve decision latitude by providing workers with the autonomy to make decisions and the opportunity to exercise and develop skill. The evidence reviewed previously suggests that a work analysis framework that satisfies these constraints will lead to the design of computer-based work that is safe, productive, and healthy. In the remainder of this book, we argue that this three-way objective can be achieved by deliberately and systematically designing information systems that support worker adaptation.

OUTLINE

The Rest of Part I—Introduction

Chapter 2 completes our introduction by providing the motivation for conducting work analysis. In particular, it shows the importance of explicitly studying work before designing a computer-based information system by illustrating the perils of not doing so.

Part II—Three Approaches to Work Analysis

Part II of the book puts our ideas in a broad intellectual context. Chapter 3 examines normative approaches to work analysis, particularly traditional task analysis. We argue that such techniques are of limited value for complex sociotechnical systems, and that more attention needs to be given to an analysis of the system being controlled. Chapter 4 examines descriptive approaches to work analysis, particularly naturalistic field studies of work in situ. We argue that such techniques are very valuable for collecting data to inform a work analysis, but are insufficient for complex sociotechnical systems. Descriptive approaches need to be complemented by work models if we are going to design systematically novel information systems that make possible new and effective ways of performing work. Chapter 5 describes a formative approach to work analysis that is well suited to the relatively unique demands of complex sociotechnical systems. The particular formative work analysis framework described in this book is based on the notion of *behavior-shaping constraints* (Rasmussen et al., 1994). As shown in Fig. 1.14, the goal is to identify the various categories of factors that shape human behavior in sociotechnical systems. The basic idea is that the behavior exhibited by workers over time is generated by, or emerges from, a confluence of behavior-shaping constraints that specify the dimensions that must be incorporated into a framework for work analysis. These dimensions can then be used to identify the requirements that a design must satisfy if it is to support work effectively.

Part III—Cognitive Work Analysis in Action

Part III describes in much more detail the framework for CWA, both conceptually and by example. Chapter 6 introduces the case study that we use as an example throughout most of Part III. We deliberately chose the case study from the "technical" domain of process control, thereby complementing the more "humanistic" case study of library information retrieval presented in Rasmussen et al. (1994). By choosing a different type of case, we hope to illustrate the breadth of the concepts described in the book.

Chapters 7–11 each discuss one of the five dimensions of work analysis corresponding to the five behavior-shaping factors illustrated in Fig. 1.14. For each dimension, a conceptual tool is presented for understanding and addressing that particular behavior-shaping factor. In addition, each dimension of analysis is illustrated by application examples, usually from the aforementioned case study.

To fully appreciate this book, it is important to emphasize that the five layers of behavior-shaping constraints are conceptual distinctions to be respected (cf. Hanson, 1958), not methodological steps to be followed in sequence. Any analysis framework

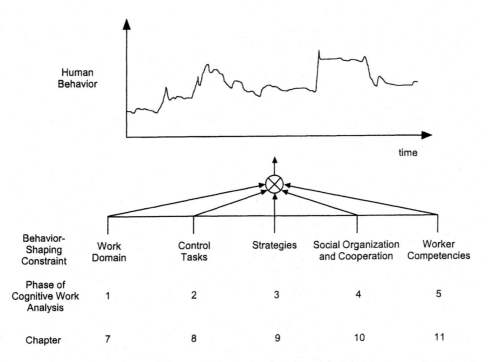

Behavior-Shaping Constraint	Work Domain	Control Tasks	Strategies	Social Organization and Cooperation	Worker Competencies
Phase of Cognitive Work Analysis	1	2	3	4	5
Chapter	7	8	9	10	11

FIG. 1.14. Overview of framework for cognitive work analysis based on layers of constraint that shape human behavior.

can be described from two complementary perspectives. One describes the types of concepts to be used for modeling and the types of representations that should be developed by the time the analysis is complete. The other describes the process and the data collection methods that an analyst should adopt in eliciting the knowledge required to perform the analysis. The former can be thought of as a *conceptual* perspective, whereas the latter can be thought of as a *methodological* perspective. If the analysis framework is to have a systematic basis, the conceptual perspective should be logical and rational (i.e., very orderly and coherent) because the concepts that are proposed for analysis should be clearly defined, internally consistent, and well integrated (cf. Beyer & Holtzblatt, 1998). In contrast, the methodological perspective is much less orderly because the methods that are chosen are eclectic and the activities that analysts actually engage in while conducting an analysis are very opportunistic, chaotic, nonlinear, and iterative (cf. Card, 1996). Although there is some structure in the methodological perspective, it would be very difficult to write a comprehensible book from this perspective without rationalizing the disorderly process actually used to conduct an analysis in practice. Thus, we have chosen to write the book from a conceptual perspective, our goal being to present certain conceptual distinctions in an orderly and understandable fashion.

You should keep in mind that the fact that the concepts in the framework are well defined and highly structured does not imply that the methodology used to actually apply the framework in practice is also orderly. It is not. Therefore, as you read the book, you should periodically return to the distinction between these two perspectives, conceptual and methodological, as this will help you understand our presentation of the framework for CWA.

Chapter 12 further illustrates CWA in action by examining its implications for both design and research. For design, the goal is to tailor the information technology to reflect, and thereby support, the characteristics of work, as identified by the work analysis. For research, the goal is to identify or manipulate the factors that can shape the behavior of research participants. We present examples that illustrate how the particular framework described in this book can be used to inform the design of a computer-based information system and to conduct research on basic issues pertinent to cognitive engineering.

Part IV—Final Words

Part IV wraps up the book by addressing some broader issues. Chapter 13 revisits the three criteria for effectiveness defined in this first chapter (see earlier discussion). We argue that the characteristics of the framework for CWA described in this book should lead to computer-based work that improves safety, productivity, and health. Furthermore, we review arguments to support the claim that the issues addressed by this framework, and thus its potential value, will only become increasingly important in the future.

Finally, the Appendix is a brief historical addendum that provides an overview of the cognitive engineering research conducted in the Electronics Department at Risø National Laboratory in Roskilde, Denmark, since the 1960s. This historical overview is important because it describes the intellectual lineage of the concepts described in this book, thereby showing how and why they were developed.

Why Work Analysis? An Ecological Perspective 2

Technology can be like a jalapeño pepper in a French sauce.
   —Negroponte (1995, p. 223)

A systematic understanding of how environments affect people can be used deliberately to create environments to suit human adaptation.
   —Carroll (1991, p. 14)

PURPOSE

In this chapter, we provide the motivation for conducting a work analysis (the argument for the specific framework that we advocate in this book is presented in chap. 5). Our objective is to convince you of two points. First, we argue that human work should be explicitly studied as part of designing a computer-based information system. We make this point with the help of several examples that illustrate the perils of not following this recommendation. Second, we argue that work analysis should be based on an ecological approach—that it should begin with, and give primary importance to, the constraints that the environment imposes on workers' actions. We make this point by discussing the limitations of the more common cognitivist approach to work analysis that begins with, and gives primary importance to, the constraints imposed by worker characteristics.

WHY CONDUCT A WORK ANALYSIS?

One of the features that distinguishes cognitive engineering from some approaches to HCI, human factors, and psychology is the great emphasis placed on studying the semantics of the work domain (Rasmussen & Goodstein, 1988; Woods & Roth, 1988b). Our claim is that a thorough work analysis of some type must be conducted if computer-based systems are to effectively support human work in complex sociotechnical systems. In this section, we argue indirectly for the value of work analysis by presenting two examples. The first example is a set of field studies of computer-based technology in hospitals in Denmark. The second example is a pair of questionnaire studies investigating novice graphic designers interacting with pencil and paper and with desktop-publishing software.

Example 1: Hospitals

The first example is taken from a research program conducted by Hovde (1990) consisting of a number of field studies examining the structure of work in various hospitals in Denmark. Some hospitals already had experience with computer-based systems to support work, whereas other hospitals had only just installed such systems. Others still did not have computer-based information systems but expressed some interest in acquiring them.

Hovde (personal communication, 1987) made several observations about the differential success of the various computer-based information systems that had been installed in the hospitals he studied. Most of these hospitals had purchased generic, "off-the-shelf" or "turn-key" application packages (e.g., database programs, scheduling programs) that purportedly would help any hospital improve the productivity of its operations. Hovde found that the hospitals that had purchased such applications were not very happy with them. In some cases, nurses and doctors even bypassed the information systems because they were so cumbersome to use. In other cases, upper management enforced the usage of the information systems, seemingly because so much money was invested in them. As a result, nurses and doctors used the information systems but only reluctantly.

One of the hospitals that Hovde investigated was unique in several respects. First, management did not have to enforce use of the information system; the nurses, doctors, and administrators used it willingly. Second, if we trust the comments obtained from workers, the computer system worked well. Third, the process that was used to develop the information system was completely different from that observed in all of the other cases. Rather than purchasing a generic, off-the-shelf application as other hospitals had done, this hospital had contracted a computer programmer (who happened to be the spouse of one of the nurses who worked at the hospital) to do the job. The specifications for the information system were developed by the programmer in conjunction with the workers who would eventually be using the information system.

An interesting phenomenon occurred during the programming phase. Being trained in logical thinking, the programmer reflected on the functional specifications as he was implementing them and came up with "more rational" ways of defining the functionality of the information system and the structure of the work flow. When he voiced his "improvements" to the future users, his ideas were almost always dismissed. Although the changes seemed logical to the programmer, they were not based on a thorough understanding of the hospital's work demands. However, this was only obvious to the workers because they were the ones who had a deep appreciation for the factors governing the context in which the information system was to be used. The programmer did as he was told, and implemented the functional specifications developed by the workers, rather than changing them to fit his own criteria.[1] As mentioned earlier, the result was an information system that was used and that seemed to achieve its goals in an effective manner.

Because of the uncontrolled nature of these field studies, we cannot be sure what caused this particular hospital's implementation to be better accepted and utilized than the others that were observed. With that caveat in mind, there is one possible explanation that highlights the importance of work analysis. The generic software packages were not designed to fit any particular hospital. In fact, in some cases, they were not even designed for hospital applications at all, but instead were intended to be computer-based tools that could be used in many application domains. As a result, they were not tailored to support the work of the hospitals in which they

---

[1]We do not mean to suggest that workers' preferences or current ways of working should provide the sole basis for systems design (see chap. 4).

a) Case where the information system is not preceded by a cognitive work analysis. There are gaps between what needs to be done and the support offered by the system.

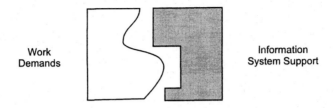

Work
Demands

Information
System Support

b) Case where the information system is tailored to the work demands based on a cognitive work analysis.

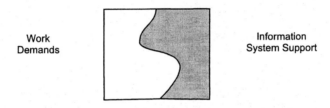

Work
Demands

Information
System Support

FIG. 2.1. Relation between work demands and information system support when design is (a) not preceded by a work analysis, and (b) informed by a work analysis.

wound up being used. As shown in Fig. 2.1a, one can think of this situation as a degree of fit between two pieces of a puzzle.[2] Because one piece was not designed with the other in mind, the fit is only approximate. It is up to the workers to use their own ingenuity to make up the remaining gap, but this takes time, effort, and knowledge. And in some cases, workers do not bother to make the extra effort required to make up this gap.

It is important to note that the pattern of behavior observed by Hovde has been frequently observed in the health care industry: "When the implementation of information systems interferes with traditional practice routines, they are not likely to be accepted by physicians" (Anderson, 1997, p. 84). For example, Anderson briefly described a case where a medical information system was introduced into a university medical center, and then was strongly opposed by physicians. Apparently, this lack of acceptance was due to a gap like the one shown in Fig. 2.1a. The new information system "lacked medical staff sponsorship, altered traditional work flow patterns, changed the relations among professional groups, and adversely affected the medical education program" (Anderson, 1997, p. 87). It is no wonder that the system was

---

[2]If you are familiar with Norman (1986), you may think that Fig. 2.1 is a diluted reincarnation of Norman's Gulfs of Execution and Evaluation (see his Figs. 3.1 and 3.2). Despite the apparent similarity, there is an important difference. Norman represented the gap between a hypothetical or prototypical workers' goals and the capabilities of a physical system (i.e., computer interface). In contrast, Fig. 2.1 shows the gap between the work demands and the support offered by a hypothetical computer-based information system. The implications of this difference should become apparent in the second half of this chapter.

received so negatively! Note that lack of use and acceptance can lead to substantial economic losses, because: "In some cases, lack of acceptance of a clinical information system has led to its discontinuation after implementation" (Anderson, 1997, p. 86).

The "home-made" application observed by Hovde provides a stark contrast to such failures, probably because it was tailored to the context in which it was going to be used. Thus, it took advantage of many of the constraints that governed the flow and structure of work in that hospital. As in Fig. 2.1b, one piece of the puzzle was specifically designed to fit the other. As a result, the fit seemed to be a very good one. The workers stated that the information system effectively supported them in their work activities. But of course, this was not an accident. A good fit can be achieved only if we systematically analyze the characteristics of the work setting. Only then can we design an information system that is effectively tailored to the demands imposed by the work (Gibbs, 1997).

Although it is consistent with the general lessons learned from experience with information systems in health care (Anderson, 1997), the evidence from Hovde's field studies in favor of work analysis is anecdotal in nature. Thus, we cannot put too much confidence in our conclusions, despite their intuitive appeal and apparent generality. The second example, presented next, goes part way toward addressing this limitation by discussing the results from an experiment evaluating desktop-publishing software for graphic designers.

Example 2: Graphic Design

Black (1990) conducted a study of desktop publishing investigating the impact of two different media (paper and pencil vs. software) on workers' activities and on product quality. This study indirectly shows the importance of work analysis by illustrating the poor fit that can exist between work demands and the support offered by an information system when that system is not based on a work analysis.

Although the task given to workers was held constant, Black (1990) found that the two working media (pencil and paper vs. computer) systematically shaped workers' activities in significant ways. For example, with pencil and paper, the interaction was direct; the marks that could be drawn were limited only by the worker's imagination and manual dexterity, neither of which is a problem for graphic designers. In contrast, with the computer-based system, the workers' actions were strongly constrained by the commands available in the software, frequently in unnatural and dysfunctional ways. Thus, there was a lack of fit between the demands of the job and the support provided by the information system, just as in Fig. 2.1a. As Black put it, with the computer medium "designers are one step removed from the marks they are making" (p. 284).

To overcome this gap between work demands and computer support, workers would have to satisfy their intentions in a less direct fashion. Typically, this would involve concatenating together a set of commands in an improvised fashion because there was no command available for the mark workers wished to draw. This type of "workaround" activity is probably familiar to anyone who has used a computer drawing program. Figure 2.2 illustrates an example, adapted from S. J. Payne (1991), that shows how cumbersome these workaround activities can be with some programs. The most direct and intuitive route to create the figure to be drawn (a bisected circle with the left half filled) would be to use the three steps shown on the left of Fig. 2.2:

Figure to be Drawn:

Intuitive Path to Draw Figure: **Actual Path Required to Draw Figure:**

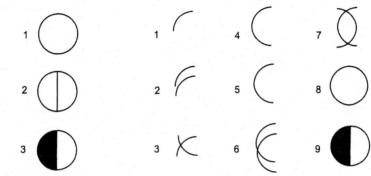

FIG. 2.2. Example showing how drawing software can cause users to engage in cumbersome work-arounds. See text for detailed description.

1. Create a circle.
2. Add a vertical bisecting line.
3. Fill the left half of the circle.

With some programs, however, enclosed spaces cannot be filled because fill-patterns are only attributes of objects, such as the entire circle. As a result, workers would have to follow the set of steps shown on the right of Fig. 2.2:

1. Draw an arc of 90° to create a quarter circle.
2. Make a duplicate of the arc.
3. Invert the copy vertically.
4. Align the copy with the original to create a half circle.
5. Group the copy and the original to create a single object.
6. Duplicate the new object.
7. Invert the copy horizontally.
8. Align the copy with the original to create a full circle.
9. Fill in the left half-circle object.

Clearly, the actual path required is much longer and far less intuitive. Whereas some existing programs support the direct path, some require an indirect path like the one just described. Although the details differ, the program examined by Black (1990) also requires workers to engage in analogous awkward and inefficient workarounds.

Another difference between pencil and paper and computer was observed for the tactics that graphic designers used early in the design process. Usually, graphic designers develop a range of design concepts and create a draft layout for each concept. Then, they rely on a visible comparison of these potential solutions to choose the concept that they are going to flesh out in more detail to create the final design. This tactic is most efficiently accomplished by viewing the initial drafts in parallel before going on with the detailed design process. But unfortunately, the software examined by Black (1990) did not directly support this parallel comparison of multiple alternatives. Only one design could be viewed at a time. Again, we have a situation, like that depicted in Fig. 2.1a, with a mismatch between the support provided by the information system and the demands imposed by the work.

Of course, there is no reason why the worker could not create several files, each with a different alternative for the same design problem. These could then be brought up individually, printed out, and then compared in parallel in paper form. By engaging in these extra steps, workers could use the computer program and still follow the design process they would use with pencil and paper. But the problem is that workers must take the initiative and devote the time and effort that is required to follow their preferred process. In this case, the computer program is an obstacle that must be overcome with improvisations and contortions. Rather than getting direct support for their work activities from the information system, workers must fight and work around the system to get the job done in the desired manner.

A third difference between pencil and paper and the computer medium is the degree to which the appearance of a draft is informative about the stage of progress. As shown in Fig. 2.3, the pencil-and-paper medium provides graphic designers with visible cues about the stage of development of the draft. These cues allow workers to focus their attention on what remains to be done. Referring to Fig. 2.3, some areas clearly have not been developed very much at all (e.g., bottom left of the page). Other areas have a rough layout specified in terms of number of columns (e.g., top right of the page). Still other areas have received more attention as evidenced by the fact that the precise content has already been specified (e.g., "National and International Implications"). Moreover, at a global level, it is clear that the document is just an early draft. Nobody would ever confuse it for the final product.

The way in which visible cues provide information about the stage of development is more directly illustrated in Fig. 2.4, which illustrates four drafts produced sequentially during the design of a particular document. Even to the untrained eye, it is obvious at a glance that each successive draft is more refined than the previous one. Compare this with the three designs illustrated in Fig. 2.5 produced with the computer-based medium. It is probably not at all apparent that these are all early drafts. Although each design has a finished appearance, the design decisions are only provisional and the text has not been properly edited. Thus, with the computer medium, the information that is available on pencil and paper about the stage of development of the design is absent. This provides yet another example of how the work demands (in this case, the requirement to keep track of the stage of development of the design) are not very well supported by the information system, creating a gap like that shown in Fig. 2.1a.

As with the previous example, this does not necessarily mean that graphic designers must change their tactics. The difference with the computer medium is that, rather than being able to rely on visible cues to focus their attention on what needs to be done, graphic designers must instead remember the decisions that are provisional and the parts of the document that need further work. Again, the lack of fit between the information system and the work demands requires people to engage in effortful and time-consuming activities if they are to do the job in the desired manner.

A final example of the way in which the medium affects the ease with which work is accomplished is the use of successive drafts as an informal history of design decisions.

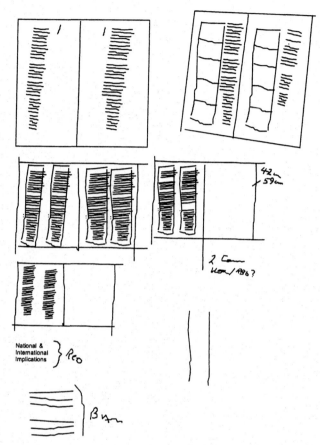

FIG. 2.3. Part of a series of rough sketches produced during the early stages of designing a leaflet. Reprinted from Black, A. (1990). Visible planning on paper and on screen: The impact of working medium on decision-making by novice graphic designers. *Behaviour and Information Technology, 9*, 283–296, published by Taylor & Francis. Copyright © 1990 Taylor & Francis Ltd. Reprinted by permission.

With pencil and paper, graphic designers would keep each draft as a record of the development of the ideas that they had generated during the course of the development of a particular document. The collection of the drafts for a particular design thereby represented a type of "design log" that could be referred to later to recall why particular decisions were made, or what alternatives had been considered but were discarded. Thus, a collection of drafts represents a valuable resource for graphic designers. Unfortunately, the creators of the software did not build in a way to easily and reliably keep a design log of previous drafts for a particular document. So yet again, we have an instance of the situation depicted in Fig. 2.1a where the computer medium was not designed to support the work demands.

As with the case of the parallel initial drafts described earlier, it is possible for graphic designers to circumvent this lack of functionality. In this particular case, this would involve remembering to save successive drafts and label them in such a way that the design problem and the version number were both easy to determine. Doing so would allow these files to be retrieved later for the purposes already described. But note, once again, that this requires some time and effort on the part of the worker. Furthermore, there is a possibility for error (e.g., if the worker accidentally overwrites an earlier draft).

In conclusion, our comparison of graphic design layout with pencil and paper versus computer repeatedly emphasizes several points. First, the work demands associated with graphic design were not supported by the information system (see Fig. 2.1a). It is very important to note that, in every instance, this lack of functionality was not due to a technological impediment. To take but one example, there is no reason why the computer software could not provide a facility for reliably and easily keeping successive drafts for subsequent efficient retrieval. There must be another reason why the information system did not support the work demands associated with graphic design. Second, because of this lack of support, workers would have to engage in workarounds in order to make up the distance between the work demands and the functionality offered by the computer system (cf. Fig. 2.1a). These workarounds, summarized in Table 2.1, have several disadvantages: They take time, they require some effort, and they are unreliable in the sense that they are error-prone. As a result, rather than helping people do their job more efficiently, the computer system sometimes became an obstacle that workers would have to struggle with to get the job done. It is important to point out that these findings are not tied to the specifics of Black's (1990) study. Instead, they seem to be generalizable across a

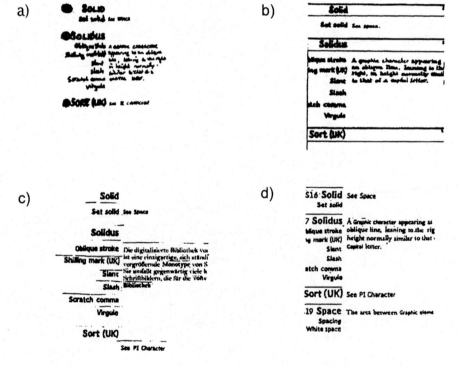

FIG. 2.4. Sections of four draft pages produced sequentially during the development of ideas for a bibliography. Reprinted from Black, A. (1990). Visible planning on paper and on screen: The impact of working medium on decision-making by novice graphic designers. *Behaviour and Information Technology, 9,* 283–296, published by Taylor & Francis. Copyright © 1990 Taylor & Francis Ltd. Reprinted by permission.

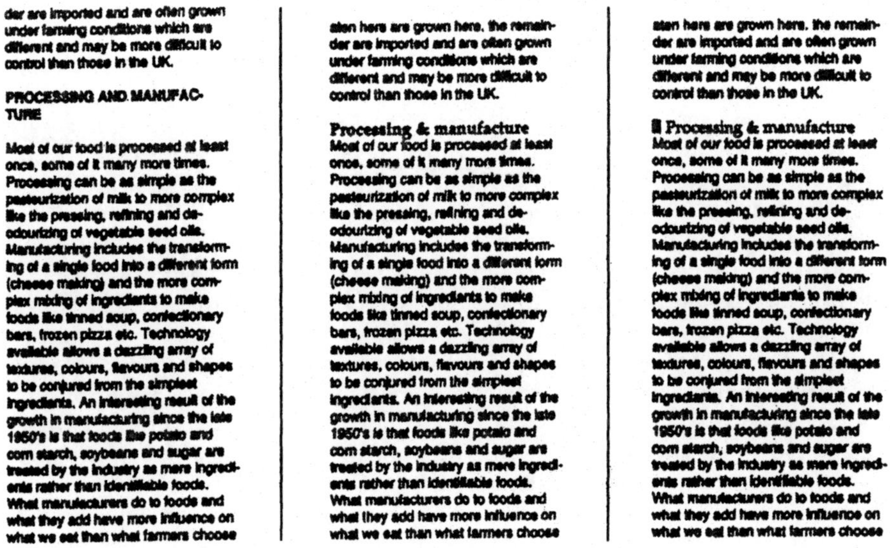

der are imported and are often grown under farming conditions which are different and may be more difficult to control than those in the UK.

PROCESSING AND MANUFAC-TURE

Most of our food is processed at least once, some of it many more times. Processing can be as simple as the pasteurization of milk to more complex like the pressing, refining and de-odourizing of vegetable seed oils. Manufacturing includes the transforming of a single food into a different form (cheese making) and the more complex mixing of ingredients to make foods like tinned soup, confectionary bars, frozen pizza etc. Technology available allows a dazzling array of textures, colours, flavours and shapes to be conjured from the simplest ingredients. An interesting result of the growth in manufacturing since the late 1950's is that foods like potato and corn starch, soybeans and sugar are treated by the industry as mere ingredients rather than identifiable foods. What manufacturers do to foods and what they add have more influence on what we eat than what farmers choose

aten here are grown here, the remainder are imported and are often grown under farming conditions which are different and may be more difficult to control than those in the UK.

Processing & manufacture
Most of our food is processed at least once, some of it many more times. Processing can be as simple as the pasteurization of milk to more complex like the pressing, refining and de-odourizing of vegetable seed oils. Manufacturing includes the transforming of a single food into a different form (cheese making) and the more complex mixing of ingredients to make foods like tinned soup, confectionary bars, frozen pizza etc. Technology available allows a dazzling array of textures, colours, flavours and shapes to be conjured from the simplest ingredients. An interesting result of the growth in manufacturing since the late 1950's is that foods like potato and corn starch, soybeans and sugar are treated by the industry as mere ingredients rather than identifiable foods. What manufacturers do to foods and what they add have more influence on what we eat than what farmers choose

aten here are grown here, the remainder are imported and are often grown under farming conditions which are different and may be more difficult to control than those in the UK.

■ **Processing & manufacture**
Most of our food is processed at least once, some of it many more times. Processing can be as simple as the pasteurization of milk to more complex like the pressing, refining and de-odourizing of vegetable seed oils. Manufacturing includes the transforming of a single food into a different form (cheese making) and the more complex mixing of ingredients to make foods like tinned soup, confectionary bars, frozen pizza etc. Technology available allows a dazzling array of textures, colours, flavours and shapes to be conjured from the simplest ingredients. An interesting result of the growth in manufacturing since the late 1950's is that foods like potato and corn starch, soybeans and sugar are treated by the industry as mere ingredients rather than identifiable foods. What manufacturers do to foods and what they add have more influence on what we eat than what farmers choose

(a) (b) (c)

FIG. 2.5. Three early drafts produced on screen. Although the drafts have a finished appearance, the text has not been properly edited and the design decisions are provisional. Reprinted from Black, A. (1990). Visible planning on paper and on screen: The impact of working medium on decision-making by novice graphic designers. *Behaviour and Information Technology, 9*, 283–296, published by Taylor & Francis. Copyright © 1990 Taylor & Francis Ltd. Reprinted by permission.

TABLE 2.1 A Summary of the Workarounds That Are Required to Deal With the Work Demands That Are Not Explicitly Supported by the Information System for Graphic Design Analyzed by Black (1990)

Work Demand	Required Workaround
Create desired marks on page	An awkward and inefficient set of commands
Compare alternative design proposals in parallel	Create separate file for each alternative design proposal; serially bring up each file; print out each file; compare all of them in paper form
Focus attention on the parts of the document that have not been completely designed yet	Remember which decisions are provisional and which parts of the document need further work
Maintain a log of previous drafts to serve as a design history	Remember to save each draft in a separate file; label the files so that the design problem and version number are easy to determine

very wide variety of application demands in which workers have to interact with computer systems that have not been tailored to the demands of the job (e.g., Bisantz, Cohen, Gravelle, & Wilson, 1996; R. I. Cook & Woods, 1996; Sachs, 1995; Vicente, Burns, Mumaw, & Roth, 1996).

An obvious question to ask is: Did workers bother to engage in these workaround activities or did they instead let the computer medium dictate, or at least shape, the practices they adopted to get the job done? Black (1990) provided some insight into this issue as well, describing the results of two questionnaire studies of the work practices used by novice graphic designers. In the first study, half of the novices worked on paper first and then moved to the computer medium, whereas the remaining students progressed in the opposite order. The novices completed the same questionnaire twice, once right after they had finished their work, and then again after they had received feedback on their designs from an experienced tutor.

Before being evaluated, both groups preferred working with the computer medium because they could create drafts that looked finished with less investment in manual labor. However, after they received technical feedback on their designs from the tutor, both groups valued working with pencil and paper more than with the computer. Black (1990) pointed out that students were initially more influenced by the ease with which they could produce finished-looking designs, which explains why the computer medium was preferred. However, with this software at least, doing what is easy does not lead to "doing the right thing." It was only when the novices received feedback about substantive structural issues that are critical to good designs that they learned this lesson. As a result, the second time they filled out the questionnaire, they paid more attention to these structural issues and their preference for the pencil-and-paper medium increased accordingly.

The second questionnaire study conducted by Black (1990) investigated how slightly more experienced novices divided their efforts between computer-based and pencil-and-paper media while solving a single design problem. The research goal was to determine if the perceived quality of the designs was influenced by the choice to work more with one medium than the other. Some novices spent much more time working with the computer medium than with pencil and paper. Interestingly, the constraints of the two media seemed to have a substantial impact on quality. More specifically, "students working predominantly on screen appeared to have compromised their initial ideas more and to have made less effort to push forward their original plans by learning about the software they were using" (Black, 1990, p. 294). This conclusion has tremendous implications, and generalizes well beyond the program that Black investigated (Norman, 1993). It suggests that, although it may be possible to overcome the lack of fit between work demands and information system support through workarounds, novices at least tend to be "sucked in" by the constraints of the information system. Because they have not yet acquired fixed working habits, novices may adopt practices that are easy rather than those that lead to good outcomes. And if the information system is not based on a sound work analysis, then what is easy is usually not what is good. As a result, quality suffers because of the lack of fit between work demands and the computer system (see Fig. 2.1a).

Interestingly, it is possible to turn around these findings in a more optimistic direction (Carroll, 1991). If novice workers tend to do what is easy, this opens up the possibility of designing the information system so that what is easy is also functional, rather than dysfunctional as in the case of the software investigated by

Black (1990). But to do this, software designers need to explicitly investigate the characteristics of the work they are designing for so that they can develop computer tools to support the end users of their products.

Summary

Hovde 's (1990) field study of information systems in hospitals, and Black's (1990) study of desktop-publishing software both show the perils of ignoring work analysis. The lesson is that a work analysis must be conducted if there is to be a good fit between the support provided by the information system and the work demands. Only then can we achieve the situation depicted in Fig. 2.1b rather than that depicted in Fig. 2.1a.

WHERE SHOULD THE WORK ANALYSIS BEGIN?

Work Demands = Cognitive Constraints + Environment Constraints

Although work analysis is not appreciated or practiced by many information system designers, it is readily accepted in the behavioral design disciplines, such as human factors and HCI. Nevertheless, we show in this section that there is a lack of consensus on where the work analysis should begin. At the risk of greatly oversimplifying matters, we can divide the concept of work demands discussed in the previous section into two subsets, cognitive constraints and environment constraints.[3] *Cognitive constraints* are work demands that originate with the human cognitive system. For example, workers' subjective preferences and current mental models place cognitive constraints on work. In contrast, *environment constraints* are work demands that originate with the context in which workers are situated. For example, the physical and social reality that serve as the context for workers' behaviors are environmental constraints because they exist independently of what any one worker might think.[4]

The concept of environment constraints may be less familiar, so we give several examples. For instance, engineers working on a collaborative design project must take into account the intentions and actions of other engineers, otherwise their efforts will not be coordinated and the project goals will be compromised. Similarly, airline pilots must take into account the positions of other aircraft and the terrain if they are to achieve their goals. Also, the decisions made by a financial analyst must take

---

[3]The dualistic distinction between cognitive and environment constraints is useful for the pedagogical purposes of this chapter. However, on close theoretical scrutiny, the distinction does not hold up because it is impossible to meaningfully consider actors and environments separately from each other (Dewey & Bentley, 1949). We address this conceptual problem in detail in subsequent chapters by showing how cognitive and environmental constraints can be seamlessly integrated into a unified framework for CWA.

[4]Note that the environment to which we are referring is different from that described in chapter 1 and depicted in Fig. 1.1. There, we described the environment for a sociotechnical system (i.e., the forces outside of that system). Here, we are describing the environment for a worker (i.e., the forces outside of that worker). The difference between these two frames of reference can be implicitly seen by comparing Fig. 1.1 with Fig. 2.6. The environment depicted in Fig. 2.6 includes all of the factors outside of the workers subsystem in Fig. 1.1.

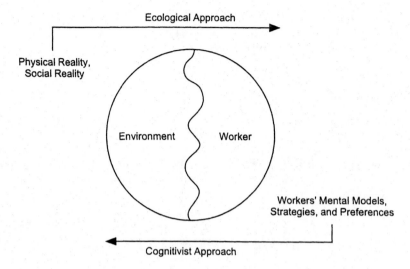

FIG. 2.6. The ecological and cognitivist approaches to work analysis.

into account the actual behavior of the stock market, otherwise the analyst's goals cannot possibly be consistently achieved. These factors are all examples of environment constraints because they must be respected by any goal-directed actor, human or otherwise, working in the respective application domain.

Because work demands are usually composed of both cognitive and environment constraints, there can be little argument that work analysis should include an investigation of both types of constraints. To overlook either would be a mistake because it would lead to a gap between work demands and information system support (like that shown in Fig. 2.1a). So why is there a lack of consensus then? The dilemma is in deciding which type of constraints should be given most importance. Should a work analysis begin with, and be primarily driven by, cognitive constraints or environment constraints? As illustrated in Fig. 2.6, two diametrically opposed approaches have been proposed to answer this question.

Ecological Approach or Cognitive Approach?

The approach we advocate is that work analysis should begin with, and give primary importance to, the constraints that the environment imposes on workers' actions. The viewpoint is based on an *ecological* perspective to human factors (Flach, Hancock, Caird, & Vicente, 1995; Hancock, Flach, Caird, & Vicente, 1995) because it gives precedence to the constraints that the work ecology imposes on goal-directed behavior (see Fig. 2.6).[5] Because this perspective provides the foundation for the framework described in this book, it is important that we make the case in favor of the primacy of environment constraints clear at the outset. Although we have

_____

[5]The ecological approach to human factors has been deeply influenced by, but is not the same as, the ecological approach to psychology (Brunswik, 1956; E. J. Gibson, 1991; J. J. Gibson, 1979). The latter has been primarily concerned with basic theories of judgment and decision making (in the Brunswikian case) or perception and action (in the Gibsonian case). The ecological approach to human factors is broader in scope and more applied in its orientation.

presented some of these arguments before (e.g., Vicente, 1990a), the importance of starting and driving work analysis from environment constraints is still not taken as seriously or as clearly as it needs to be.

One of the primary reasons for this lack of appreciation is that the dominant viewpoint in psychology, human factors, and HCI has been based on a *cognitivist* (or information-processing) perspective (e.g., Wickens, 1992) that gives primary importance to the constraints that the human cognitive system imposes on goal-directed behavior (cf. Bannon & Bødker, 1991; Flach, 1990b). From this viewpoint, the work analysis should begin with and be driven by, not an analysis of environment constraints, but an analysis of workers' cognitive characteristics (see Fig. 2.6). An example of the cognitivist viewpoint is the widely held belief that it is always important to identify a worker's mental model of the work domain and then design a human-computer interface to be compatible with that model. This specific design principle is used in the remainder of this section as an exemplar of the cognitivist perspective, thereby providing us with a focus to show some of the differences between the ecological and cognitivist approaches to work analysis. Again, it is important to emphasize that what is under discussion is not whether we should analyze only environment constraints or only cognitive constraints. Both are crucial. As illustrated in Fig. 2.6, the point of debate is which type of constraint should be the starting point and driver for work analysis.

Before comparing these two approaches, we must introduce the distinction between correspondence-driven work domains and coherence-driven work domains (Vicente, 1990a). As shown in Fig. 2.7, *correspondence-driven* work domains impose dynamic, environmental constraints on the goal-directed behavior of workers. Comparing Figs. 2.6 and 2.7, we see that correspondence-driven domains have a physical or social reality that serves as a context for worker behavior. The examples of environment constraints given earlier (i.e., engineering design, commercial aviation, finance) all come from correspondence-driven work domains precisely because such domains impose environment constraints on workers' actions. As shown in Fig. 2.8, *coherence-driven* work domains do not have any dynamic, environmental constraints that must be respected. Comparing Figs. 2.6 and 2.8, we see that coherence-driven

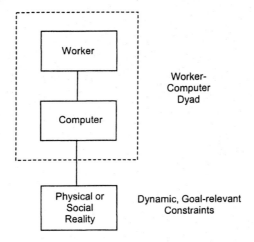

FIG. 2.7. Correspondence-driven work domains (adapted from Vicente, 1990).

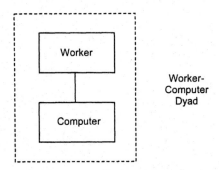

FIG. 2.8. Coherence-driven work domains (adapted from Vicente, 1990).

domains do not have a physical or social reality, outside of the worker and the computer, that must be taken into account by workers in order for goals to be achieved. A simple example is a person working with a word processor. In these cases, the focus of attention is solely on the worker-computer dyad without reference to an external environment.

It should be clear from Figs. 2.6 to 2.8 that the cognitivist perspective is the most effective approach if we are designing for coherence-driven work domains. Because there are no dynamic, environmental constraints, the work analysis can begin with, and give primary importance to, worker cognitive characteristics since these human factors should dictate how the resulting interface is to be designed (see Vicente, 1990a, for a more detailed justification of this point). However, in this book we are specifically concerned with complex sociotechnical systems that, as discussed in chapter 1, pose a distinct and challenging set of demands on both workers and designers. It should be equally clear from the examples listed previously that such systems are correspondence-driven. There are always dynamic, environmental constraints that need to be taken into account, whether they be imposed by a social reality (as in the case of engineering design or the stock market), or by a physical reality (as in the case of aviation or medicine). Accordingly, in this section, we examine the usefulness of the cognitivist and ecological approaches to work analysis specifically for the case of correspondence-driven work domains.

The examples we present in this section are all drawn from nuclear power. We deliberately chose this biased approach because the point we are trying to make should be especially obvious in the nuclear domain, where physical reality is of paramount importance. But as our examples show, the primacy of environment constraints is not widely appreciated even in this industry.

The Cognitivist Approach

Example 1. In a review of advanced alarm and diagnostic systems for nuclear power plants, Kim (1994) provided a clear statement advocating the cognitivist approach: "The computerized systems for on-line management of plant anomalies must provide the operators with representations of plant functions that are compatible with their mental image of the plant, the so-called mental model or conceptual model" (p. 293). This may seem like a logical and defensible claim, but it overlooks an important factor. What if the worker's mental model is incomplete, or even worse, incorrect in the sense that it does not faithfully capture how the plant actually works? Given the human factors design inadequacies in existing nuclear power plant control rooms—and many other application domains, for that matter—and the fact that the control room design can shape workers' mental models, we cannot take it for granted that workers' mental models are accurate. How useful would it be, then, to design an interface that is consistent with a worker's mental model but that describes the plant in a way that is incorrect, incomplete, or ultimately misleading? Or in other words, how useful would it be to begin a work analysis with the worker's existing mental model rather than with the actual environment constraints? The answer to this question should be self-evident because the laws of physics are unforgiving.

But because the primacy of environment constraints is not widely appreciated, it is worthwhile to illustrate the consequences of the cognitivist approach for correspondence-driven work domains. We use the classic example of the Three Mile Island (TMI) nuclear power plant accident to illustrate our point (Rubinstein, 1979). There were many factors that contributed to this accident, but we focus on just one particular issue that is pertinent to the present discussion. To introduce the issue, we have to briefly describe how part of the TMI nuclear power plant was intended to function.

Figure 2.9 shows a very simplified diagram of what is known as the primary coolant loop. For the sake of clarity, only three components are shown: the nuclear reactor, the pressurizer, and the primary pump. These components are connected by pipes, allowing water to circulate from the reactor to the pressurizer and then back again. Under normal circumstances, the primary loop is kept under high pressure so that, despite the fact that the water in the loop is very hot, it is kept in a liquid state (see Fig. 2.9a). This is important because if the water begins to boil, then it may be difficult to circulate and remove heat from the reactor. This could cause the reactor temperature to increase, which can eventually cause a core meltdown—a severe event that can significantly impact public safety. The pressurizer is intended to act as a buffer, with the level increasing or decreasing as a function of the amount of water in the primary loop.

During the TMI accident, the pressurizer relief valve opened automatically to reduce the pressure in the primary loop. When the primary pressure came back into its normal range, the pressurizer relief valve was supposed to close, but it remained stuck open instead (see Fig. 2.9b). This caused the primary pressure to decrease to the point where the water in the reactor started to boil. Thus, instead of being in a liquid state as it was supposed to be (Fig. 2.9a), the water was now in a two-phase mixture of steam and liquid (Fig. 2.9b). This, in turn, caused the level in the pressurizer to increase.

At this point, the workers' mental models come into play. The TMI operators were trained to believe that pressurizer level is an indication of how much liquid water there is in the primary loop. As shown in Fig. 2.9a, this is true when the plant is in its normal state. However, in the abnormal conditions that had arisen, the workers' belief was no longer valid. Although pressurizer level had increased, the amount of liquid water in the primary loop had actually decreased due to the boiling in the reactor and the water lost through the stuck open pressurizer relief valve. Because of their inaccurate mental model, operators incorrectly inferred from the high pressurizer level that there was too much liquid water in the primary loop and acted in ways that turned out, in retrospect, to be highly inappropriate and detrimental to

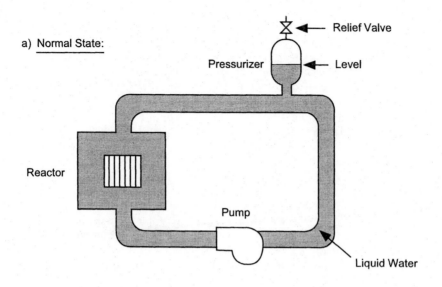

a) Normal State:

Relief Valve

Pressurizer — Level

Reactor

Pump

Liquid Water

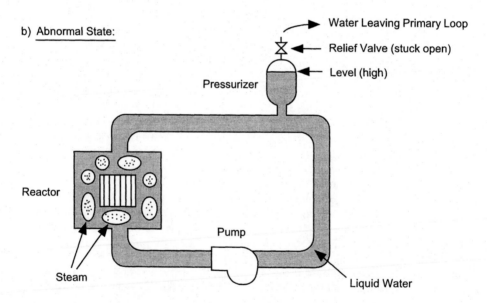

b) Abnormal State:

Water Leaving Primary Loop

Relief Valve (stuck open)

Level (high)

Pressurizer

Reactor

Pump

Steam

Liquid Water

FIG. 2.9. A highly simplified diagram of the primary coolant loop in the TMI nuclear power plant during (a) normal state, (b) abnormal state.

the safety of the plant (Rubinstein, 1979). Note, however, that the incorrect mental model that they held had served them well enough for years before the accident. The specific situation that revealed the deficiencies in their mental model had never been experienced before. We revisit this point later.

This example shows that the cognitivist approach to work analysis can lead to dangerous design recommendations. If we base the design of an interface on an analysis of workers' current mental models, then we will reinforce any existing misconceptions that workers already have. In the case of TMI, this would correspond to reinforcing the belief that pressurizer level is always a valid indication of the amount of liquid water in the primary coolant loop—a belief that is just plain wrong. And as the TMI accident showed, reinforcing this belief would be a tragic mistake because it could lead workers to act inappropriately in emergency situations that, in retrospect, uncover a discrepancy between their mental models and the way the plant actually works. Instead, we should be creating designs that reinforce accurate beliefs about how the plant actually works so that workers can act appropriately, especially in life-threatening emergencies.

Example 2. The TMI accident occurred almost two decades ago, so it is possible that the lesson just described has been learned and incorporated by the nuclear industry. A much more recent anecdote shows that this is not the case, and at the same time, further illustrates the advantages of the ecological approach of giving primary importance to environment rather than cognitive constraints. One nuclear power plant vendor chose an exceptionally skilled operator to participate in the design of a new control room.[6] This practice is consistent with existing theories of "participatory design" or "user-centered design," the goal being to get worker input into the design process (see chap. 4). This is a laudable goal and represents a distinct improvement over traditional design practices that essentially leave the worker out of the design process. Nevertheless, this anecdote shows that it is possible to take a user-centered perspective too literally.

When the new control room design was shown to other operators, the designers quickly realized that the design process they had adopted was faulty. The first valuable lesson they learned was that other operators did not think of the plant in the same way as the exceptional operator who was part of the design team. In fact, no two operators seemed to have the same mental model of the plant. This finding caused the designers to reflect more deeply on the relationship between operators' mental models and the engineering laws and principles governing plant behavior. This reflective exercise revealed another valuable lesson, namely that all of the operator's mental models were limited in the sense that they contained misconceptions, omissions, or both. As a result, the control room in question had to be redesigned to better reflect the way in which the plant actually worked. Needless to say, this was a very costly and time-consuming process.

This anecdote from the nuclear industry shows that adopting the cognitivist approach of starting and driving work analysis from workers' current mental models can be a costly mistake in correspondence-driven work domains. If designers had adopted an ecological perspective and had started the work analysis by determining, in detail, how the plant really works and then used that analysis to drive the design process, then there would have been no need to redesign the control room interface at a late stage of the design process.

Example 3. We might be tempted to think that the two examples we have described so far are mere outliers. That is, the cognitivist perspective cannot possibly be widely held in the nuclear industry where the primacy of actual work environment constraints (e.g., physical conservation laws) are so obvious, and where the potential consequences of ignoring those constraints (e.g., leaking radiation into the atmosphere) are so enormous. Surprising as it may be, there is compelling evidence to the contrary.

---

[6]Due to the embarrassing nature of the events, the identity of the company has been kept anonymous.

The U.S. Nuclear Regulatory Commission (USNRC), as its name indicates, is the government agency that is responsible for regulating the nuclear industry in the United States. After the TMI accident, the USNRC published a technical report (NUREG-0700) that contains guidelines for how to improve the human factors design of nuclear power plant control rooms (USNRC, 1981). With the more recent trend from traditional analog hard-wired to advanced computer-based control room designs, the guidelines in NUREG-0700 have become somewhat dated. As a result, the USNRC decided to update the report to add guidance that is more relevant to advanced control room design (USNRC, 1996). There is a great deal of very useful information in the new version of NUREG-0700, including an exemplary design process that tries to ensure that a thorough task analysis is conducted early in the design life cycle. Despite this, however, the new, revised version of NUREG-0700 clearly includes a design recommendation based on the cognitivist perspective: "[Computer-based support systems] should be consistent in content and format with the cognitive strategies and mental models employed by the user" (USNRC, 1996, p. 327).

What can we conclude from such a statement? There is no doubt that progress has been made in the nuclear industry since the TMI accident. Nevertheless, the benefits of starting and driving work analysis by examining environment constraints rather than by analyzing workers' existing mental models do not seem to be widely, or at least consistently, appreciated, even by a government nuclear regulatory agency.

Example 4. Although designers in industry or government regulators may not be aware of the limitations of the cognitivist approach to work analysis, we might think that competent researchers are quite aware that work analysis should be initiated and driven by environment constraints. If so, then we would expect that all prominent human factors and HCI theories would reflect an appreciation for the ecological approach. To show that this is in fact not the case, we examine briefly the well-known proximity compatibility principle (PCP) (Wickens & Carswell, 1995). In a nutshell, the PCP states that the perceptual characteristics of displays should be designed to be compatible with the cognitive processes used by workers to perform a particular task. For example, if workers use two sources of information to complete a task, then the display should somehow integrate these data for the workers. Conversely, if workers focus on a single source of information to complete a task, then that datum should be presented separately from other data.

It is evident from the applications that have been chosen to evaluate the PCP (e.g., aviation, process control) that it is intended to be applied to correspondence-driven work domains. Nevertheless, Wickens and Carswell (1995) were clearly advocating the cognitivist perspective. For example, in defining various forms of task proximity (which, according to PCP, dictate what type of displays should be designed), Wickens and Carswell discussed the concept of functional similarity, which "refers to the similarity of the units or objects being measured, *as represented in the operator's semantic space*" (p. 476, emphasis added). They then went on to state that "functional similarity could be derived from multidimensional scaling techniques eliciting the structure of the *operator's* semantic space or mental model of the displayed system" (p. 476, emphasis added). These unambiguous statements advocate that the design process be driven by an analysis of workers' mental model of the system, the hallmark of the cognitivist approach.

Therefore, the recommendations made by the PCP are subject to the limitations already identified previously. Basing a work analysis on a worker's mental model overlooks the fact that such models may be incorrect or incomplete. Furthermore, basing a display on a "buggy" mental model can be a costly mistake, as the previous anecdote from the nuclear industry made clear. For correspondence-driven domains, a different approach must be adopted.

The Ecological Approach

The approach we advocate is based on an ecological approach that suggests that work analysis should begin with, and be driven by, an explicit analysis of the constraints that the environment imposes on action. Only by adopting such a perspective can designers ensure that the content and structure of the interface is compatible with environment constraints (Vicente, 1990a). The desired objective of this approach is to ensure that workers will acquire a *veridical* mental model of the environment, so that their understanding corresponds, as closely as possible, to the actual behavior of the context with which they interact (e.g., the nuclear power plant).

Our argument for the ecological perspective is not specific to the nuclear industry. It generalizes meaningfully to any application domain where there is an external physical or social reality—outside of the person and the computer—that imposes dynamic, goal-relevant constraints on action (Fig. 2.7). For these correspondence-driven work domains, workers' mental models should correspond with this external reality. The only way to accomplish this objective efficiently is to adopt an ecological approach, giving primacy to environment constraints.

Another example of a correspondence-driven work domain where the applicability of the ecological approach can be readily appreciated is aviation. In this domain, it is important that pilots' mental models of how the aircraft functions, of geometric constraints in three-dimensional space, and of the flight management system (FMS) be accurate. Thus, interfaces displaying the status of aircraft systems, air traffic, and the FMS should be based on the constraints that describe how these entities actually work, not on how pilots might think that they work.

This point is illustrated by the findings of Sarter and Woods (1994), who conducted a simulator study investigating how experienced commercial airline pilots interacted with the FMS on a Boeing 737-300. One might think that because these pilots were well trained and had a great deal of experience, they would have a near-perfect understanding of how the FMS works. But the data from this study show that, on the contrary, pilots have gaps in their understanding of the functional structure of the FMS. Moreover, it seems that the pilots are not aware of the gaps in their knowledge. For example, after the experiment, one pilot volunteered: "I never knew I did not know this. I just never thought about this situation" (p. 25). This finding has important implications. If workers are generally not aware of the deficiencies in their mental models, and if designers use these models as the basis for interface design, then these deficiencies are almost sure to be transferred to the resulting interface. And as with the TMI example discussed earlier, this practice would eventually lead to a rude awakening when the pilot's assessment of a situation does not correspond to what is actually going on. Furthermore, these discrepancies are most likely to be revealed and to lead to a breakdown in performance only during rare, abnormal, time-critical situations, where the consequences of failure are the greatest (Sarter & Woods, 1994), just like what was observed at TMI.

In conclusion, the primacy of environment constraints advocated by the ecological approach is not just relevant to the nuclear power domain. It is equally relevant to aviation, medicine (e.g., R. I. Cook, Potter, Woods, & McDonald, 1991), and other correspondence-driven work domains.

Potential Misunderstandings

It is important that we to address several additional points in order to avoid any potential misunderstandings. First, some of you might think that we have presented a straw man. After all, how could anyone possibly advocate designing an interface based on a grossly incorrect worker mental model? The only way to determine if we have misrepresented others' views is to look closely at the words they use. As Dewey and Bentley (1949) observed half a century ago: "What the writers 'really mean' is much less important logically than what they say (what they are able to say under their manner of approach) when they are manifestly *doing their best to say what they mean*" (p. 40, emphasis in original). It is certainly true that in none of the examples cited previously did anyone explicitly advocate that interfaces should be designed according to mental models that are incorrect or incomplete. However, it does not follow that the actions and words described in those examples reflect a deep appreciation for the advantages of the ecological approach. On the contrary, Wickens and Carswell (1995), for instance, never explicitly stated that a display should be designed to be compatible with the way the work domain actually functions. Moreover, the words that they did use (see earlier discussion) indirectly conflict with such an assertion.

Second, you might think that workers' mental models do not have to be accurate. As long as they are "close enough," then performance may not suffer. This is probably true for the set of situations that workers encounter frequently and to which they have adapted. However, accident analyses and simulator studies have repeatedly shown that when unfamiliar, abnormal events that are outside of workers' adaptation boundaries occur, the divergence between workers' mental models and the actual behavior of the work domain becomes practically significant and can greatly jeopardize system safety (Rasmussen, 1969; Rubinstein, 1979; Sarter & Woods, 1994). Thus, close enough may be good enough most of the time, but in complex sociotechnical systems where the consequences of error can have enormous financial, safety, and environmental implications, a more cautious and thorough approach is required.

Third, some of you will have realized that there are several different correct mental models for any system with objective characteristics. This is absolutely correct but it does not invalidate our point. Just because there may be more than one equally faithful mental model does not mean that designers can assume that workers currently have one of those models. There are literally an infinite number of mental models that are incorrect. In correspondence-driven domains, it is always important to make sure that the model serving as the basis for interface design captures the way in which the work environment actually functions. The best way to ensure that this occurs is by basing work analysis on the ecological approach.

Integrating Cognitivist and Environment Constraints

Our discussion so far has focused primarily on contrasting the cognitivist and ecological approaches to work analysis. This tack was deliberately chosen to point out the comparatively unappreciated importance, indeed primacy, of environment constraints. But as we pointed out at the start, if any information system is to be effective, it must integrate environment and cognitive constraints. How can we retain the value of both types of constraints, while still acknowledging the primacy of environment constraints?

Our answer to this question can be explained with reference to Fig. 2.6. The ecological approach to work analysis begins by analyzing physical reality (e.g., nuclear physics) and social reality (e.g., the intentions of a remote collaborator), whereas the cognitivist perspective begins by analyzing workers' cognitive characteristics (e.g., mental models, strategies, preferences). At the risk of being overly repetitive, we reemphasize that what is under consideration is which set of constraints should be given precedence, not which set should be the subject of work analysis—both are important. Having said this, the basic point of this section is that cognitive compatibility is of little use unless ecological compatibility has already been established. That is, making an interface compatible with a worker's mental model will not do much good if that model does not correspond with the way the environment actually functions. It is for this reason that environment constraints need to be given priority.

However, this does not mean that there is no room for cognitive constraints. On the contrary, attention to human characteristics is of the utmost importance and should also be given an important place in work analysis. Once it has been established that environment constraints have been respected, knowledge of human characteristics provides an important basis for making the remaining design decisions. In this way, work analysis can provide a way of making sure that physical and social reality are not ignored, while simultaneously ensuring that cognitive characteristics are not overlooked. The end result should be computer-based systems that will present workers with the information they need to develop accurate mental models in a form that is compatible with existing knowledge of human cognition. The framework that we present later in this book embodies this integrative approach, synthesizing environment and cognitive considerations while giving primacy to the environment.

Summary

In this section, we have presented four examples from the nuclear industry to show that, even in this patently correspondence-driven domain, researchers, designers, and regulators have not taken the primacy of environment constraints as strongly as they should. A number of important limitations arising from this oversight were discussed. We concluded that a work analysis framework for complex sociotechnical systems must integrate environment and cognitive constraints while giving primacy to the environment.

CONCLUSION

This book is primarily about analysis, not design. But as we pointed out in chapter 1, if analysis has any pragmatic value at all, it must be in the implications it has for design. The first part of this chapter showed that when computer-based information systems are designed without taking into account work demands, problems arise.

The cognitive burden imposed on workers is increased. Also, workers have to engage in workarounds to get the job done. Not only that, but these workarounds can lead to errors. And in some cases, the workarounds may be so costly in terms of time and effort that workers may just give in and do what is easy with the interface that they have, rather than what is required to get the job done effectively. As a result, the quality of the work is sacrificed. To avoid these problems, the design of information systems for complex sociotechnical systems should be based on some kind of work analysis. If the work demands are identified and then built into the information system, then many of the problems identified above should be resolved, or at the very least, greatly reduced.

In the second part of the chapter, we argued that work analysis should begin with environment constraints. The value of this perspective was illustrated by contrasting the widely held cognitivist approach with the less prevalent ecological approach. The limitations of the cognitivist approach were illustrated in some detail, showing that it is impotent without ecological compatibility. Nevertheless, both types of constraints need to be considered in work analysis. The approach that we advocated, and that is embedded in the remainder of the book, is to first analyze the constraints imposed by the environment. These constraints must be dealt with if there is to be any hope of achieving effective performance. Analyzing cognitive constraints provides a valuable way of closing the remaining degrees of freedom. This fundamental idea will be developed further in subsequent chapters.

THREE APPROACHES TO WORK ANALYSIS II

Normative Approaches to Work Analysis: "The One Best Way?" 3

Each employee of your establishment should receive every day clear-cut, definite instructions as to just what he is to do and how he is to do it, and these instructions should be exactly carried out, whether they are right or wrong.
— Frederick Winslow Taylor (cited in Kanigel, 1997, p. 377)

How is it possible to write a procedure for absolutely every possible situation, especially in a world filled with unexpected events? Answer: it's impossible. [Yet] procedures and rule books dominate industry.
— Norman (1998, p. 156)

PURPOSE

In chapter 2, we argued that it is important to conduct some form of work analysis, and that such an analysis should be based on an ecological approach. As we pointed out in chapter 1, various types of work analysis techniques have been developed since the beginning of this century. In this chapter and the next, we critically examine some of these techniques to see how well they stand up to the challenges imposed by complex sociotechnical systems (see chap. 1). Normative approaches to work analysis, particularly task analysis, are examined in this chapter. To anticipate, we conclude that some form of task analysis is indispensable in complex sociotechnical systems, but that a work domain analysis is also needed to overcome the limitations of task analysis.[1]

THREE GENERATIONS OF WORK ANALYSIS TECHNIQUES

Because many different types of work analysis techniques have been proposed, it is useful to categorize them in a way that highlights important similarities and significant differences as well. Rasmussen (1997a) distinguished between three generic categories of models that can be adopted for the specific purpose of grouping work analysis techniques. *Normative models* prescribe how a system should behave. In contrast, *descriptive models* describe how a system actually behaves in practice. Finally, *formative models* specify the requirements that must be satisfied so that the

---

[1]As mentioned in chapter 1 (see Footnote 2), task analysis and work domain analysis are two different types of work analysis. One of the main purposes of this chapter is to explain the difference between these two techniques.

system could behave in a new, desired way.[2] Rasmussen used these three model categories to examine the evolution of theories in a number of diverse areas of research, including engineering design, organizational behavior, decision making, and human error. His review revealed a progression over time in each of these areas, from normative first-generation models, to descriptive second-generation models, and only very recently, to formative third-generation models.

The same progression can be seen in the evolution of approaches to work analysis. Seminal Tayloristic work methods analyses (Taylor, 1911), traditional human factors task analyses (R. B. Miller, 1953), and early HCI Goals, Operators, Methods, Selection Rules (GOMS) analyses (Card, Moran, & A. Newell, 1983) are all exemplars of normative approaches to work analysis. Each of these approaches prescribes a normative, rational benchmark for how workers should behave in different situations. More recently, researchers from the anthropological (Hutchins, 1995a; Suchman, 1987), activity theory (Bødker, 1991; Nardi, 1996b), and naturalistic decision-making research communities (Klein, Orasanu, Calderwood, & Zsambok, 1993; Zsambok & Klein, 1997) have all pointed out that workers' actions frequently do not—and indeed, should not—always follow these normative prescriptions. These researchers have come to this conclusion by studying how workers actually behave in situ and by developing descriptive models of current practice. To be clear, the problem with normative models is not that they are normative per se, but that the assumptions they make about human work are not realistic and thus not very useful (Sheridan, in press). But as we show in chapter 4, there are problems in deriving implications for design from descriptive work analysis techniques too, at least for complex sociotechnical systems. These problems have been recognized by researchers from a diverse set of communities, including HCI (Carroll, Kellogg, & Rosson, 1991), sociology (Schmidt, 1991b), anthropology (Button & Dourish, 1996), and naturalistic decision making (T. E. Miller & Woods, 1997). Nevertheless, a solution to these problems has remained elusive.

What we need is a third-generation, formative work analysis framework that helps us specify the design attributes that computer-based information systems should have to satisfy safety, productivity, and health goals.[3] Chapter 5 provides a sketch of the formative approach to work analysis that is the subject of this book. But first, the limitations and implications of normative approaches to work analysis are described in this chapter. Descriptive approaches are described in chapter 4.

Outline

We begin by describing one type of normative approach to work analysis—task analysis. We then show that task analysis techniques actually fall into two qualitatively different categories, constraint and instruction based, with instruction-based techniques being far more prevalent. Our review reveals that instruction-based ap-

---

[2]Rasmussen (1997a) used the term *predictive* rather than *formative* for this construct. We changed the label because the term *predictive* has connotations that could easily lead to misunderstandings (see Footnote 3).

[3]Note that the requirements that emerge from a formative work analysis do not uniquely specify a new design. That is, they rule out many design alternatives, but additional insight (e.g., human factors and HCI design principles) are required to flesh out a detailed design. This is one of the reasons why Rasmussen's (1997a) label *predictive* can be misleading (see Footnote 2). Formative work analyses are predictive in the qualitative sense used in dynamical systems theory (Port & van Gelder, 1995) but not in the formal, quantitative sense that most of us usually associate with the concept of prediction.

proaches to task analysis are limited in their ability to identify comprehensively the information requirements in complex sociotechnical systems. Fortunately, however, constraint-based task analyses provide a viable alternative that offers some important insights. Their weakness is that they are not well suited to the demands associated with unanticipated events. Work domain analyses are proposed as a complementary technique that can help analysts identify the information support workers need to deal with unanticipated events.

THE NORMATIVE APPROACH: TASK ANALYSIS

Scope

As we mentioned, there are different exemplars of normative approaches to work analysis (e.g., Tayloristic work methods, task analysis, GOMS). Rather than provide an exhaustive and detailed review of each of these individual techniques, we use task analysis as a prototypical exemplar with which to examine critically normative approaches to work analysis.[4]

In this section, we first describe various types of task analysis techniques and provide simple examples of each type. Second, we show how these techniques fall into two qualitatively distinct categories. Third, several important implications for complex sociotechnical systems arising from this distinction are discussed. To anticipate, we show that there is a conflict between constraint- and instruction-based task analysis techniques. This conflict is unraveled in the next section of this chapter.

What Is Task Analysis?

Even the more specific category of task analysis actually represents a diverse set of approaches. In what is perhaps the most comprehensive and clearly-written guide to task analysis produced to date, Kirwan and Ainsworth (1992) reviewed many, if not all, of the major techniques for performing a task analysis. These methods can be used to identify and examine "the tasks that *must* be performed by users when they interact with systems" (Kirwan & Ainsworth, 1992, p. vii, emphasis added). Or in slightly different terms, "Task analysis can be defined as the study of what an operator (or team of operators) is *required* to do, in terms of actions and/or cognitive processes to achieve a system goal" (Kirwan & Ainsworth, 1992, p. 1, emphasis added). These two definitions unambiguously put task analysis techniques into the class of normative approaches to work analysis. The emphasis is on identifying what workers should be doing to get the job done, reminiscent of Taylor's quest for the "one best way" (Kanigel, 1997). Our story in the Preface provides an example: The regulators were looking for "procedural compliance" when evaluating the nuclear power plant operators' performance in the simulator. Despite their many differences, task analysis techniques generally share this rational quest for identifying the ideal way(s) in which the job should be performed (e.g., Kirwan, 1992).

---

[4]Two caveats are in order. First, we use the term *task analysis* to refer to the end product (i.e., the model or representation) generated by the process of studying a task. This is a typical convention in North America, although in some cases the term *task description* is used for this purpose instead (Singleton, 1974). Second, task analysis can be used in either a descriptive or a normative fashion, but we only focus on the latter in this chapter. Descriptive forms of work analysis are the subject of chapter 4.

To be sure, the precision with which this objective is sought differs considerably among task analysis techniques. Three different levels can be identified (for examples of task analysis techniques at these levels, see Kirwan & Ainsworth, 1992):

1. *Input–Output*—This first level identifies the inputs that are required to perform the task, the outputs that are achieved after it is completed, and the constraints that must be taken into account in selecting the actions that are required to achieve the task. An example is illustrated in Fig. 3.1 for the task of determining the rate of gasoline consumption in km/liter for an automobile. The two inputs and the single output for the task are shown, but the steps that are used to achieve the task are not specified. Note that the inputs are specified in miles and gallons, so a conversion process is required to perform the task correctly. Thus, certain constraints must be obeyed to reach the goal state. For the example in Fig. 3.1, two relevant constraints are the number of kilometers in a mile and the number of liters in a gallon. Even for a task as simple as this one, there are several different sets of actions that we could adopt to perform the same task. For example, we could use a calculator, in which case the main steps are reading off numbers from the odometer and the gas pump and then typing those numbers into the calculator. Or alternatively, we could use mental arithmetic, in which case some cognitive steps are also required to arrive at the correct answer. The input–output level deliberately ignores these process details. It only provides a very high level product description of the task.

There are several points worth noting here. First, the constraints on action can be of different types. In the example presented in Fig. 3.1, the constraints were relationships between variables. However, they could also be sequential in nature (e.g., there must be water in the kettle before you turn the stove on). Second, it is not possible to perform the task correctly without factoring in these constraints. They must be taken into account, independent of how the task is actually performed. Third, the constraints serve the function of "problem state space reduction" (Sheridan, in press) but they usually do not specify a unique action flow sequence. They merely limit the options (degrees of freedom) on viable action sequences (i.e., all sequences that do not factor in the correct conversion factors are ruled out). In the words of Kugler, Kelso, and Turvey (1980), "it is not that actions are caused by constraints, it is rather that some actions are excluded by them" (p. 9).

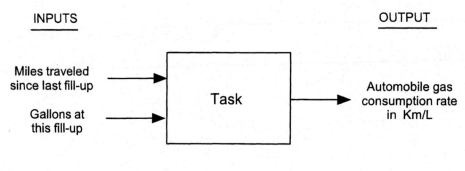

FIG. 3.1. Example of input/output task analysis for determining the rate of gasoline consumption in an automobile.

2. *Sequential Flow*—This level identifies a temporally ordered sequence of actions that is required to complete the task. Such a description is equivalent to a flowchart of the process that workers should follow to perform the task, much like the specification of a computer algorithm. Traditionally, the steps in the flow sequence have been overt actions, although cognitive steps can also be specified. The steps themselves are usually taken from a behavior taxonomy such as the classic Berliner taxonomy, shown in Table 3.1. The column on the extreme right provides a "periodic table" of elemental action verbs that can be used to compose a flow sequence for a particular task. For example, one specific sequence of actions for performing part of the task in Fig. 3.1 could be: (a) *Read* current odometer value, (b) *read* odometer value at last fill-up, and (c) *calculate* the difference to obtain miles traveled since last fill-up. The result would be an algorithm for performing the task in a particular way. Note that, because it is more detailed, this level of task analysis is usually dependent on the device workers currently have available to perform the task. For example, if you had a trip odometer in your car, the action sequence just specified would be unnecessary, as long as you reset the odometer the last time you filled your tank. Instead of calculating the miles traveled since the last fill-up, you could simply read off this value from the trip odometer.

Like computer program flowcharts, the flow sequence can include branch points to represent alternative paths in the process sequence (Meister, 1985). In practice, however, this branching is avoided to the extent possible because it makes for a more complicated task representation (Meister, 1995a). Another example of a task analysis conducted at the sequential flow level is shown in Fig. 3.2. Note the single sequence of overt actions.

3. *Timeline*—This final level identifies a temporally ordered sequence of actions that is required to achieve the task, with duration estimates for each action. This level is the most detailed of all, and was traditionally used in forced assembly line operations. Sticking with the example we used in the previous level, we might obtain the following task description (the time estimates are purely hypothetical): (a) 0–1 s: *Read* current odometer value, (b) 1–2 s: *read* odometer value at last fill-up, and (c) 2–3 s: *calculate* the difference to obtain miles traveled since last fill-up. This type of task analysis should be familiar to industrial engineers who have been exposed to time and motion studies (Maynard, 1971), to HCI professionals who have been exposed to GOMS and the Model Human Processor (Card et al., 1983), and to human factors professionals who have been exposed to time-based measures of mental work load (Meister, 1985). Not only is the sequence of actions that should be performed specified, but so is the duration of each of those actions. A more complex example is illustrated in Fig. 3.3, involving time sharing of multiple tasks shown by the overlap in the timeline. If we are to believe this analysis, there is only one right way to perform this task at this level of resolution. All of the discretion has essentially been eliminated, leaving only a single, very well defined procedure for workers to follow.

Constraints Versus Instructions

These three levels of task analysis, summarized graphically in Fig. 3.4, differ in their specificity. If we adopt a multidimensional state space as a frame of reference, the first level of task analysis (input–output) shown in Fig. 3.4a merely identifies a *point* in the

| TABLE 3.1 | Classification of Task Behaviors |

Perceptual processess	Searching for and receiving information	Detects Inspects Observes Reads Receives Scans Surveys
	Identifying objects, actions, events	Discriminates Identifies Locates
Mediational processes	Information processing	Categorizes Calculates Codes Computes Interpolates Itemizes Tabulates Translates
	Problem solving and decision-making	Analyzes Calculates Chooses Compares Computes Estimates Plans
Communication processes		Advises Answers Communicates Directs Indicates Informs Instructs Requests Transmits
Motor processes	Simple/Discrete	Activates Closes Connects Disconnects Joins Moves Presses Sets
	Complex/Continuous	Adjusts Aligns Regulates Synchronizes Tracks

Note. From Meister (1985). Copyright 1985 by John Wiley & Sons, Inc. Reprinted by permission.

RECEIPT AND HANDLING OF A LOADED IF-300 CASK

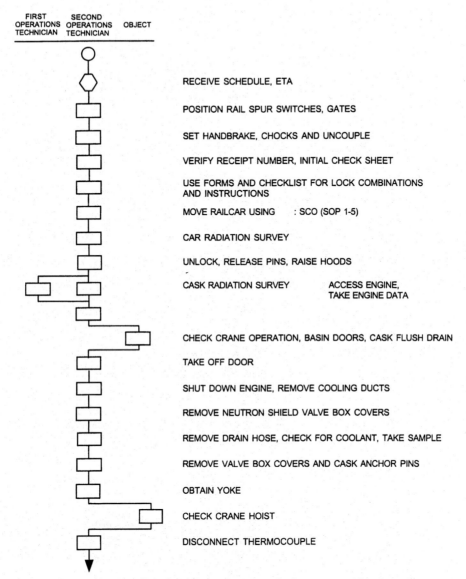

FIRST
OPERATIONS
TECHNICIAN

SECOND
OPERATIONS
TECHNICIAN

OBJECT

RECEIVE SCHEDULE, ETA

POSITION RAIL SPUR SWITCHES, GATES

SET HANDBRAKE, CHOCKS AND UNCOUPLE

VERIFY RECEIPT NUMBER, INITIAL CHECK SHEET

USE FORMS AND CHECKLIST FOR LOCK COMBINATIONS
AND INSTRUCTIONS

MOVE RAILCAR USING   : SCO (SOP 1-5)

CAR RADIATION SURVEY

UNLOCK, RELEASE PINS, RAISE HOODS

CASK RADIATION SURVEY   ACCESS ENGINE,
                                 TAKE ENGINE DATA

CHECK CRANE OPERATION, BASIN DOORS, CASK FLUSH DRAIN

TAKE OFF DOOR

SHUT DOWN ENGINE, REMOVE COOLING DUCTS

REMOVE NEUTRON SHIELD VALVE BOX COVERS

REMOVE DRAIN HOSE, CHECK FOR COOLANT, TAKE SAMPLE

REMOVE VALVE BOX COVERS AND CASK ANCHOR PINS

OBTAIN YOKE

CHECK CRANE HOIST

DISCONNECT THERMOCOUPLE

FIG. 3.2. Example of sequential flow task analysis for the receipt and handling of a loaded cask.

space (i.e., the goal state), the inputs that are required to get to that point (not shown in the figure), and the constraints that must be respected (i.e., the areas that must be avoided) if workers are to get to the goal state. The second level of task analysis (flow sequence), shown in Fig. 3.4b, represents a marked increase in specificity. Not only is the goal state identified, but so is a sequence of actions for getting to the goal state. There is still some leeway, however, because the timing of the actions is not specified. Therefore, in terms of a state space representation, this level of task analysis corre-

TIMELINE ANALYSIS

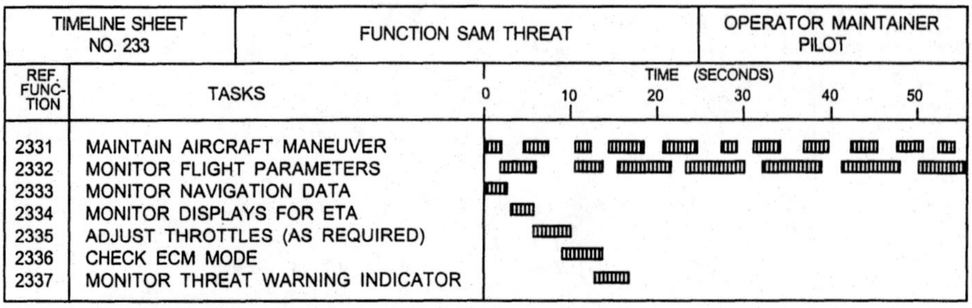

TIMELINE SHEET NO. 233		FUNCTION SAM THREAT						OPERATOR MAINTAINER PILOT		
REF. FUNC- TION	TASKS	TIME (SECONDS)								
		0	10	20	30	40	50			
2331	MAINTAIN AIRCRAFT MANEUVER									
2332	MONITOR FLIGHT PARAMETERS									
2333	MONITOR NAVIGATION DATA									
2334	MONITOR DISPLAYS FOR ETA									
2335	ADJUST THROTTLES (AS REQUIRED)									
2336	CHECK ECM MODE									
2337	MONITOR THREAT WARNING INDICATOR									

FIG. 3.3. Example of a timeline task analysis for a military aviation task (Meister, 1985). From *Behavioral analysis and measurement methods* by David Meister, Copyright © (1985, John Wiley & Sons, Inc). Reprinted by permission of John Wiley & Sons, Inc.

sponds to a *path* that must be followed (see Fig. 3.4b). Finally, the third level of task analysis (timeline) is the most specific of all. It prescribes a sequence of actions, each with a particular duration. This level of analysis can be represented as a unique *trajectory* in the state space for reaching the goal (see Fig. 3.4c).[5]

The graphical comparison in Fig. 3.4 clearly reveals an otherwise subtle qualitative distinction between the three levels of task analysis. The first level, being based on *constraints,* specifies what should not be done, whereas the next two levels, being based on *instructions,* specify what should be done (cf. Kugler et al., 1980). This qualitative difference, although perhaps unfamiliar to us, is of absolutely fundamental importance to the approach taken in this book. Therefore, it is discussed at length in chapter 5 and revisited several times in later chapters.

It is important to add that instruction-based techniques are, by far, the more prevalent of the two types of task analysis. For example, almost all of the techniques described in Kirwan and Ainsworth's (1992) guide to task analysis are instruction based. This observation has important implications that are highlighted later in this chapter.

Implications

The point of doing any kind of work analysis is to derive implications for systems design. Different forms of work analysis make different assumptions about the nature of work, so they lead to different designs, which, in turn, lead to different types of guidance for workers. For example, constraint-based and instruction-based approaches to task analysis have different implications for workers. First, instruction-based approaches are more detailed, so they leave less discretion to workers (cf. Leplat, 1989). At the lowest level (timeline, see Fig. 3.4c), the analyst must make decisions about what actions should be performed to accomplish the task, how they should be sequenced together, and how long each will take. There is very little left

---

[5]Note that the distinction between a path and a trajectory is a relative one. For example, if we adopt a more fine-grained level of analysis (i.e., if we "zoom in"), what looked like a trajectory becomes a path. Conversely, if we adopt a more coarse level of analysis (i.e., if we "zoom out"), what looked like a path can look like a trajectory.

a) <u>Level 1 - Input–Output:</u>

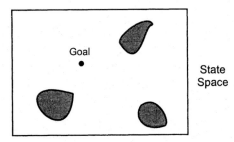

b) <u>Level 2 - Flow Sequence:</u>

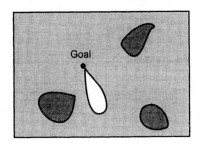

c) <u>Level 3 - Timeline:</u>

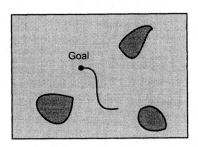

FIG. 3.4. A qualitative graphical comparison of the specificity associated with the three levels of task analysis described in the text. Shaded regions represent hypothetical areas that should be avoided, according to the analysis.

for the worker to do except follow the resulting procedure or work flow, reliably and consistently (Reed, 1996). With constraint-based approaches, on the other hand, none of these decisions are made up front by the analyst. Only guidance about the goal state and the constraints on action are provided, not about how the task should be accomplished (Fig. 3.4a). Consequently, workers are given some discretion to make decisions online about how to perform the task. This contrast shows that constraint-based approaches to task analysis provide more worker discretion than do instruction-based approaches. And as we mentioned in chapter 1, decision latitude is a key factor in determining worker health.

Of course, we could just as easily turn these observations around by examining, not the discretion, but the guidance offered to workers by the two types of task analysis. We could argue that instruction-based approaches provide more detailed guidance and thus are less likely to lead to human error. In contrast, constraint-based approaches offer minimal guidance and thus are more likely to lead to human error. Thus, whether instruction-based or constraint-based approaches to task analysis are better depends on the view that we hold of people (cf. Reed, 1996; Sachs, 1995). If we view people in an optimistic light, we may want to adopt the constraint-based approach—provide a computer system with information about the goal to be achieved, the current state of affairs, and the constraints to be obeyed, and then let workers decide how to achieve the goal. If we view people in a pessimistic light, then we may want to adopt the instruction-based approach—identify up front the "correct" way to perform the task, and then build the results of that analysis into a computer-based work flow that leads workers along and enforces that "proper" way of doing the job (e.g., Andognini, 1997; Marsden, 1996).

Second, this difference between levels of analysis can also be viewed from the perspective of accommodating performance variability (cf. T. J. Smith, Henning, & K. U. Smith, 1994). As shown in Fig. 3.4, constraint-based analyses accommodate a great deal of variability in worker action (i.e., many ways of doing to the task), whereas instruction-based analyses accommodate very little variability in worker action (i.e., in the extreme, only one way of doing the task). This observation has implications for two very important factors, namely the extent to which workers encounter situations that foster learning and the ability that workers are given to deal with unusual circumstances.

With regard to opportunities for learning, in the subarea of control theory known as system identification (e.g., Johannson, 1993), it is well established that it is necessary to "stimulate" a system by varying its control inputs widely to obtain a comprehensive understanding of the characteristics of that system. Also, in the subarea of control theory known as adaptive control (e.g., Narendra, 1986), it is well known that action variability provides the "persistent excitation" that is required for adaptive control. Furthermore, in psychology it is well established that providing opportunities for people to engage in exploratory behavior fosters learning (E. J. Gibson, 1991). In motor control, researchers have found that the inherent variability in the human motor system serves an educational function because it allows people to experience the varying set of conditions that are required for improved motor skill (Bernstein, 1996). These converging findings from disparate research areas show that there is a very robust relationship between action variability and learning opportunities (cf. Norros, 1996).

As for the ability to deal with novelty, if the conditions under which a task is to be performed vary from one situation to the next (e.g., starting to the right of the goal in Fig. 3.4 rather than to the left), then the actions that are required to get to the goal must differ for the two cases (cf. Bartlett, 1958; Bernstein, 1996). This relationship is explored in detail in the next section. For now, we merely observe that constraint-based approaches to task analysis provide workers with more flexibility and better opportunities to learn than do instruction-based approaches.

Third, recall that the definitions provided earlier indicate that task analysis identifies the tasks that *workers* are required to perform. But, at a detailed level, what workers are required to do depends on the device they have available to them. Note that, in our terms (see Glossary), *device* refers to the interface and automation but not to the work domain. Thus, any detailed account of a task is likely to make assumptions about the device that workers have available to perform the task. Instruction-based analyses are subject to this limitation. We saw this in the very simple case of figuring out the gas consumption for an automobile. If our car has a trip odometer, then the actions that we must perform to complete the task are different from those that are required if a trip odometer is not available. Similarly, if a calculator is available, the set of actions that we need to perform to complete the task will also change, in part because some steps have been automated and are thus not under workers' purview. Not only that, but even the type of calculator makes a difference. If automatic conversion functions are available, then we do not have to know what the multipliers are; otherwise we do. Therefore, instruction-based analyses are more *device-dependent* (Benyon, 1992a) because their content and form changes as a function of the interface and automation that workers have available to perform the task.

In contrast, because constraint-based analyses merely provide an envelope on task achievement, they are more *device-independent*. For example, the requirements shown in Fig. 3.1 must be taken into account if the task is to be accomplished correctly and consistently. If the fact that there are 3.785 liters in a gallon is not taken into account somehow, whether it be by the worker or by the device (e.g., a calculator), then the correct answer cannot possibly be consistently obtained. It is worthwhile emphasizing that this statement is true, even if the actor performing the task is not a person. An ant, for instance, would have to deal with the same constraints. In general terms, some relationships are properties of a task and quite independent of any particular device (Diaper & Addison, 1992). By constraining rather than instructing workers, constraint-based analyses describe properties of the task, rather than properties of how to do the task with a particular device. This difference may be subtle, but it is quite important. Of course, the device-dependent/independent distinction lies on a continuum rather than in the two discrete, bipolar categories we have been describing, but this does not change the main point of our argument. The level of detail at which a task analysis is conducted is proportional to the device-dependence of that analysis. Therefore, constraint-based analyses are generally less device-dependent than are instruction-based analyses.

Consequently, instruction-based analyses are plagued by a fundamental logical problem that has very important design implications. If an instruction-based task analysis is going to be performed, then assumptions about the design of the device (i.e., computer interface and automation) available to workers must be made. The reason for this is that the particulars of the timeline or flow sequence will change as a function of the device that is available (see earlier discussion). But, the reason we want to do a task analysis in the first place is to figure out how the device should be designed. We are thus faced with an infinite regress that is apparent, albeit implicit, in the practical guide to task analysis by Kirwan and Ainsworth (1992):

- Task analysis defines what a *worker* is required to do.
- Task analysis can be used by designers at an early stage of design to decide how to allocate functions between workers and machines.
- Some tasks are better performed by workers whereas other tasks are better performed by machines.

What is left implicit is the fact that different function allocation policies lead to different worker responsibilities. For instance, if a task is completely automated, then there may be no worker activities for that task. At the other extreme, if there is no automation, then all of the activities may have to be performed by the worker. The very same logic applies to interface design. Returning to the trip odometer example, with one interface a worker may have to perform one action (calculate the difference between current and previous km), whereas with a different interface a worker may have to perform a very different action (read the trip odometer) to achieve *the very same task goal.* So we are left with a conundrum where "what the worker is required to do" is both an input to, and an output of, the design process. Note that it is not possible to escape this dilemma completely by iteration because any design decision can result in a change in worker activities or responsibilities. The implications of this task-artifact cycle (Carroll et al., 1991) are discussed in more detail in chapter 4.

Conclusions

What can we conclude from this review of task analysis techniques? The primary advantage of the instruction-based approach seems to be that more guidance is provided. By following the steps laid out in a flow sequence or timeline, workers may be less likely to forget a step, perform the wrong step, or perform the right steps in the wrong order. In short, by providing workers with a "precompiled" solution to task demands, there may be less chance of human errors arising from human information-processing limitations.

The constraint-based approach appears to have several advantages. First, more discretion is given to workers to decide exactly how the task should be performed. Based on Karasek and Theorell's (1990) demand-control model (see chap. 1), we would therefore expect better worker health. Second, constraint-based approaches also accommodate greater variability in action. As a result, workers have more opportunities to learn and better chances of coping with unusual or changing circumstances. Thus, constraint-based approaches foster learning and support greater flexibility. Third, constraint-based approaches make fewer assumptions about the properties of the device (i.e., interface + automation) available to workers. Consequently, it is more likely that the new design will result in new functional possibilities, rather than being constrained by designers' current assumptions about functionality. Given Landauer's (1995) arguments and evidence (see chap. 1), we would thereby expect better productivity.

When taken together, these arguments seem to point in two conflicting directions. On the one hand, it seems that we should adopt instruction-based approaches to task analysis to provide workers with detailed guidance that they can use to perform tasks with minimal errors. On the other hand, it seems that we should adopt constraint-based approaches to give workers more discretion, accommodate more variability in action, and identify new possibilities for doing the job, thereby resulting in better health, flexibility, and productivity, respectively. Given that instruction-based approaches are, by far, the most prevalent, we might think that their benefits outweigh those of constraint-based techniques. This possibility is explored next.

CONSTRAINTS OR INSTRUCTIONS? THE VIEW FROM CONTROL THEORY

Control theory provides a generic language for understanding action (Flach, 1990a; Powers, 1973a, 1973b; T. J. Smith et al., 1994), and so we use it here to resolve the apparent conflict between constraint-based and instruction-based approaches to task analysis. Although control theorists have developed sophisticated mathematical formalisms for analyzing and representing systems, we only use qualitative concepts from control theory in an informal way to simplify the discussion. In this subsection (based on the work of Marken, 1986; Powers, 1973a, 1973b, 1978), we present two models of goal-oriented behavior, the second being more complex than the first. The basic assumption behind each model is that people are adaptive, goal-oriented agents. Although they are not realistic representations of human action in complex systems, the models can nevertheless be used to uncover a number of important insights.

Simple Negative Feedback Loop

Figure 3.5 shows the simplest possible model of goal-oriented behavior, a negative feedback loop. The variables in the loop are:

Goal *(g)*—the desired state to be achieved and maintained.
Output *(o)*—the output of the system (i.e., the current state of the plant).
Error *(e)*—the difference between the desired state and the current state.
Action *(a)*—the action of the worker.

Although our notation does not make this explicit, each of these variables is dynamic. In addition to these four variables, the negative feedback loop is also characterized by two parameters:

W—the control strategy used by the worker.
P—the dynamics of the plant (the control-theoretic term for the system being controlled).

Given these definitions, a number of deductions can be made. First, the action is determined by the error signal and the strategy used by the worker:

$$a = W \ e. \tag{1}$$

The error signal, in turn, is the difference between the goal state and the output state:

$$e = g - o. \tag{2}$$

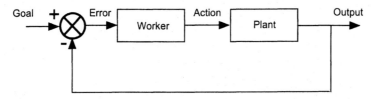

FIG. 3.5. A very simple model of goal-directed behavior.

Substituting Equation 2 into Equation 1 leads to the following:

$$a = W\,(g - o). \tag{3}$$

That is, the actions taken by workers are determined by the goal state, the state of the plant, and the control strategy used by workers. If we know these three terms, then we can predict the sequence of actions that workers should take. This is precisely the approach adopted by instruction-based task analyses (see earlier discussion). The analyst specifies the goal (g), *assumes* a set of initial conditions (o at time t = 0) and the strategy that the worker will use (W), and then derives a flow sequence or timeline of actions (a) that will satisfy the goal (e.g., Kirwan, 1992).

However, in complex sociotechnical systems, the initial conditions frequently cannot be known with certainty. For example, when a particular task is to be performed on a flexible manufacturing system, the initial configuration of that system may vary from one situation to the next. Similarly, the initial state of a petrochemical plant may also vary over time when performing say a start-up task. What happens in these types of situations? Because o at time t = 0 is unknown, a cannot be predicted by the analyst. This simple lesson is familiar to anyone who has ever tried to follow a set of instructions for using say a home appliance after having made an error. The instructions (basically a list of actions) are based on certain assumptions about the initial state of the world. When we make a mistake, those assumptions are no longer valid. As a result, we can no longer use the instructions to perform the task, unless we can find a way to resynchronize ourselves with the states that are assumed in the instructions. The very same conclusion applies when conducting an instruction-based task analysis—if the initial conditions are unknown, then we cannot predict the actions that workers need to take to perform the task.

There is a similar problem arising from uncertainty associated with W, the worker's strategy. In complex sociotechnical systems with many degrees of freedom, there can frequently be more than one strategy to achieve the very same task goal (see chap. 9). For example, if we emphasize speed, then one strategy may be viable; whereas if we emphasize accuracy, a different strategy may be preferable. Thus, we may not be able to predict the actual strategy used by workers, particularly if different strategies are used by different workers or by the same worker on different occasions. Under these conditions, a will again be unpredictable because of its dependence on W. Thus, there are two reasons why we may not be able to predict the precise actions required to perform a particular task, both causing difficulties for instruction-based task analysis techniques.

We can gain additional insights into the relationship between the predictability of conditions, the variability in action, and the feasibility of constraint- and instruction-based forms of task analysis by adopting a slightly more complex model of goal-directed behavior.

Negative Feedback Loop With a Disturbance

Figure 3.6 illustrates a negative feedback loop like the one in Fig. 3.5, except that a source of disturbances (d) has been added to the output of the plant. Disturbances are factors or events that affect the state of the plant in ways that have not been, or cannot be, anticipated by system designers. A very simple example is the wind that can push your car when you are driving. Although designers can expect that wind

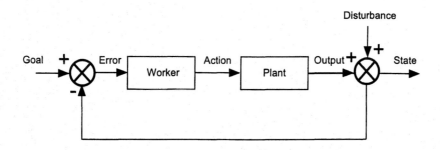

FIG. 3.6. A slightly more complex model of goal-directed behavior.

gusts will occur, they cannot anticipate their magnitude and timing. The wind gusts *(d)* are added to the output *(o)* of your car and thereby cause a different state of the world *(s)* (i.e., the car is in a different position than it would have been in without the wind). In the simpler previous model (Fig. 3.5), the output *(o)* and the state of the world *(s)* were the same thing because there was no disturbance *(d)*. However, in this more complex model, the output *(o)* and the state of the world *(s)* are different because of the presence of the disturbance *(d)*. Many kinds of disturbances can be found in complex sociotechnical systems, some relatively minor and some more complex: an unanticipated change in the economy in the financial industry, a rush order in the manufacturing industry, an unanticipated network failure in the telecommunications industry, an unexpected allergic reaction of a patient to an antibiotic in the medical domain, and multiple, interacting equipment failures in a nuclear power plant (cf. Norros, 1996). The severity of these examples should make it clear that it is important to understand the implications of disturbances for the control of complex sociotechnical systems. Although there can be many different sources of disturbance, for now, we do not discriminate between them. All of them are lumped together in the variable *d*.

As with the previous case, a number of deductions can be made from this very simple model. The actions are still a function of the error signal and the strategy used by the worker:

$$a = W\ e. \tag{4}$$

In this case, the error signal is the difference between the goal state and the state of the world:

$$e = g - s. \tag{5}$$

The output of the plant is determined by the actions performed by the worker and the dynamics of the plant itself:

$$o = P\ a. \tag{6}$$

Also, the state of the world is a simple function of two variables, the disturbance and the output from the plant:

$$s = o + d. \tag{7}$$

Substituting Equation 7 into Equation 5 leads to the following:

$$e = g - (o + d). \tag{8}$$

If we substitute Equation 6 into Equation 8, we get:

$$e = g - (P\ a + d). \tag{9}$$

For the goal to be satisfied, the error signal must go to zero. If we let $e = 0$ in Equation 9, we can see what relationship must be true for the goal to be achieved:

$$g = a\,P + d. \tag{10}$$

Solving for the actions that the worker must perform, we get:

$$a = (g - d)\ /\ P. \tag{11}$$

Equation 11 has a straightforward, yet very important, qualitative interpretation. The actions that workers must perform (a) to achieve the goal are a function of the goal itself (g), the dynamics of the plant (P), and the disturbance acting on the system (d). But as we mentioned before, the disturbance—by definition—cannot be predicted. It follows that the actions that workers need to take to satisfy the goal also cannot be predicted.[6]

The example mentioned earlier of driving a car under windy conditions gives an intuitive appreciation for this fact. If the goal (g) is to be satisfied (e.g., stay in your lane), the actions (a) performed by the driver must oppose the disturbances (d) caused by the wind. If the wind pushes the car to the left, the driver must respond by steering to the right, and vice versa. But because nobody (including the driver) can predict what the pattern of disturbances will be, the precise actions that the driver must take, as well as their timing, cannot be planned up front. Instead, they must be generated online, in real time, as the wind acts on the car.

There are several points that follow from this observation. First, the very same set of actions will have different effects at different times because the pattern of disturbances will rarely be repeated. Each situation is a unique context characterized by a different set of contingencies. Second, as a result of this situated context, if the same goal is to be achieved on different occasions, then a different set of actions will be required to achieve the very same task goal. Thus, the actions have to vary if the outcome is to remain the same. This phenomenon has been referred to as *context-conditioned variability* in the motor control literature (Turvey, Shaw, & Mace, 1978), *unanticipated variability* in the cognitive engineering literature (Roth, Bennett, & Woods, 1987), and *situated action* in the cognitive science literature (Suchman,

---

[6]Some of you might find it difficult to reconcile the fact that the workers' actions cannot be predicted in these two negative feedback models with the fact that negative feedback control systems work so well in many applications. In fact, there is no conflict because feedback control systems do not predict their actions either. Instead, these are generated online, in real time, as a function of the current context.

1987). We adopt the term context-conditioned variability because it has historical precedence and because it seems to be the most transparent label for this important phenomenon.

Resolution

We are now in a position to try to resolve the apparent conflict between constraint- and instruction-based approaches to task analysis described earlier. The feasibility of the two types of analysis can be understood by distinguishing between closed and open systems. *Closed systems* are completely isolated from their environment. From the viewpoint of the analyst, the behavior of the system can be well understood by examining influences that are internal to the system itself. Conversely, *open systems* are subject to influences (i.e., unpredictable disturbances) that are external to the system.[7] Note that the closed–open system distinction lies on a continuum. Some systems may be more open (i.e., be affected more by disturbances) than others.

With this distinction in hand, the apparent conflict between conducting an instruction-based analysis to reduce human error and a constraint-based analysis to support safety, flexibility, and health may be resolved in a systematic way. The more closed a system is, the more amenable it is to instruction-based forms of task analysis. More concretely, in cases where the analyst can anticipate the conditions under which the work will be done, a realistic and appropriate flow sequence or timeline may be specified. Such conditions are usually obtained in highly proceduralized tasks where the initial conditions are predictable and there are not that many correct ways to perform the task (i.e., the degrees of freedom are few or none). Conversely, the more open a system is, the less amenable it is to an instruction-based form of task analysis. Because there are unpredictable external disturbances acting on the system, it will not be possible to accurately preidentify the different flow sequences or timelines that lead to the satisfaction of the goal. Open systems give rise to context-conditioned variability. Workers must adapt online in real time to disturbances that cannot possibly be foreseen by analysts (Hirschhorn, 1984). Such conditions are usually obtained in cognitive tasks where there are many ways of achieving the same end (e.g., creative problem-solving tasks), or in dynamic systems whose demands change in an unpredictable manner (e.g., Norros, 1996).

Given this insight, we can understand why instruction-based task analysis techniques have been used effectively by system designers for many years, and why they are much more prevalent than constraint-based approaches. In the past, the demands on workers have been comparatively procedural in nature (see chap. 1), the prototypical example being forced assembly line operations. Consequently, analysts could treat such systems as being essentially closed for practical purposes.

---

[7]The system is defined by the analyst (i.e., it is an epistemological entity, not an ontological entity). As a result, if the model adopted by the analyst has gross deficiencies, then the system will be open because it is being affected by factors that are not accounted for in the analyst's model. Therefore, at any one point in time, it is generally not possible to distinguish between the following two situations: (a) External factors are affecting the behavior of the system, and (b) internal factors not accounted for by the analyst's model are governing the behavior of the system. Only when better models are developed can it be established that a situation that looked like case (a) actually turned out to be case (b). For our purposes, it is sufficient to include both cases under the definition of open system because each presents the analyst with sources of disturbance that—by definition—cannot be anticipated.

In these cases, instruction-based analyses were feasible and captured task demands in a way that was sufficiently accurate to derive useful design implications. But more recently, as the demands on workers have become more complex, systems are becoming more and more open (see chap. 1). Specifically, complex sociotechnical systems are subject to external disturbances (e.g., faults, changes in demands) that must be compensated for; they are subject to various forms of uncertainty that must be accommodated; their demands are primarily cognitive and social in nature, so there is no one right way of getting the job done; their behavior is dynamic, requiring workers to adapt to moment-by-moment changes in context. Collectively, these factors create a need for context-conditioned variability that is substantial enough in magnitude that it cannot be meaningfully ignored (cf. Norros, 1996). Our story in the Preface is a vivid indication of this fact. Operators sometimes *had* to deviate from the "one best way" embedded in their procedures, otherwise the desired goal state would not be achieved. As the quote from Norman (1998) at the start of this chapter indicates, the general point is that, in complex sociotechnical systems, it is not possible to identify a flow sequence or timeline that is appropriate for every possible situation. In other words, instruction-based task analysis techniques do not do justice to the richness of the set of actions that are required to cope with the entire set of job demands. The flow sequence and timing of the actions that will satisfy even anticipated goals are not fixed, are too numerous to enumerate, and are subject to factors that cannot be identified a priori (for an example, see Ujita, Kawano, & Yoshimura, 1995).

These limitations have been identified previously, not just by critics of instruction-based task analysis (e.g., Bannon & Bødker, 1991; Greif, 1991; Hirschhorn, 1984), but even more important, by its proponents as well:

- "Workload assessment methods (such as timeline analysis) are relatively crude and measure the operator time involved at a console but not the cognitive demands or 'burden' actually levied on the operator in non-routine situations. More sophisticated technology needs to be developed" (Kirwan, 1992, p. 386).

- "As the cognitive (i.e., mental) content increases, [flow sequence] representations become less satisfactory. In particular there may be many potential internal cognitive mechanisms and the real cognitive structures used by the operators may be 'opportunistic', rather than clear decisions made on strict criteria. Modeling such activity can be difficult" (Kirwan & Ainsworth, 1992, p. 94).

- "As with timeline analysis, [task analytic] simulation techniques are currently poor at covering mental aspects of workload" (McLeod & Sherwood-Jones, 1992, p. 308).

- "Decision making and problem solving . . . activities can be solved successfully by several different strategies and the individual choice will depend on very subjective criteria" (Meister, 1995a, p. 120).

Although these limitations are frequently ignored in practice (e.g., see some of the case studies in Kirwan & Ainsworth, 1992), there seems to be little doubt that instruction-based forms of task analysis are strongly limited for open systems, such as those described in chapter 1, because they tend to ignore the variability and richness of action in such systems. This limitation becomes especially significant when we consider that the vast majority of existing task analysis techniques are instruction based, and thus not particularly well suited for our purposes.[8]

This criticism is not a call to completely abandon procedures. On the contrary, procedural guidance of some type has a useful role to play in complex sociotechnical systems (see chap. 8). The point is instead that instruction-based approaches are rigid and that we must find some way of supporting the context-conditioned variability needed for open systems.

Who Cares? Design Implications

What are the practical implications of this criticism for design? What happens if we design a computer-based aid for an open system using an instruction-based task analysis? Will it be "good enough," or will it lead to significant problems that merit our attention?

Bisantz et al. (1996) provided a cogent example of what happens when an instruction-based form of task analysis is used to design a computer aid to support workers in an open system requiring context-conditioned variability. They conducted a field study of a computer information system that was intended to help cooks in fast-food restaurants make decisions about what type of food to cook, how much to cook, and when to cook it. There are conflicting criteria governing these decisions: (a) the cooks need to have enough food already prepared to meet customer demand in a timely fashion, and (b) the cooks should not prepare too much food that will not be purchased, otherwise it will go to waste.

The restaurants that were observed in the field study had installed an information system that provided minute-by-minute cooking instructions to assist cooks in making the decisions identified above. The prescriptions made by the aid were projections of appropriate actions, based on predicted sales patterns, actual current sales, and the amount of each product currently on hand. Most of these inputs had already been stored in the computer system by its designers, but the amount of each product currently on hand had to be entered in manually by the cooks. In other words, the information system prescribed to workers what actions were to be performed, in what order, and at precisely what time, based on preidentified variables and relationships. We can infer from this that the computer system was based on some type of instruction-based analysis.

An important characteristic of this work environment that was not taken into account in the design of the aid is that there are several sources of disturbances. For example, customer demand is not completely predictable. To be sure, there are certain trends as a function of the time of day (e.g., lunch, dinner). However, unexpected rushes can and do occur due to local community happenings or due to busloads of travelers coming off of the highway. There is also a disturbance (or source of uncertainty) introduced by the fact that cooks would sometimes forget to input the amount of product on hand, enter it incorrectly, or enter it only after a

---

[8]It is possible, of course, that some tasks in some complex sociotechnical systems may be so constrained that there is only one way to perform the task safely or efficiently. In such rare cases, a constraint-based task analysis would reveal that there are no degrees of freedom for action, and that a detailed procedure would in fact be appropriate (see chap. 5). The result in this case would be similar to that obtained from an instruction-based task analysis.

substantial delay. The amount of product currently available is one of the inputs into the algorithm governing the recommendations made by the aid, so if there are errors, missing data, or old data in the input, the recommendations made by the aid will not be appropriate. This possibility was not taken into account by the rationalized, ideal assumptions built into the design of the computer aid. In summary, there are external disturbances making this an open system.

The control-theoretic analysis presented earlier allows us to predict how well this computer aid should function in such an environment. The aid is essentially trying to enforce a particular way of doing the task that is rational given the assumptions built into the aid. We would thereby expect that the computer system should make it very difficult, or perhaps even impossible, to accomplish the task in any way other than the rationalized set of actions prescribed by the aid. And because this environment is subject to external disturbances, it should be difficult for cooks to use the computer aid and still adapt to the contingencies of the moment. In other words, because the aid seems to have been designed according to an instruction-based task analysis, it should not be flexible enough to support the context-conditioned variability that is required to achieve task goals in this open system.

The empirical observations of Bisantz et al. (1996) are entirely consistent with this prediction. Because the aid was too rigid, it went essentially unused, despite the fact that there was considerable pressure imposed on workers by management to obey the aid's cooking instructions. Note that this finding was obtained in a comparatively simple fast-food restaurant, not a nuclear power plant! Although it does not rate highly on many of the dimensions of complexity described in chapter 1, the fast-food environment is sufficiently open to disturbances that instruction-based task analyses lead to the design of ineffective information systems. Moreover, this finding is not an isolated one. There are other documented cases showing that computer information systems based on prescriptive rationalizations of how tasks should be conducted either go unused or are ineffective in open systems with unpredictable disturbances (e.g., Guerlain, 1995; Hirschhorn, 1984; Roth et al., 1987; Sachs, 1995).

The bottom line is that instruction-based forms of task analysis are not very useful for appreciably open systems. If this limitation is ignored, then the resulting design will range from being not as effective as it could be to unusable, as in the case of the fast-food example. And as we mentioned, most existing task analysis techniques are of the instruction-based type and thus not ideal for our purposes. What about constraint-based forms of task analysis? Can they meet the challenges of open systems?

Constraint-Based Approaches: A Partial Savior

Fortunately, constraint-based approaches to task analysis can accommodate some of the context-conditioned variability required to achieve goals in open systems. As we discussed earlier, that is one of the primary advantages of constraint-based analyses—they are flexible enough to accommodate variability in worker action. Because they specify only the goal to be achieved and the constraints on achieving that goal, such analyses are indifferent to variations in precisely how the goal is to be achieved. In fact, both of the negative feedback models discussed earlier (see Figs. 3.5 and 3.6) are actually constraint-based structures, not instruction-based. There is no pre-planned set of actions built into the control loops. On the contrary, the actions that are required to achieve the goal *emerge* on the spot in real time as a function of the initial conditions and disturbances of the moment. In other words, the negative feedback models are capable of a very elementary form of context-conditioned variability (or situated action; Suchman, 1987).

The only factors that remain constant in the control loop are the goal to be achieved, and the constraints on goal achievement. Constraint-based approaches provide only this level of guidance. For closed systems, this vagueness could be a disadvantage. It might be preferable to minimize errors caused by human informa-tion-processing limitations by providing more detailed guidance (e.g., a flow se-quence or a timeline). However, the preceding analysis shows that this vagueness is actually an advantage for open systems. The only way to adapt to disturbances is to leave some decisions about how the task should be performed to the worker, rather than to specify all of the details up front based on idealized (and unrealistic) assumptions about what should be done.

It is extremely important to note, however, that *discretion is not the same as complete freedom.* This is a point that is frequently overlooked, and therefore, a common cause of misunderstanding of the ideas presented in this book. We are not advocating that workers be allowed to do whatever they want. This would be irresponsible because it would include unintentional errors ("it seemed like a good thing to do at the time"), not to mention deliberate acts of sabotage. The discretion and flexibility that we are advocating is bounded by constraints. Errors and acts of sabotage fall into the areas of the state space that workers are required to avoid. Thus, the position we are advocating is to give workers discretion *within the boundaries of safe and effective operation.* Does this mean that constraint-based task analyses are the work analysis technique of choice for complex sociotechnical systems? Unfortunately not.

Taking Stock and Looking Forward: An Unresolved Problem

Recall that we began this chapter with the goal of determining if normative work analyses could meet the challenges imposed by complex sociotechnical systems. Task analysis was chosen as a prototypical example of such normative methods, and several different techniques were described. We realized that these different techniques could be organized into two categories, constraint based and instruction based. Instruction-based techniques are, by far, the most numerous, so we might expect that they would be more useful. We examined the viability of each of these categories of task analysis for complex sociotechnical systems by adopting a con-trol-theoretic perspective. Our conclusion was that instruction-based analyses are not very useful for systems that are substantially open to unpredictable disturbances. Complex sociotechnical systems clearly fall into this category, as do many other simpler work environments.

These conclusions were reinforced by the case study of an information system installed in fast-food restaurants. The computer aid observed there seemed to be designed according to an instruction-based analysis of the task. Because the work environment was subject to unpredictable disturbances, workers did not use the aid for the purposes for which it was intended. Empirical data from other studies cited earlier reinforce the same point—instruction-based task analyses lead to brittle computer systems that cannot accommodate the context-conditioned variability that is required for efficient and successful performance in open systems. And because most existing task analysis techniques are of the instruction-based variety, they are not well suited to our purposes.

Constraint-based forms of task analysis provide an attractive alternative. They give workers more discretion, thereby fostering better worker health (Karasek & Theorell, 1990). They can lead to new design functionality, making it more likely that productivity will be improved (Landauer, 1995). They also accommodate more variability in worker action, providing the flexibility required to adapt to unpredictable disturbances and increased opportunities for learning. Thus, constraint-based forms of task analysis are well suited to the demands of complex sociotechnical systems, and go partway toward addressing the challenges imposed by such systems.

However, they are not capable of meeting all of those challenges. This important qualification can be recognized if we uncover an implicit assumption that is buried in the two control-theoretic models presented earlier and the case study from the fast-food industry. In all three of these cases, it has been assumed that the goal to be achieved is well defined and can be established ahead of time, even though the sequence and timing of actions that are required to achieve the goal cannot be predicted. But there are some situations in complex sociotechnical systems where even the particular goal to be achieved may not be identifiable beforehand. The prototypical example is an unanticipated emergency in a nuclear power plant (e.g., the TMI accident discussed in chap. 2). If the event that triggers the emergency cannot be known beforehand, how can we know what the goal should be? Although it is not widely recognized, this problem can exist in many comparatively more mundane situations. A good example was provided by Shepherd (1992) in an application of task analysis to maintenance training for mechanical fitters in a chemical company: "A major problem that had to be dealt with at the outset was that fitters, like most craftsmen, are supposed to do anything they are called upon to do: there is apparently no one task to analyze" (p. 328).

One way to deal with this dilemma is to specify the task goal at a high (i.e., vague) level. For example, Shepherd (1992) identified one task goal as requiring workers to "Note any problems" (p. 335). This tactic tries to overcome ignorance about the precise goal to be pursued by specifying a very broad goal that subsumes all of the more specific goals that might be associated with unanticipated events. The problem with this approach is that it merely provides a place holder for what workers are supposed to do. It does very little to identify the information or knowledge that workers require to cope with the novelty imposed by unanticipated events.

The crux of the problem seems to be that all task analysis methods are *event-dependent* (Vicente & Tanabe, 1993). They require a specification (or assumption) of at least a class of initiating events before the analysis can even get off the ground. Otherwise, the precise goal to be pursued cannot be identified. Therefore, even constraint-based forms of task analysis do not provide a very satisfactory basis for dealing with unanticipated events.

This is an unacceptable situation for complex sociotechnical systems because, as we mentioned in chapter 1, unanticipated events pose the greatest risk to safety in such systems. We need some other form of work analysis if we are to support workers in these challenging situations. More specifically, we need *event-independent* work analysis techniques (Vicente & Tanabe, 1993) whose relevance and utility are not tied to a specific, finite class of anticipated events. In the next section, we develop a better idea of how this need can be met.

TABLE 3.2 Relative Advantages and Disadvantages of Directions and Maps		
	Directions	**Maps**
Mental economy	efficient	effortful
Ability to adapt to unforeseen contingencies	brittle	flexible
Scope of applicability	narrow	broad
Ability to recover from errors	limited	great

DEALING WITH UNANTICIPATED EVENTS: THE VIEW FROM SPATIAL NAVIGATION

In this section, we derive some insights about how to conduct an event-independent work analysis by drawing on a problem with which everyone is familiar, namely spatial navigation. This analogy reveals the power of *work domain analysis* (see Footnote 1), a complementary form of work analysis that overcomes the weaknesses of task analysis.

Directions Versus Maps

Thorndyke and Goldin (1983) investigated how people learned to find their way in their everyday environments (e.g., while walking or driving in a city). Among the types of spatial knowledge they identified were procedural knowledge and survey knowledge.[9] *Procedural knowledge* represents the sequence of actions that people are required to take to follow a particular route. An example would be a set of directions to get from your home to a friend's home by rote (e.g., Go up highway 427 northbound, take the Rathburn Road exit west, keep going straight for about 2 km, turn left at Mill Road, etc.). *Survey knowledge* represents the spatial relationships between locations and routes in an environment. An example would be someone who has bird's-eye-view knowledge of the relative location of the streets in their neighborhood. Thus, procedural knowledge tells you what to do, whereas survey knowledge just tells you about the layout of the land. These two types of knowledge lead to the design of two complementary types of navigation aids. Procedural knowledge can be embedded in *directions,* and survey knowledge can be embedded in a *map.*

Strengths and Weaknesses. These two types of spatial aids have complementary advantages and disadvantages, as shown in Table 3.2. Instructions are more efficient because they tell you what you have to do. This mental economy can be achieved only because someone (e.g., the person who gave you the directions) was able to derive an appropriate set of actions to get from a particular starting point to a particular destination. They had to conduct this "analysis" beforehand and only then could they have built in the insights gained from that analysis into a set of

---

[9] The third form of spatial representation identified by Thorndyke and Goldin (1983) was landmark knowledge.

directions. Because they have already done most of the thinking for us, all we have to do is follow the actions in the directions. In contrast, with a map we have to do the thinking ourselves to derive a particular route from where we are to where we want to be. As a result, navigating with a map is less efficient than navigating with directions in situations for which directions are available.

However, this decrease in efficiency is compensated for by an increase in both flexibility and generality. Maps are more flexible because they allow us to derive different routes to get to the same location. This is particularly useful when an unforeseen event occurs. For example, if we are driving home and find that there is an accident on our regular route, causing a tremendous traffic jam, we can use our map to adapt online by deriving a new route to our destination that avoids the traffic jam. In contrast, if we only had directions, we would not be able to adapt to this unforeseen contingency. All we would have is a procedure for how to get from one point to another using a particular route. As soon as we deviate from that route, our directions would be of little use because they are tailored to that one sequence of actions. Thus, although they may be less economic, maps are more flexible (or conversely, less brittle) than directions.

Maps also have another related advantage, namely their generality. Because they are analogical representations (Woods, 1998), maps contain all of the possible destinations and starting points for a given geographical area (assuming they are comprehensive and accurate).[10] This property provides a tremendous amount of generality. It allows us to derive a route to get from any one point to any other point. In other words, maps are event-independent in the sense that their applicability is not tied to the particular starting or destination points that comprise a particular navigation event. This is an obvious advantage, especially if we frequently have to deal with novel travel plans. For example, if we are salespeople and we have to get to the same destination (e.g., our home) from a new starting point at the end of each day (e.g., a different customer's place of business), then maps provide us with the type of representation that we need to accommodate this variability in task demands. Similarly, if we leave from the same starting point at the start of each day (our home again), but have to drive to different destinations every day, then maps offer the same powerful generality. This provides a stark contrast to the limited possibilities made available by directions. Because they are tailored to one starting point, one route, and one destination, directions have a much more limited scope of applicability.

This observation also has important implications for the ability to recover from errors. If all we have is a set of directions, as soon as we make a mistake our navigational aid is useless, unless we can get back onto the route embedded in our directions or unless the particular error has been anticipated and built into the directions (e.g., "if you get to the bridge, you have gone too far"). In contrast, if we have an accurate and comprehensive map, recovery from errors is effortful, but possible. It is effortful because it requires some replanning to compensate for the error and get back on track. It is possible, however, because the utility of the map

---

[10]You may be wondering what happens if a map is out of date or contains an error. One of the benefits of analogical representations, of which maps are an example, is that they contain a great deal of redundancy (e.g., the position of any one object can be cross-referenced from many different locations). This redundancy can actually allow people to discover errors in the representation through a process of converging operations.

is independent of our current location (as long as the map covers the area of interest). Thus, we can determine how to get from where we mistakenly wound up to where we really want to go to.

Relevance to Work Analysis

The distinction between these two forms of navigational aids—directions and maps—has an analog in two forms of work analysis. Like directions, *task* representations tell workers what goals they should be trying to achieve, or also how they should be achieving them (see previous discussion). Like maps, *work domain* representations merely describe the structure of the controlled system (see chap. 7 for more details). More intuitively, a task is *what* workers do, whereas a work domain is what workers do it *on* (i.e., the object of action). For those familiar with computer programming, this distinction is similar to that between a control structure and a data structure, because the control structure operates on the data structure.

Table 3.3 shows that the comparative advantages and disadvantages of directions and maps are analogous to those between task and work domain analyses. Task analyses are efficient because they identify what needs to be done (in the case of constraint-based analyses), and perhaps even how it should be done (in the case of instruction-based analyses). But, as a result of this specificity, task analyses do not provide the support required to adapt to unanticipated events (see earlier discussion). Task analyses are also narrow in their generality because they are applicable only to the tasks that have been identified up front (as in the case of constraint-based analyses), or even more narrowly, to the particular ways of doing the task that have been identified up front (as in the case of instruction-based analyses). Finally, and as a result, task analyses are also limited in their ability to support recovery from errors.

As shown in Table 3.3, work domain analyses have a complementary set of strengths and weaknesses. Their primary disadvantage is that they do not tell workers what to do. They merely describe the capabilities of the system that workers will be acting on. As a result, work domain analyses put greater demands on workers for situations in which appropriate task descriptions are available. The good news is that work analyses are flexible because they provide workers with the information they need to generate an appropriate response, online in real time, to events that have not been anticipated by designers. Moreover, work domain analyses also have a broader scope of applicability. Because they merely show what the work domain is capable of doing—independent of any particular event—they provide workers

TABLE 3.3   Relative Advantages and Disadvantages of Task Analysis and Work Domain Analysis

	Task Analysis	Work Domain Analysis
Mental economy	efficient	effortful
Ability to adapt to unforeseen contingencies	brittle	flexible
Scope of applicability	narrow	broad
Ability to recover from errors	limited	great

with the discretion to meet the demands of the job in a variety of ways that suit their preferences or the particular needs of the moment. Finally, for the reasons already discussed, work domain analyses also provide workers with the support they need to recover from errors.

Conclusions

Two points are worth highlighting from this comparison. First, work domain analyses are absolutely essential for complex sociotechnical systems. They provide a way of supporting worker adaptation to novelty, thereby addressing the criterion of safety, and they provide discretion—not complete freedom—to workers, thereby addressing the criterion of health. Second, because they have complementary strengths and weaknesses, it would be useful to include both work domain analysis and constraint-based task analysis techniques in a single, integrated framework for work analysis. The framework described in the remainder of this book achieves this goal by including a work domain analysis phase (see chap. 7) and a constraint-based task analysis phase (see chap. 8) in one integrated, overarching framework.

SUMMARY

What type of work analysis is appropriate for complex sociotechnical systems? In this chapter, we evaluated the suitability of normative approaches to work analysis in the form of task analysis. We learned that the vast majority of existing task analysis techniques are instruction based. Yet, such techniques are not well suited for complex sociotechnical systems because they underestimate context-conditioned variability, and if used, can lead to unusable or ineffective computer information systems. Thus, most existing task analysis techniques are not very useful for our purposes. Constraint-based task analysis techniques are better suited for complex sociotechnical systems. Moreover, they are more likely to lead to improvements in flexibility, productivity, worker health, and on-the-job learning. However, they are not capable of dealing with the demands imposed by unanticipated events. Fortunately, work domain analyses provide a basis for dealing with such events. Thus, a work analysis framework for complex sociotechnical systems should include both work domain analysis and constraint-based task analysis techniques. The bottom line is that constraint-based task analysis techniques are necessary, but they are far from sufficient. In the next chapter, we see if additional insights can be garnered from descriptive approaches to work analysis.

Descriptive Approaches to Work Analysis: "What Workers Really Do" 4

The principle of flexibility creates a conception of work in which the worker's capacity to learn, to adapt, and to regulate the evolving [automation] becomes central to the machine system's developmental potential.
—Hirschhorn (1984, p. 58)

Innovation = imagination of what could be based on knowledge of what is.
—Lucy A. Suchman (personal communication, April 1997)

PURPOSE

In chapter 3, we reviewed normative approaches to work analysis to see how well they stand up to the unique challenges imposed by complex sociotechnical systems. In this chapter, we do the same for descriptive approaches to work analysis. The important contributions of descriptive approaches are illustrated through four case studies. The convergent findings from these cases help us identify additional dimensions that must be considered in work analysis. We argue that these descriptive techniques are very important and useful in understanding what workers really do and what they would like to do. Nevertheless, there are limitations in extracting design implications from descriptive approaches to work analysis. Our conclusion is that the descriptive analysis of current practice should be viewed as one of several possible means to investigate intrinsic work constraints, rather than an end in itself. Computer-based information systems should not be designed based solely on studies of current practice, nor should they be designed to support just the practices in which workers are currently engaged. These insights directly motivate a need for the formative approach to work analysis introduced in the following chapter.

DESCRIPTIVE APPROACHES: CURRENT PRACTICE

Scope

Descriptive approaches to work analysis are qualitatively different from normative approaches. Rather than postulating "rational" benchmarks for how workers should behave, descriptive approaches seek to understand how workers actually behave in practice. This goal is accomplished by conducting field studies that document the (usually quite dynamic) practical challenges that workers actually face on the job, and the (usually quite ingenious) practices that workers have developed to cope with those challenges. Thus, descriptive approaches to work analysis are called descriptive because they seek to document and understand current practice in order to suggest ideas for new designs.

In North America, descriptive work analyses, particularly field observations in naturalistic settings, are frequently seen as a relatively recent innovation. However, there is a rich tradition of field study research in Europe going back at least 30 years (e.g., see some of the selections in Edwards & Lees, 1974, as well as Pejtersen, 1973; Rasmussen & Jensen, 1973). The Francophone ergonomics community, in particular, has placed great emphasis on phenomenological descriptions of current practice in naturalistic work settings (for overviews in English, see De Keyser, 1991; De Keyser, Decortis, & Van Daele, 1988). This research tradition makes a strong distinction between task and activity (e.g., Leplat, 1989, 1990) that is equivalent to the distinction between normative and descriptive approaches to work analysis. The term *task* refers to the official actions that are prescribed to workers, and is thus representative of the normative approach. The term *activity*, on the other hand, refers to the informal actions that workers actually perform in practice, and is thus representative of the descriptive approach. This distinction is at the very heart of Francophone ergonomics, and shows that descriptive approaches to work analysis have a relatively long history, at least in Europe.

As with normative approaches, there are many different perspectives that can be categorized as being descriptive. A comprehensive review of this body of research would require a book of its own. Therefore, we limit ourselves to describing four prominent case studies of descriptive work analyses conducted by researchers from different disciplines. Despite the diversity in the background of the analysts and in the application domains, the findings from these studies exhibit a surprising degree of convergence.

What Is Current Practice?

A Case Study From Situated Action. Suchman (1987), an anthropologist, conducted a representative study (cf. Brunswik, 1956) examining how pairs of office workers cooperated and interacted with a prototype expert help system for a photocopying machine. She used conversation analysis—a perspective for understanding face-to-face communication between people—as a basis for understanding communication between workers and computers. A key insight from conversation analysis is that

> Communication succeeds in the face of . . . disturbances not because we predict reliably what will happen and thereby avoid problems, or even because we encounter problems that we have anticipated in advance, but because we work, moment by moment, to identify inevitable troubles that arise. (p. 83)

These were the same insights that we derived in generic form from the negative feedback models in chapter 3. In an open system with external disturbances, workers must exhibit context-conditioned variability if they are to accomplish task goals.

The theoretical foil (i.e., antagonist) that Suchman (1987) was arguing against in her study was the artificial intelligence approach to planning that had been frequently adopted by cognitive science researchers. This view assumes that detailed plans can be derived, a priori, to specify what actions must be taken to accomplish particular tasks. Essentially, this is the same perspective that is embodied in the instruction-based approaches to task analysis discussed in chapter 3.

Suchman (1987) compared these two views of human activity by studying how workers interacted with a computer system that was designed according to this very detailed instruction-based view of action. She videotaped office workers' first encounters with a computerized expert help system that was intended to help them operate a large and complex photocopier. In each session, two workers, neither of whom had used the expert system before, collaborated in pairs to perform representative photocopying tasks. The interactions between the workers and the expert system, and between the workers themselves, were logged, transcribed, and analyzed.

As we would suspect from chapter 3, Suchman (1987) observed that the success of an expert help system designed according to an instruction-based approach is "constrained by limitations on the designers' ability to predict any user's actions" (p. 120). The results of her study showed that this condition cannot be satisfied, even in the comparatively simple situation of two office workers interacting for the first time with a photocopier. Workers found alternate, equally plausible interpretations of the messages presented by both the help system and the photocopier, interpretations that unfortunately were not aligned with those that the designers intended. Furthermore, workers encountered contingencies that were not included in the detailed plans built into the expert help system. Consequently, the advice offered by the system was no longer appropriate for the unanticipated situation in which the workers found themselves, much like the computer "aid" for fast-food restaurants described in chapter 3. Needless to say, workers became confused and performance was far from efficient.

Suchman's (1987) conclusions are very similar to the ones we expressed in chapter 3. In open systems, context-conditioned variability is required to get the job done. In Suchman's own words, "purposeful actions are inevitably *situated actions*" (p. viii, emphasis in original) that are responding to "local interactions contingent on the actor's particular circumstances" (p. 28). Thus, the actual behavior of workers in situ is quite different from the rationalized ideal set forth by normative techniques for work analysis.

A Case Study From Naturalistic Decision Making. Klein (1989), an experimental psychologist, described several retrospective naturalistic studies of current practice that he conducted with his colleagues. Their first project investigated how fire-fighting commanders reported making decisions under stress. Klein and colleagues adopted a phenomenological approach, interviewing highly experienced commanders to learn how they made decisions in dynamic situations involving time pressure, risk, and personal accountability. In one study, the interviewees had an average of 23 years of job experience. During the interviews, they were asked to describe critical, nonroutine incidents that presented them with particularly challenging decision problems.

The theoretical foil for Klein's (1989) study was classic decision theory, which claims that workers should follow a thorough and "rational" approach if they are to make good decisions. For example, multiattribute utility theory models (e.g., Raiffa, 1968) recommend that workers: generate the set of options that are available, identify an appropriate set of evaluation dimensions, attach weights to each of these dimensions, rate each option on each dimension, and then select the option with the highest utility score. How well do these prescriptive approaches account for the decision-making behavior reported by the fire-fighting commanders interviewed by Klein?

Consistent with Simon's (1956) much earlier observations, Klein (1989) found that the normative canons prescribed by the classic approach to decision making did not do a very good job of accounting for the retrospective interview data he collected. This led him to propose *recognition-primed decisionmaking* (RPD), a descriptive model of how experts frequently make decisions under time pressure and risk (see Klein, 1989, for a complete description). The RPD model provides a strong contrast to normative decision models. Rather than generating many alternatives in parallel, assigning weights, and comparing options, the fire-fighting commanders reported that they were instead acting and reacting to the evolving situation on the basis of their experience. Why did experts "go with the flow" rather than follow the prescriptions of the normative approach?

Just as there were good reasons for the nuclear power plant operators in our story to deviate from their procedures (see Preface), there were also several good reasons why the fire fighters did not follow the canons of normative decision models. First, there was little time available to choose a course of action, so it would be very difficult to engage in the laborious mental deliberations recommended by prescriptive, analytical approaches. As a result, the fire-fighting commanders did not try to find an optimal action, but instead focused on identifying a course of action that was feasible, timely, and cost-effective. In Simon's (1956) terms, they satisficed rather than optimized. Second, and more important, the commanders were not novices at the job but instead had a great deal of experience that they could exploit. Consequently, they were able to rely on their well-tuned perceptual capabilities to recognize informative cues in the environment. Given these cues, the commanders could selectively and efficiently recognize a particular situation as belonging to a category with which they were familiar. And because they were able to recognize the situation, they also knew what actions were appropriate, based on what had worked in the past under similar circumstances. By relying on their experience, the first option that fire-fighting commanders identified frequently turned out to be a workable action. Rather than have to generate all possible action alternatives up front—including many that, on the basis of experience, would be clearly inappropriate for the current situation—commanders could quickly identify a single viable option that frequently allowed them to get on with doing their job. As one fire fighter put it, "We don't make decisions. We put out fires."

In summary, Klein found that expert fire-fighting commanders were able to exploit their experience base to recognize situations in an efficient and timely manner, and to directly associate that situation assessment with a relevant action. Subsequent studies in other domains have often led to similar results (see Klein, 1989, 1997, for reviews). These findings thereby lend some generalizability to RPD, "a descriptive model" of expert decision making (Klein, 1989, p. 85). They show that the way in which workers actually made decisions was quite different from the normative approach prescribed by classic decision theory.

A Case Study From Activity Theory. Bødker (1991), a computer scientist, was involved in a number of projects involving the participatory design of computer-based systems. The largest of these was the UTOPIA project, "a Scandinavian research project on trade union based development of and training in computer technology and work organization, especially text and image processing in the graphic industries" (Bødker, 1991, p. 7). As part of this project, descriptive work analyses were conducted to understand how workers currently achieved task goals. Based on these analyses, requirement specifications for future devices were developed and then prototypes were created and tested. Workers actively participated in each of these phases.

According to Bannon and Bødker (1991), the theoretical foil for Bødker's (1991) study was the Anglo-American approach to HCI. Although things have changed somewhat in the last 5 years or so, this research tradition had focused primarily on the artifacts being designed, with comparatively little regard for the context in which the artifacts were going to be used. For example, much more attention was paid to the syntactic and lexical features of the interface (e.g., Should windows be tiled or overlapping? How broad and deep should a menu be?) than to the semantics of the application domain. This emphasis could also be seen in the importance placed on usability—designing computer-based systems that are easy for workers to use—and the comparative lack of attention to usefulness—designing the functionality that is required to do a particular job effectively (cf. Landauer, 1995). Referring back to chapter 2, we can say that Anglo-American HCI was much more influenced by the cognitivist perspective than by the ecological perspective.

While conducting descriptive analyses of work during the UTOPIA project, Bødker (1991) concluded that the Anglo-American approach to HCI did not provide an accurate or comprehensive account of her findings. The traditional paradigm seemed too narrow and superficial to account for the richness and complexity of the practices exhibited by workers. This led Bødker to search for an alternative theoretical framework that was capable of explaining her descriptive findings. Note that Bødker was not content to merely document what she observed. She also wanted to identify a systematic framework consisting of well-defined concepts that could be meaningfully applied in naturalistic settings (cf. the motivation for our Glossary).

This quest led Bødker (1991) to *activity theory,* a psychological theory that had been developed by basic researchers in the Soviet Union. Bødker found that activity theory could be adapted to the practical needs of HCI, and more important, that its concepts provided a good descriptive understanding of the data collected during UTOPIA and other projects. Although a comprehensive account of activity theory is well beyond the scope of this chapter, we can briefly describe some of its characteristics to illustrate the general nature of the theory (see Bedny & Meister, 1997; Nardi, 1996b, for more theoretical details and sample applications). The core focus of activity theory is the study of goal-directed activity. As simple as this idea may appear, in the context of HCI it results in a profound shift in emphasis. Instead of focusing primarily on the device, activity theory focuses primarily on the goals that workers are trying to satisfy with the device. In this view, the device is merely a means, not an end in itself. Thus, rather than viewing HCI as working with computers, activity theory views HCI as working through computers.

This change in emphasis has several important implications. First, it requires a deep understanding of the semantics of the domain (see chap. 1). Rather than just focusing on superficial surface features of the device, analysts have to spend much more time understanding the application domain. Otherwise, they will never be able to understand workers' activities. Second, goal-directed activity is usually conducted with tools in a social context that has a cultural history, at least in naturalistic settings. Consequently, the scope of HCI must be considerably broadened to accommodate these additional factors. To understand activity, we must understand how tools influence current practices, a relationship we address in more detail later. We must also understand how people interact with, and learn from, one another. Thus, social and organizational issues come to the fore. In addition, we must understand how current practice has been influenced by cultural history. Decisions made in the past, and influences of the past, shape current practice. Thus, the development of these changes over time can itself become the object of inquiry. Third, because of its emphasis on studying activity in situ, activity theory puts a premium on understanding human expertise and skill. This characteristic is valuable in HCI applications where we want to construct computer-based systems that enhance and foster worker competence. Finally, as a result of all of these factors, activity theory views action as situated. Workers' actions are adapted to the current context. Thus, once again, we find the importance of context-conditioned variability in understanding workers' actions in open systems.

These general characteristics are represented in activity theory by a number of systematic concepts, such as: activity, action, operation, object, tool, subject, and community. According to Bødker (1991) and others (see Nardi, 1996b) these concepts, with their focus on practice, provide a much broader frame of reference for HCI than the traditional Anglo-American approach.

A Case Study From Distributed Cognition. Hutchins (1995a), an anthropologist, conducted a field study of current practice in the domain of ship navigation. He collected his data as an observer, primarily on the navigation bridge of a number of U.S. Navy ships at sea. During this time, he made detailed records of workers' actions and communications using written notes, audiotapes, and eventually, a video camera. The recordings were subsequently transcribed and analyzed based, in part, on Hutchins' extensive experience with, and knowledge of, the navigation task. The goal of these investigations was to better understand the nature of human cognition in naturalistic settings.

The theoretical foil for Hutchins' (1995a) work was the physical symbol system hypothesis (PSSH) put forth by A. Newell and Simon (1972). The PSSH is based primarily on findings from experiments with inexperienced people performing laboratory puzzles and games (e.g., cryptarithmetic, chess, and Tower of Hanoi). Note that these are closed systems (see chap. 3). Furthermore, in these experiments, people are typically faced with novel and thus, mentally challenging, situations. Nevertheless, people are usually forced to work alone and are usually not allowed to use any aids (e.g., pencil and paper) to solve the task at hand. The view that has developed from these studies is that human cognition involves mental information processing driven by well-defined rules and representations stored in human memory. In essence, human cognition is likened to a digital computer.

Hutchins' (1995a) naturalistic observations of current practice led him to a very different view of human cognition. First, he found that knowledge and information processing are not confined to the brain, but are instead distributed spatially across individuals and artifacts and temporally as a function of the history of a particular culture. Workers were rarely observed in the isolated, pensive activities that we would expect based on the PSSH. Instead, they frequently relied on external artifacts to reduce the burden on their limited cognitive resources, thereby allowing them to achieve task goals in a more economic fashion. Second, workers frequently accomplished task goals, not in isolation through mental information processing, but as a functional team through mutual coordination of their actions. As a result, team communication played a very important role. Furthermore, the redundancy achieved by having multiple people involved in performing a task also had the added benefit of creating a robust mechanism for error detection and error recovery. When one worker makes an error, it is frequently noticed by another because team members have access to each other's actions and communications. Third, cognition "in the wild" is an emergent activity that is not completely specified ahead of time, whether it be by written procedures or by mental rules and representations. Instead, the pattern of coordinated action exhibited by teams are generated anew each time "without there being a representation of that overall pattern anywhere in the system" (Hutchins, 1995a, p. 200). Using the terms we introduced in chapter 3, cognition is situated and exhibits context-conditioned variability. Finally, Hutchins also found that the historical influence of culture was very important. Many of the useful artifacts that workers rely on (e.g., navigation charts) and many of the practices in which workers are engaged (e.g., simplifying rules for calculation) were adaptive products of hundreds of years of navigational experience. In short, cognition in naturalistic settings is not a strictly mental activity mechanically performed by an individual in isolation using mental rules and representations. Instead, it is an emergent, distributed activity that is performed by people with tools, within the context of a team or organization, in a cultural context that has evolved over years of experience.

It is important to note that not all of the work practices that Hutchins (1995a) observed were documented as official practice. Although there were procedures that were said to prescribe how the navigation task is supposed to be performed, "the normative procedures are an ideal that is seldom achieved, or seldom achieved as described" (Hutchins, 1995a, p. 28). Thus, like the other descriptive analyses reviewed earlier, Hutchins found that normative descriptions of work (i.e., the task, in Francophone ergonomics) do not correspond to the practices in which workers are actually engaged (i.e., the activity, in Francophone ergonomics). Thus, yet again, we see the value added by descriptive approaches to work analysis.

The Importance of Studying Current Practice

Table 4.1 summarizes these four case studies of descriptive work analysis. There are several differences between the four cases. First, they were conducted by researchers from largely disparate backgrounds. Second, they focused on very distinct application domains. Third, their respective theoretical foils also differed, at least at a superficial level. Despite these substantial dissimilarities, the cases are united by two overarching themes.

Common Themes. The first common thread is the enormous value of conducting descriptive studies of work in representative or naturalistic settings. As shown in Table 4.1, each of the four studies led to important insights that have influenced entire areas of research. Suchman's (1987) research is largely responsible for the increase in attention to anthropological methods and theories in the HCI community. Klein's (1989) research has influenced many psychologists and human factors researchers to study and better understand workers' expertise under naturalistic conditions. Bødker's (1991) work has greatly increased the profile of Soviet activity theory in the applied world of HCI. Finally, Hutchins' (1995a) research has had a

TABLE 4.1 Comparison of Four Case Studies of Descriptive Studies of Work Analysis

Case Study	Analyst's Background	Application Domain	Theoretical Foil	Major Contribution
Suchman (1987)	Anthropology	Photocopying	Artificial Intelligence	Situated Action
Klein (1989)	Experimental Psychology	Fire Fighting	Classical Decision Making	RPD Model
Bødker (1991)	Computer Science	Graphic Design	Anglo-American HCI	Activity Theory-Based HCI
Hutchins (1995a)	Anthropology	Ship Navigation	Physical Symbol System Hypothesis	Distributed Cognition

significant impact on basic and applied researchers in cognitive science by showing the value of distributed cognition for studying and shaping cognition. In each of these four cases, the researchers would not have been able to develop the same insights had they not ventured out into the field and meticulously analyzed and documented the varied and complex demands imposed on workers, and the ingenious adaptations that workers have developed to make those demands manageable. The picture of human work that emerges from these studies is very different from the one painted by the unrealistic assumptions made by the normative approaches to work analysis that we reviewed in chapter 3. Therefore, these four case studies collectively illustrate the "value added" by studying what workers really do.

The second unifying theme is the converging characterization of human work. First, although there are differences between the approaches, all four cases illustrate the importance of context-conditioned variability. Cognitive work in naturalistic settings takes place in an open system, and so people must frequently adapt to the contingencies of the moment. Second, work has a strong social component and rarely takes place in isolation. Workers are usually part of a team that must communicate and coordinate effectively to achieve their goals. Third, work is also seldom solely focused on internal mental processing because workers create tools (e.g., external representations) that are tailored to the work demands, thereby reducing the burden on scarce cognitive resources. Fourth, current work practices are shaped by historical cultural factors that have been ignored by many traditional approaches. Fifth, workers are frequently under time pressure and other constraints, so they must develop the expertise required to get the job done in the face of such pressures. Frequently, such expertise has a strong perceptual component, thereby reducing the load on cognitive resources. Based on all of these insights, we can make the following strong claim for open systems: *Workers do not, cannot, and should not consistently follow the detailed prescriptions of normative approaches.* Our story of "malicious procedural compliance" in the Preface is a vivid indicator of this point.

A set of articles in a recent special section of the *Communication of the ACM* devoted to representations of work emphasizes this conclusion (Suchman, 1995). These articles show that normative approaches to work analysis provide an inadequate basis for design because their assumptions ignore the richness and variability of actual work practices (see chap. 3). This point is an important one to make because normative approaches still dominate the design of computer information systems. In a more recent article, Simonsen and Kensing (1997) also advocated an examination of current work practices as part of systems design. They explicitly stated that the approach they have adopted "develops descriptive understanding in contrast to prescriptive" (p. 82).

Implications for Work Analysis. The characterization of work emerging from these studies has important implications for a viable framework for work analysis in complex sociotechnical systems. If we include the insights from the previous chapter, we can conclude that such a framework must include at least five dimensions of work:

1. A *work domain analysis* is needed to identify the information requirements that will allow workers to deal with unfamiliar and unanticipated events (see chap. 3).

2. A *constraint-based task analysis* is needed to identify the information requirements that will help workers achieve anticipated task goals in a flexible, situated manner (see chap. 3).

3. An analysis of effective *strategies* is required to identify the mechanisms that can generate the practices that have emerged historically in the culture of a particular application domain. If our work analysis explicitly identifies such strategies, then designers can build information systems to support them.

4. Perhaps the most important lesson emerging from descriptive analyses of work is that we need to analyze the *social and organizational* factors that govern work. Although we speak of creating a computer-based information system, we are also inducing an organizational structure when we introduce information technology into the workplace. Thus, our work analysis should explicitly address the issue of how work can be allocated across individuals, how those individuals can be organized into groups or teams, and how individuals, groups, and teams can communicate with each other. By explicitly analyzing these social factors, we can try to coordinate technological change and organizational change.

5. Finally, it would also be useful to explicitly identify the various demands that the application domain imposes on individual *workers' competencies* (i.e., expertise). Identifying those demands would allow us to design computer-based information systems that facilitate skill acquisition and that support expert action.

WHY DESCRIPTIVE APPROACHES ARE NOT ENOUGH: THE TASK-ARTIFACT CYCLE

Although much can be gained by studying workers' current practices, it does not follow that computer-based information systems should be designed based only on descriptive studies, nor should they be designed to support only current practices. This is a subtle distinction that we believe has not received the attention it deserves. The point is a very important one, however, because it illustrates the limitations of descriptive approaches to work analysis and motivates a need for formative approaches.

Limitations of Basing Design Solely on Current Practice

The rationale behind our claim is illustrated in Fig. 4.1, which depicts two idealized, intersecting sets. The set on the left is defined by the constraints on achieving work goals, independent of any particular device. We have labeled this set *intrinsic work*

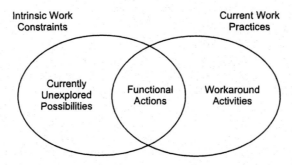

FIG. 4.1. Why computer-based information systems should not necessarily be designed to support current work practices. See text for details.

constraints to emphasize the fact that these constraints are an inherent part of work in a particular domain. This concept may seem like a Platonic ideal, so it is important to point out some examples: the position of mountains in air traffic control, foreign exchange rates in international financial markets, the laws of physics in a power plant, and the maneuvering capabilities of various enemy aircraft in military command and control. In each case, the constraints in question must be taken into account by any actor if task goals are to be reliably and consistently achieved, regardless of the device available. Thus, intrinsic work constraints are, in principle at least, device-independent because they delimit the actions that are required to get the job done, not the actions that are required to get the job done with a particular device. In subsequent chapters, we develop this concept in a more concrete and detailed manner.

The significance of intrinsic work constraints can be made clear by contrasting this set with the one on the right in Fig. 4.1, which is the set of actions that comprise *current work practices.* When we conduct a descriptive work analysis, we are observing a subset of the set on the right of Fig. 4.1. A crucial point, whose significance is discussed later, is that current practice is device-dependent because it represents the strategies and tasks that workers are using to get the job done with whatever tools they currently have at their disposal.

One important implication to emerge from this graphical representation is that these two sets are overlapping but not equivalent.[1] There are three qualitatively different regions that must be distinguished in Fig. 4.1. The first region is the intersection of the two sets, which represents the subset of actions that are intrinsic to getting the job done and that are a part of current work practice. These are *functional actions,* and their identification is a primary reason for conducting field studies of work under naturalistic conditions.

The second region in Fig. 4.1 is the subset on the far right, which is composed of actions that are not directly part of getting the job done, but that are nevertheless part of current practice. These *workaround activities* are a frequent source of inefficiencies in HCI. You may recall that Table 2.1 in chapter 2 provides a number of instances from the graphic design software investigated by Black (1990). For example, printing out one alternative draft at a time is not an inherent part of graphic design. It is an activity that workers have to engage in when the information system does not support the

---

[1]The relative size of, and the amount of overlap between, these two sets is unknown and surely varies from case to case. The depiction in the figure is intended to illustrate only qualitative distinctions, not quantitative differences.

demands of the job (in this case, comparing several initial design proposals in parallel before proceeding with the detailed design work). There are many other examples we could cite as belonging to this category (e.g., R. I. Cook & Woods, 1996). One omnipresent example in today's world of graphical user interfaces is the set of overhead activities associated with interface navigation and management. In some application programs, workers spend a great deal of their time and energy managing windows and other interface objects, such as dialogue boxes, messages, menus, and palettes (e.g., Harrison, Ishii, Vicente, & Buxton, 1995). In fact, when we videotaped workers interacting with such programs and then replayed the tape in fast-forward mode, what we saw was a furious level of activity with workers frequently moving interface objects, resizing them, bringing them up, and making them go away. All of these are workaround activities. They are not directed at accomplishing work objectives, at least not in a direct way. Instead, they are "overhead" activities that workers have to perform because of the impoverished interface they currently have.

Finally, the third region in Fig. 4.1 is the subset on the far left, which is composed of the actions that are an intrinsic part of work but that are not a part of current work practices. These are *currently unexplored possibilities* for getting the job done. They may be potentially very efficient and productive ways of accomplishing work. However, these actions are usually not a part of current practice because it is not feasible for workers to perform them with the current inadequate level of support with which they have been provided. For instance, currently unexplored actions may require too much time, computational effort, memory demands, or knowledge with the existing device. Returning to the domain of graphic design described by Black (1990), a hypothetical example could be the practice of comparing several alternative design proposals in parallel before going on to develop a detailed design. If this activity is cumbersome to accomplish with the existing information system (as it was with the software analyzed by Black), then workers may simply decide to omit this task. The result would be a reduction in product quality because of poor design practices. The general point is that workers sometimes do not exploit certain work domain possibilities, perform certain tasks, or adopt particular strategies because it would be too effortful or time consuming to do so with the tools they currently have available to do the job. However, these unexplored possibilities could very well become a productive part of workers' practices, if the requisite computer support were provided. In chapters 9 and 12, we provide detailed examples illustrating this very point.

Implications. This discussion is relevant to both design and work analysis. In terms of design, the implication is that computer-based information systems should ideally be designed to support intrinsic work constraints, not just current work practices.[2] As suggested by Fig. 4.1, there are two reasons to justify this claim. First, we do not want to base our design on the workaround activities that are vestiges of poor device design. And as Benyon (1992a) pointed out: "Current practice is *always* tied to existing technology. . . . [Thus,] embodying current practice in future systems is a fundamental error" (p. 114, emphasis in original). Thus, there is no

---

[2]This claim is valid only for the revolutionary design conditions we described in chapter 1 (cf. Alexander, 1964). There are many situations in industry where designers are forced to work strongly within the context of previous designs. In these cases, making assumptions about the existing device may be a good thing. However, it is important to realize that the resulting design will not be nearly as effective as it could be in these cases, because it is inheriting some of the deficiencies or limitations of the previous device.

reason why workaround activities should be supported by the new device (barring technological constraints). Second, we do want to support currently unexplored possibilities. There is no reason to exclude ways of doing the job that could be effective but that are too effortful or time consuming to adopt with the existing device (barring technological constraints). After all, such new functionality can lead to marked improvements in productivity (Landauer, 1995).

In terms of analysis, the conclusion of this discussion is that work analysis should focus on identifying intrinsic work constraints rather than just current practice (see Fig. 4.1). More specifically, the technique we choose for work analysis should try to explicitly "peel away" the subset of current practices that originate from poor device design, and it should explicitly try to uncover productive practices that can be used, not just those that are being used (Rasmussen et al., 1994). This does not mean that descriptive studies of current practice are not valuable. As the four cases reviewed earlier show, they are essential. Instead, the point is that future designs should go beyond current practice by removing unwanted inefficiencies and by adding new functional possibilities. The most direct way of achieving this goal is to adopt a work analysis technique that explicitly tries to identify intrinsic work constraints.

Are These Limitations Recognized?

The limitations we have identified have been explicitly recognized by a number of researchers from different backgrounds, including some who are advocates of descriptive approaches to work analysis. One example that we have already mentioned is that of Benyon (1992a), who stated that "embodying current practice in future systems is a fundamental error" (p. 114). More recently, Beyer and Holtzblatt (1998) echoed this opinion, stating that designing a new device to match existing work practices exactly "would be a sure path to failure" (p. 7). Holmqvist and Andersen (1991), who are strong advocates of descriptive work analyses, also pointed out that they "do not believe that a system designer should design the computer-based . . . system as a copy of the existing work" (pp. 91–92). T. E. Miller and Woods (1997), coming from a cognitive engineering perspective, voiced a similar opinion.

Thus, there are a number of researchers who believe that the design of future devices should not be based solely on current work practices. This is not to say that it is not valuable to study current work practices. On the contrary, as the overlap in the center of Fig. 4.1 shows, studying current practice can shed a great deal of light on intrinsic work constraints. However, the analysis of current practice should be viewed as one of several possible means to investigate work constraints, rather than an end in itself. This point can be better appreciated by examining how difficult it has been to move effectively from descriptive work analyses to implications for the design of computer-based information systems.

From Descriptive Analysis to Design Implications: The Track Record

Social Science. In a recent article, Button and Dourish (1996) observed that the attempt to incorporate the insights from anthropological analyses of current practice (specifically, those using an approach known as ethnomethodology) into the design of computer-based information systems has been "problematic" (p. 20). Descriptive work analysis techniques have certainly been very effective in understanding how existing information systems fail to support human work. However, Button and Dourish stated that the attempt to design new information systems using these techniques "is fraught with methodological dangers" (p. 19). In some ways, this should not be surprising because ethnomethodology was not designed for such purposes. That "tradition is in analysing practice, rather than 'inventing the future' " (p. 21).

Similar, if less specific, concerns have been voiced by other social scientists. For example, Heath and Luff (1996) stated that:

> It is being increasingly argued that the requirements for complex systems need to be derived from a deeper understanding of real-world, technologically supported cooperative work, which in turn might lead to a distinctive, more social scientific, approach to user-centered design. *Whether or not such developments can be drawn from work in the social sciences is unclear.* (p. 98, emphasis added)

In summary, social scientists have contributed greatly by providing us with a better descriptive understanding of how work is conducted in situ (e.g., Engeström & Middleton, 1996). However, an additional body of knowledge and set of techniques are needed to design a better device.

Activity Theory. The same point can be illustrated in a more specific form by referring to the contributions to HCI based on activity theory documented in Nardi (1996b). There are few examples of novel designs based on an activity theory analysis (see Bødker, 1991, for an exception). There are at least two probable reasons for this. First, the application of activity theory to HCI is relatively recent, so there may not have been enough time for examples to emerge. Second, it is possible that the descriptive nature of activity theory makes it difficult to develop a novel design. This interpretation is supported by juxtaposing quotations from several chapters in Nardi's edited volume:

1. "Activity theory . . . proposes a very specific notion of context: the activity itself is the context" (Nardi, 1996c, p. 76).
2. "The activity is the way the subject sees the practice" (Christiansen, 1996, p. 176).
3. "The main strategy for research is simply to ask the persons to tell us about their action plans and goals" (Raeithel & Velichkovsky, 1996, p. 219).
4a. "Activities are always changing and developing" (Kuutti, 1996, p. 33).
4b. "Activity is, according to the theory, . . . a prism that moves and changes all the time" (Christiansen, 1996, p. 195).

These quotes illustrate that the focus of the work analysis is a moving target, making stable generalizations to new products difficult.

Like ethnomethodology (see previous discussion), activity theory was also not originally developed for design purposes. According to Nardi (1996a): "Activity theory focuses on *practice* . . . [and] is a powerful and clarifying *descriptive* tool rather than a strongly predictive theory. The object of activity theory is to *understand* the unity of consciousness and activity" (p. 7, emphasis added). Descriptive understanding is essential in determining why existing devices are ineffective or how workers' activities have evolved over time. Our point is that this knowledge is useful, but not sufficient, for designing better information systems. Thus, despite its virtues, activity theory is subject to the limitations identified earlier. As a result, the claim that "activity theory . . . holds the best conceptual potential for studies of human-computer interaction" (Kaptelinin, 1996, p. 107) seems somewhat exaggerated. Different methods have different strength and weaknesses (cf. Weinberg, 1982), so no framework—including the one described in this book—can be considered to be "the best" overall.

Francophone Ergonomics. It is very important to point out that the concerns that we and others have voiced are not specific to work analysis techniques from the social sciences or from activity theory. They are instead deeply rooted in a fundamental limitation of descriptive approaches to work analysis. This conclusion is supported by the fact that other researchers, from independent traditions, have encountered identical problems in using this class of methods.

The Francophone ergonomics community, for example, has experienced this very same problem, despite having the benefit of over 30 years of experience with descriptive work analyses:

> One fact, however, seems certain. A wealth of data has been collected but the data are still in their raw state. It has not yet been possible to extract any formal models or generalizations from them. Hence, the difficulty of creating an ergonomics of design, which would enable us to proceed by methods other than empirical testing, trial and error or successive corrections. (De Keyser et al., 1988, p. 151)

Human Factors. The deep-rooted limitations of descriptive approaches to work analysis can also be observed in the human factors community. In chapter 3, we described how task analysis can be used in a normative fashion. It is also possible to use the same techniques to describe, not what workers should do, but what they are currently doing. To take but one example, the task analysis technique known as *link analysis* (Kirwan & Ainsworth, 1992) can be used to determine empirically the frequency with which workers interact with particular instruments or locations (e.g., the probability of visually scanning from one display to another, or the probability of walking from one area of a control room to another). Such task analyses suffer from the same problems as other descriptive techniques. For instance, though link analysis can help analysts identify bottlenecks in current practice, it is not nearly as effective in formatively specifying the attributes that a new device should possess if it is to eliminate or reduce those bottlenecks. Kirwan and Ainsworth provided a clear statement of the problem: "As the [link] analysis is very context-specific [i.e., device-dependent], it is not normally possible to add information, or make inferences, if the system changes . . . any changes will probably result in the analysis having to be totally restarted" (p. 123). The very same conclusion can be made for other descriptive task analysis techniques, such as activity sampling (Kirwan & Ainsworth, 1992). Thus, the limitations associated with descriptive approaches are not limited to work analysis techniques from any one discipline.

People Are Adaptive: The Task-Artifact Cycle

The problem is that the introduction of new technology is, to use Schmidt's (1991b) felicitous phrase, "like writing on water" (p. 76). Analyses of current practice lead to design ideas for supporting that practice. However, workers are adaptive, so the

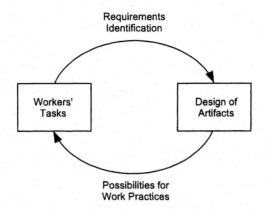

FIG. 4.2. The task-artifact cycle (adapted from Carroll et al., 1991).

introduction of a new design results in new practices that, more often than not, are not fully or well supported by the new device. The result is a new set of unforeseen problems (see Vicente & Williges, 1988, for an example). We have already encountered the tip of this iceberg in chapter 3 when we discussed the regress caused by the device-dependence of instruction-based approaches to task analysis.

This interdependence between current practice and the design of a device has been concisely captured in Carroll et al.'s (1991) task-artifact cycle, illustrated in Fig. 4.2. The circular nature of technological intervention is clearly shown. If we conduct a descriptive work analysis to understand workers' current tasks, we will identify requirements that could be used to design a new artifact. However, once this artifact is introduced into the workplace, new possibilities for work practices are created, thereby shaping workers' practices. In the words of Carroll et al.: "A task implicitly sets requirements for the development of artifacts to support it; an artifact suggests possibilities and introduces constraints that often radically redefine the task for which the artifact was originally developed" (p. 79). Thus, by basing new designs on work analyses of current practice, designers will always be one step behind their interventions.

Summary

In this section, we have tried to show why descriptive approaches to work analysis are valuable but limited in their ability to inform the design of computer-based information systems. Current practice reflects inadequacies and limitations in the existing device and, at the same time, hides potentially valuable but unexplored ways of working. These problems have been recognized by some researchers. Moreover, the difficulties in going from descriptive analysis to design implications have been encountered by a number of distinct research traditions that use descriptive methods for work analysis, including social sciences, activity theory, Francophone ergonomics, and traditional human factors. The root cause of these difficulties lies in the regress imposed by human adaptation, and described by the task-artifact cycle.

Our understanding of this problem suggests that work analysis techniques should seek to identify the set of intrinsic work constraints rather than just the set of current work practices. This alternative approach should make it easier to move from analysis to design. Before we explore this view, however, we examine the techniques that other researchers have already developed to try to overcome the task-artifact cycle.

EXISTING TECHNIQUES FOR GETTING AROUND THE TASK-ARTIFACT CYCLE

The problems imposed by the task-artifact cycle have been recognized by at least some researchers for quite some time (e.g., Greenbaum & Kyng, 1991). Thus, it should not be surprising to find that at least two major techniques have been developed, and used, to try to get around the task-artifact cycle: rapid prototyping and scenario-based design. In this section, we review the valuable contributions of each of these techniques.

Rapid Prototyping and Iterative User Testing

One way to "go beyond the present interface" (Kyng, 1995, p. 49) is to create prototypes of new designs, evaluate them by having workers use them to perform representative tasks, and then use the evaluation findings to iterate on the design. The Scandinavian school of participatory design has made particularly effective use of this technique (e.g., Bødker, 1991; Greenbaum & Kyng, 1991). The rationale behind prototyping is straightforward, and can be explained with reference to Fig. 4.1. Instead of basing a design solely on descriptive work analyses of current practice, building and testing prototypes provides an opportunity to "dig into" the subset of currently unexplored possibilities. Each round of prototyping and testing can be used to evaluate hypotheses about what new functions might be useful to workers. Ideally, with each iteration, the subset of currently unexplored possibilities should become smaller and smaller, as valuable functionality is added (or changed) in each generation of prototypes. Similarly, building and testing prototypes also provides an opportunity to "dig out of" the subset of workaround activities. Each round of prototyping and testing can be used to evaluate hypotheses about deficiencies in the device that induce workarounds. Ideally, with each iteration, the subset of workaround activities should become smaller and smaller, as device deficiencies are removed in each generation of prototypes. Thus, the end goal is to iteratively maximize the overlap between the two sets in Fig. 4.1 through an active learning process involving both analysts and workers. At the end of the design life cycle, there should ideally be little distance between the intrinsic work constraints and the work practices induced by the new design. Bødker and Grønbœk (1996) provided a nice case study illustrating how prototypes can be used to try to overcome the inertia of current practice.

Scenario-Based Design

Whereas prototyping and testing provide an empirical way of trying get around the task-artifact cycle, scenario-based design (Carroll et al., 1991; Carroll & Rosson, 1992) provides an analytical technique for trying to achieve the same objective: "Scenarios have the important property that they can be generated and developed even before the situation they describe has been created" (Carroll & Rosson, 1992, p. 190). Thus, rather than—or in addition to—building prototypes, it is possible to develop scenarios that envision what it might be like for workers to interact with a device that has yet to be designed. These scenarios can then be analyzed to identify the psychological consequences of particular device features in the situations defined by the scenarios generated by the analyst. These consequences can, in turn, be analyzed to determine how well the envisioned design supports workers in achieving task goals. Those insights can be iteratively used to build new scenarios and new envisioned artifacts. As with prototypes, the goal is to minimize the workaround activities induced by the new device and to incorporate new possibilities for functionality. In short, scenarios provide a means for analytically evaluating "simulated future work" (Kyng, 1995, p. 55). Carroll and Rosson provided several examples of how scenario-based design can be used to try to overcome the inertia of current practice.

Limitations: The Problems of Device-Dependence and Incompleteness

There can be no question that both prototypes and scenarios are valuable tools that can be effectively used in conjunction with the framework that we are advocating in this book. For us, the important question is: Are these techniques sufficient to overcome the task-artifact cycle when designing computer-based support for complex sociotechnical systems?

As valuable as they are, both prototyping and scenarios suffer from two limitations: strong device-dependence and incompleteness.[3] Regarding the former, prototyping and testing actually do not overcome the regress in the task-artifact cycle (see Fig. 4.2). Instead of escaping from the regress, these techniques actually involve iterating through the task-artifact cycle numerous times, with the expectation that the design will be improved with each iteration. Figure 4.3 illustrates this relationship graphically. Each prototype constitutes an artifact (e.g., A_1) that introduces a new set of possibilities for doing the task (e.g., T_1). Analysis of this task then identifies new requirements that, if adopted, lead to the design of a new artifact (e.g., A_2). Iteration through the cycle continues in this manner until the analyst is satisfied or (more typically) runs out of resources. Thus, iterating through the task-artifact cycle in this way is akin to a dog chasing its tail. Some advocates of these techniques acknowledge this point: "Generally, our approach should be thought of as an iterative one, where the introduction of the new computer application in the work setting may cause unforeseen changes in work, in turn leading to demands for new or changed computer applications" (Bødker, Greenbaum, & Kyng, 1991, p. 151). Note that Bødker et al. were referring to iteration in the workplace after a design has been introduced, not to the testing iterations that take place before a design is introduced into the workplace.

Because prototyping and testing do not avoid the task-artifact cycle, the requirements that are identified by analysts are limited by the choice of prototypes that are tested. However, there is no systematic way to go from results of testing to prototype

---

[3]Other techniques that have been proposed (Greenbaum & Kyng, 1991) for trying to get around the task-artifact cycle (e.g., mock-ups, Future Workshops, and metaphorical design), though useful, also suffer from the same limitations.

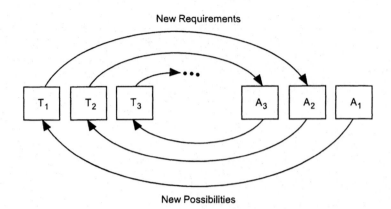

FIG. 4.3. Iterating through the task-artifact cycle. Compare with Figure 4.2 (T_i = Task$_i$; A_i = Artifact$_i$).

attributes. This latter step is largely limited by the ingenuity and creativity of the designer. Of course, creativity can never be eliminated from design (Ferguson, 1977), but the smaller we can make the gap between analysis and design, the less dependent we are on the skills of any one designer. Furthermore, any design change, even one occurring in a very late iteration, can lead to new, unforeseen problems. Thus, the aforementioned regress is never really avoided because the testing results are strongly device-dependent. Beyer and Holtzblatt (1998) encountered this problem in their experiences in the HCI industry: "Prototyping and usability testing could iterate an existing system, but couldn't suggest wholly new directions" (p. 20). Like us, they thereby felt a need to seek "a process that would lead to new kinds of systems rather than iterating existing systems" (p. 20).

Scenario-based design is limited for analogous reasons. Close inspection of the examples generated by Carroll and Rosson (1992) shows that scenarios are also device-dependent. For example, the scenario depicted in their Figure 2 (p. 187) refers to attributes of the interface (e.g., default settings in a menu). As a result, the insights that can be garnered from the analysis are limited by the interface attributes that have been specified in the scenarios.

Ideally, we would like to have a work analysis technique that makes minimal assumptions about the device (i.e., interface + automation) up front. As mentioned in chapter 3, the functionality and characteristics of the device should be an output of the work analysis. That is, the reason why we are doing an analysis in the first place is to determine how the device should be designed. The limitations of prototyping and scenarios are that they require an interface to even get started. Rather than being solely an output of the work analysis, the device specifications are needed as inputs to bootstrap the building of prototypes and scenarios. Consequently, as shown in Fig. 4.3, the task-artifact cycle is never really broken.

As for incompleteness, the insights gained from prototyping are limited by the number and range of tasks that workers are asked to perform during testing. As we discussed in chapter 3, task demands in open systems cannot be fully anticipated or enumerated in detail, so the representative set of conditions used during testing is bound to be incomplete, perhaps drastically so. Moreover, because prototyping and testing are bottom-up activities, there is no way of knowing how much of the state space of possibilities has been covered. For instance, the operators at TMI had experienced many different situations in training and operations before the accident we described in chapter 2 occurred. However, none of those situations had been sufficient to reveal the profound deficiencies in systems design uncovered by the unique event prompting the accident. Similarly, the insights gained from scenarios are limited by the number and range of scenarios that are considered in the analysis. But as Carroll and Rosson (1992) observed: "For any reasonably complex system, the scenario representation is necessarily incomplete (there is an infinity of possible use-scenarios)" (p. 185). And again, because scenarios are generated in a bottom-up fashion by the analyst, there is no way of knowing how much of the state space of possibilities has been covered.

On the one hand, it is certainly true that breadth of coverage is likely (although not guaranteed) to improve as a function of the number of prototype and test iterations we conduct, or the number of scenarios we generate and analyze. On the other hand, incompleteness means that there will be situations for which the device has not been tailored. If these situations are encountered by workers, then the likelihood of an error is likely to be significantly greater than in the situations that have been explicitly anticipated. For work domains where the consequences of errors due to incompleteness are not severe (e.g., word processing), this situation may be acceptable. However, in many complex sociotechnical systems, the potential hazards involved are very large (see chap. 1). Thus, incompleteness becomes a much more significant concern, particularly because the accident data show that disasters occur in situations that have not been anticipated by designers.

Consequently, it is important to look for work analysis techniques that offer greater device-independence and breadth of coverage. Whatever method we choose will not offer any complete guarantees, but we must make more of an effort to overcome the limitations of descriptive work analyses and the companion techniques of prototyping and scenario building. The potential consequences of not doing so are too great to ignore, because they can be measured in enormous amounts of monetary losses, substantial damage to the environment, or tragic loss of human life.

AN ALTERNATIVE WAY OF GETTING AROUND THE TASK-ARTIFACT CYCLE: MODELING INTRINSIC WORK CONSTRAINTS

The only viable way we know of breaking free from the task-artifact cycle, and its associated limitations, is to escape from the language of tasks and artifacts. Tasks and artifacts are intimately bound to current practice, which inevitably leads to the deep-rooted problems we have discussed previously. In this section, we argue that modeling intrinsic work constraints will allow us to cut through the Gordian knot of the task-artifact cycle by identifying functional design possibilities, independently of current work practices.

Completeness: The Need for Models

As we mentioned earlier, one of the key limitations of relying solely on scenarios or on prototyping and testing is that the insights that we gain are strongly bound by the number and type of scenarios and prototypes that we can afford, or have the ingenuity, to include. Prototype testing and scenario building are inductive, bottom-up activities and will always be incomplete because we can never test all possible prototypes under all possible conditions or analyze every possible scenario. Thus, we need to find some way of generalizing our insights to the prototypes and scenarios that we have not considered. Perhaps even more important, however, is the fact that it is not possible to determine the limits on our knowledge. With prototype testing and scenario building, we cannot systematically know what factors we have included and what factors we have missed. Each iteration is a data-driven step, and we have no knowledge of the overall space of possibilities in which we are searching.

Modeling provides a way of reducing the impact of these limitations. First, models provide an explicit, top-down basis for generalization. They generalize beyond instances by representing classes rather than exemplars. This feature of models was well captured by Ahl and Allen (1996):

> Although they are a critical part of science, data are not the purpose of science. Science is about predictability, and predictability derives from models. Data are limited to the special case of what happened when the measurements were made. Models, on the other hand, subsume data. Only through models can data be used to say what will happen again, before subsequent measurements are made. Data alone predict nothing. (p. 45)

As a result, models provide a broader basis for work analysis. Second, although they will always be incomplete, models also make explicit what attributes have been included. In fact, the process of model building is essentially that—the specification of the equivalence classes that define the model (see Glossary). Thus, models provide a more systematic and explicit basis for work analysis. Note that these features are particularly important—perhaps even essential—for complex sociotechnical systems. As Beyer and Holtzblatt (1998) pointed out, "the more complex the work, the more critical it is to maintain a coherent representation [of work]" (p. 212).

Device-Independence: Focusing on Intrinsic Work Constraints

This discussion naturally leads to the question: Model what? After all, there are many different entities that could be modeled. As this chapter has shown, the focus of attention in HCI in the past has been on artifacts and tasks. This focus leads to the problems of strong device-dependence that are inevitably caused by human adaptation. To reduce these problems, we need to find a way of escaping from the resulting regress illustrated by the task-artifact cycle in Fig. 4.3.

As shown in Fig. 4.1, one way to do this is to try to identify the set of intrinsic work constraints. By focusing on the constraints that are an inherent part of doing the job, we can try to get away from the details of how to do the job with a particular device. The primary goal in doing so is to let the work analysis suggest how the device should be designed, rather than to make assumptions about the device before the analysis even begins. By pursuing this goal, we can try to avoid including workaround activities that are a part of current practice. At the same time, we can systematically and explicitly seek out effective possibilities for accomplishing work that are not a part of current practice.

In some ways, this approach represents a radical change in work analysis and, by implication, design as well. Let us be concrete about what it really means. Work analysis methods should not prespecify:

- The existing set of sensors that are used to obtain data from the environment.
- The content and structure of the database that organizes all of the information in the information system.
- The functionality of the automation currently in place.
- The allocation of functions between computers and workers.
- The allocation of job responsibilities to individuals or groups.
- The appearance and structure of the interface.
- Workers' competencies.

The reason why work analysis should not inherit these decisions is because each of these issues is a point of design leverage. As a result, they should not be inputs into the work analysis, otherwise we would be inheriting the vestiges of the old, and likely inadequate, design. Instead, the decisions concerning what information should be gathered, how it should be organized, how to automate, what to automate, how to organize work, how to display information, and how to train operators should all be made based on the findings (i.e., the outputs) obtained from the work analysis. The very point of conducting a work analysis is to provide an informed basis for making decisions about these points of design leverage. To adopt the existing, or a priori, solutions to these issues is to miss important opportunities for creating novel and more effective ways of supporting human work. In the words of Beyer and Holtzblatt (1998), this approach "makes deciding how customers will work in the future the core design problem and uses those decisions to drive the use of technology" (p. 3).

As we have pointed out repeatedly in this chapter, this is not to say that it is not valuable to study current work practices. On the contrary, as the overlap in the center of Fig. 4.1 shows, studying current practice can shed a great deal of light on intrinsic work constraints. However, the analysis of current practice should be viewed as one of several possible means to investigate work constraints, rather than an end in itself. It is the work constraints themselves that should take precedence, if the pitfalls identified earlier are to be avoided.

There are several other ways of identifying intrinsic work constraints. For example, workaround activities are frequently indirect pointers to functionality that could be supported by the device (e.g., Vicente, Burns et al., 1996). In addition, sometimes workers try to use certain strategies but fail to carry them out effectively, not because the strategies are not useful but because the support to implement them reliably is not currently available. These strategies represent currently unexplored possibilities for doing work (see chap. 9). Also, by studying the structure of the work domain, it may be possible to identify tasks that should be performed but that are not currently being performed. Similarly, by studying the structure of the work domain and the tasks that need to be performed, it may be possible to identify novel strategies that could be used by workers. Finally, in some application domains, it may be possible to use analytical models (e.g., based on operations research techniques) to identify potentially very effective but currently unexplored possibilities for accomplishing work (e.g., Dessouky, Moray, & Kijowski, 1995).

SUMMARY

What type of work analysis is appropriate for complex sociotechnical systems? In this chapter, we evaluated the suitability of descriptive approaches in the form of field studies of current practice in naturalistic settings. Such studies have convincingly shown the strong limitations of normative approaches to work analysis discussed in chapter 3. Descriptive approaches have also led to important implications for work analysis. In addition to the work domain and constraint-based task analyses reviewed in the previous chapter, a framework for work analysis must also address strategies, social-organizational factors, and competencies for expertise.

Despite these valuable insights, descriptive approaches have limitations of their own, stemming from the regress captured by the task-artifact cycle. To bootstrap a descriptive work analysis, an existing, simulated, or envisioned device is required. As a result, descriptive approaches inherit the deficiencies of current practice, namely workarounds and currently unexplored possibilities (see Fig. 4.1). Thus, it is not generally useful to design future devices strictly to support current work practices.

Proponents of descriptive approaches to work analysis have proposed several techniques to get around the task-artifact cycle, the most prominent being prototypes and scenarios. Although certainly very useful—and perhaps even sufficient for some types of design problems—these techniques are not sufficient to satisfy fully the needs of complex sociotechnical systems. Prototypes and scenarios are both limited by problems of strong device-dependence and incompleteness. These limitations can be directly addressed by explicitly modeling intrinsic work constraints. With this approach, the various characteristics of the device and the organizational structure of work emerge as outputs of the work analysis rather than being required as inputs to get the analysis started. The bottom line is that descriptive studies of current practice can provide a useful window into intrinsic work constraints, but a different perspective is required to cut through the task-artifact cycle. In short, a formative approach to work analysis is needed to deal with the unique needs of complex sociotechnical systems.

Toward a Formative Approach to Work Analysis: "Workers Finish the Design" 5

> We must design jobs in such a way that workers can effectively control the [automated] controls, modifying them and regulating them to prevent failures and errors anticipated by the engineers. To do so, we must transcend our Taylorist inheritance and develop a new theory and practice of job design.
> —Hirschhorn (1984, p. 158)

> A framework for cognitive [work] analysis attempts to formulate a model which may serve to bridge the gap between engineers' analyses of control requirements and psychologists' analyses of human capabilities and preferences.
> —Rasmussen (1986a, p. 180)

PURPOSE

In the last two chapters, we identified the limitations of normative and descriptive approaches to work analysis. In this chapter, we begin by describing the generic, defining characteristics of formative approaches. This class of techniques explicitly addresses the limitations we identified in chapters 3 and 4. Then, we outline one particular formative framework, CWA, that is specifically tailored to the unique demands of complex sociotechnical systems. We demonstrate how CWA can provide a systematic basis for designing information systems with the autonomy and support that workers need to engage productively in flexible, adaptive behavior. By deliberately creating the conditions for productive adaptation, we can give workers some responsibility to "finish the design" locally as a function of the situated context, thereby improving safety, productivity, and health.

Before we go on, there are two pedagogical issues we need to address. First, this may be a good time for you to review the terms defined in our Glossary. Some of the statements we make in this and subsequent chapters may seem either unclear or counterintuitive because we are using certain terms in an unorthodox fashion. Having you consult the Glossary will hopefully help explain the rationale behind our statements. Second, recall from chapter 1 that we have written the bulk of this book from a conceptual rather than a methodological perspective. This choice becomes particularly clear beginning in this chapter. You will see that we have tried to present certain conceptual distinctions in an orderly and understandable fashion. However, this coherence of explanation should not be confused with the turmoil of process. The framework we advocate is well defined and highly structured, but—as is typical of most real-world systems design projects—the methodology that would be used to actually apply the framework in practice is much more opportunistic, chaotic, nonlinear, and iterative (cf. Card, 1996). For the sake of comprehensibility, we decided to adopt the former perspective in writing the book, rather than the latter.

THE FORMATIVE APPROACH: DESIGNING A FUTURE PRACTICE

The main conclusion from chapter 4 was that work analysis for complex sociotechnical systems should focus on identifying and modeling intrinsic work constraints. How does this basic insight lead to a class of formative approaches to work analysis? In this section, we answer this question by describing the defining features of such approaches. It is important to note that formative frameworks are a relatively recent innovation, so there are very few exemplars. Therefore, our characterization is only a tentative one.

What Makes Formative Approaches Unique?

Normative approaches focus on legislating work. Descriptive approaches focus on portraying work. In contrast, formative approaches focus on identifying requirements—both technological and organizational—that need to be satisfied if a device is going to support work effectively. Although these requirements do not uniquely specify a new design, they are still highly informative and thus very valuable because they can be used to rule out many design alternatives.

Figure 5.1 illustrates the basic structure of formative approaches.[1] Beginning on the far left of Fig. 5.1, the first step (Rasmussen, 1986b) is to identify a set of conceptual distinctions that can be directly linked to particular types of systems design decisions (e.g., sensor design, database design, automation design, organizational structure design, training program design). As Beyer and Holtzblatt (1998) observed: "A language of work for design will represent those aspects of work that matter *for design*" (p. 83, emphasis in original). By way of example, the first column in Fig. 5.1 illustrates place holders for the five conceptual distinctions that constitute the CWA framework described in this book. Each of these is identified later in this chapter and described in much more detail in chapters 7–11. By directly linking conceptual distinctions with classes of design interventions, the language we use ensures that our work analysis findings have direct and obvious implications for design.

The second step in the rationale, shown in the second column in Fig. 5.1, is to develop a corresponding set of modeling tools.[2] For each conceptual dimension, we should provide some tool that can be used to model that particular aspect of work. These tools should provide a logical, coherent structure with which to identify relevant design issues. A different modeling tool is required for each conceptual distinction because each distinction represents a qualitatively different aspect of work. Thus, the modeling tools provide a structured way of realizing the aforementioned conceptual distinctions. As illustrated in Fig. 5.1, the conceptual distinctions and their corresponding modeling tools comprise a framework for work analysis. Any one particular formative work analysis framework will be comprised of different

---

[1] Note that our description is very abstract in this subsection because we want to illustrate the defining characteristics of formative approaches. We make up for this abstractness later by adding concrete details that flesh out the generic description provided here.

[2] In this case, modeling tools refer to conceptual structures, not computational artifacts. It would certainly be useful to embed these conceptual structures in computer-based aids to support analysts in conducting work analyses in a more efficient way. Though very important, this issue is beyond the scope of this book (see Preface).

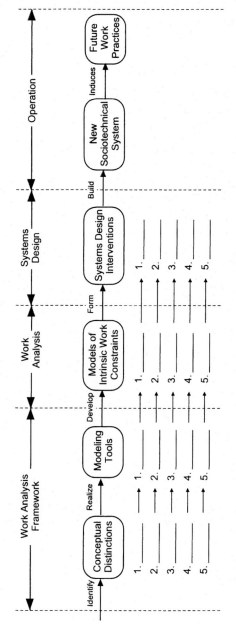

FIG. 5.1. The basic rationale behind formative approaches to work analysis. See text for description.

conceptual distinctions and modeling tools, presumably as a function of the type of problems for which it is intended.

As shown in the third column in Fig. 5.1, the third step in the rationale behind formative approaches involves using the aforementioned tools to develop models of intrinsic work constraints for a particular application. The focus is on intrinsic work constraints rather than tasks and devices so that the problems with the task-artifact cycle—strong device-dependence and incompleteness—can be overcome. As we described in detail in chapter 4, this focus will allow the properties of the device and the resulting practice to emerge as outputs of the work analysis, rather than being assumed a priori as unalterable givens. Whereas the modeling tools provide generic conceptual *structures* that can be applied consistently across many different application domains, the *content* identified by those tools will vary considerably across domains. Different application domains have different semantics, and thus, different design requirements. Discovering what these requirements are and modeling them using the tools described earlier are the goals of this third step. This modeling process is the heart of the work analysis effort.

The fourth step in Fig. 5.1 involves going from work analysis to systems design. Because each of the conceptual distinctions is linked to a particular category of design interventions (see previous discussion), each of the work models should have formative implications for design. For example, some models will help determine the structure of the information system database, whereas others will help determine worker training requirements. As shown in Fig. 5.1, these design interventions result in a new sociotechnical system consisting of, not only a new device, but also a new, corresponding organizational structure. Once these changes are implemented in operation, a future set of work practices will be induced. In the ideal case, this set of work practices should correspond to the set of actions defined by the intrinsic work constraints represented in Fig. 4.1, thereby eliminating workarounds and unexplored work possibilities (see chap. 4).

Several important points emerge from this discussion. First, whereas some approaches to work analysis focus more on devices, formative approaches focus on modeling work constraints. The primary goal of the chain in Fig. 5.1 is to design a future work practice rather than to design the details of the device. After all, the latter is merely a means to achieve the former. As Beyer and Holtzblatt (1998) observed, "The real invention of a design team is a new way for people to work" (p. 215). Second, rather than focusing on the way things should be (the normative approach), or on the way things are (the descriptive approach), formative approaches focus on the way things could be by identifying novel possibilities for productive work. The explicit modeling of intrinsic work constraints makes this feasible by escaping from the limitations imposed by the task-artifact cycle (see chap. 4). Again, Beyer and Holtzblatt captured the rationale well: "Understanding the structure of work leads to supporting work at the level of structure, which rarely changes, and suggests structural changes that radically improve the work" (p. 148). Third, because formative approaches are model based, they generalize beyond particular ways of doing work. Rather than building in a fixed and narrow work flow, the models of intrinsic work constraints introduce boundaries on productive work that still allow for flexibility and continuing evolution of work practices. Thus, the goal is "to build a structure into the [device] that supports the user's natural movement through her work *and* is flexible enough to enable the invention of new ways of working" (Beyer & Holtzblatt, 1998, p. 302, emphasis added). This point is elaborated on later in this chapter when we discuss how it is possible to design deliberately to support flexible, adaptive action.

COGNITIVE WORK ANALYSIS: MODELING BEHAVIOR-SHAPING CONSTRAINTS

The previous section has described a class of approaches to work analysis. Although there are currently very few exemplars, many could conceivably be developed under this class, each with different properties. For instance, the *contextual design* approach to work analysis proposed by Beyer and Holtzblatt (1998) is one example of a formative framework because it satisfies the rationale depicted in Fig. 5.1. In this section, we outline the framework that is the central focus of this book—the CWA framework proposed by Rasmussen et al. (1994), which is a formative approach to work analysis that is particularly well suited to the demands of complex sociotechnical systems.

Conceptual Distinctions

As shown in Fig. 5.1, the first step in laying out a particular formative framework for work analysis is to identify a set of conceptual distinctions that can be subsequently realized in a corresponding set of modeling tools. Based on the insights gained in the last two chapters, we can identify five different aspects of work that need to be represented for complex sociotechnical systems (cf. Rasmussen, 1986c; Rasmussen et al., 1994): (a) work domain, (b) control tasks, (c) strategies, (d) social organization and cooperation, and (e) worker competencies. These five concepts fully instantiate the first column in Fig. 5.1 for the CWA framework. Each concept represents a different layer of intrinsic work constraint.[3]

The first concept, *work domain,* represents the system being controlled, independent of any particular worker, automation, event, task, goal, or interface. As we discussed in chapter 3, a work domain analysis is like a map in that it shows the "lay of the land" independently of any particular activity on that land. That is, it shows the possibilities for action. For example, in the nuclear domain, the plant itself could be the work domain. Work domain analysis is described in detail in chapter 7.

The second concept, *control tasks,* are the goals that need to be achieved, independently of how they are to be achieved or by whom. As we discussed in chapter 3, a constraint-based task analysis is most appropriate for representing the control tasks in a complex sociotechnical system. Such an analysis identifies the product

---

[3]It is fascinating to observe that a very similar theoretical approach has been developed in a completely different research area, human motor control. We might think that operating a complex sociotechnical system and controlling your body are very different phenomena and thus would require completely different theoretical explanations. Despite the many differences between the two problems, this appears not to be the case. Motor control researchers within the ecological psychology perspective have adopted the notion of behavior-shaping constraints and have identified three sources of constraints that are relevant to motor control: environment, task, and organism (K. M. Newell, 1986). This theoretical similarity further illustrates the strong meta-theoretical parallels between ecological psychology and cognitive engineering (Vicente, 1995b). In this case, these parallels lead to viewing the control of the human motor system and the control of complex sociotechnical systems as both involving the coordination of many degrees of freedom by a resource-limited agent (or agents) (Vicente & Rasmussen, 1990). As a result, remarkably similar theoretical perspectives emerge for what seem to be completely different phenomena.

constraints that govern activity on the work domain (as opposed to the constraints that govern the work domain itself). In other words, the focus is on identifying what needs to done, independently of the strategy (how) or actor (who). Control task analysis is described in detail in chapter 8.

The third concept, *strategies,* are the generative mechanisms by which particular control tasks can be achieved, independently of who is executing them. Whereas control tasks are product representations (descriptions of what is to be accomplished), strategies are process representations (descriptions of how it can be accomplished). They describe how control task goals can be effectively achieved, independently of any particular actors. For example, there may be different processes that could be used to perform the activities associated with starting up the plant. Strategy analysis is described in detail in chapter 9.

The fourth concept, *social organization and cooperation,* deals with the relationships between actors, whether they be workers or automation. This representation describes how responsibility for different areas of the work domain may be allocated among actors, how control tasks may be allocated among actors, and how strategies may be distributed across actors. Thus, a social-organizational analysis describes how actors may be organized into groups or teams, how they may communicate and cooperate with each other, and what authority relationships may govern their cooperation. For example, the start-up of a plant could be conducted in a supervisory control mode with a crew of workers monitoring the steps being carried out by the automation. Social organization and cooperation analysis is described in detail in chapter 10.

Finally, the fifth concept, *worker competencies,* represents the set of constraints associated with the workers themselves. In addition to considering generic human capabilities and limitations, this analysis also identifies the particular competencies that various workers should exhibit if they are to function effectively. Different jobs require different competencies. Thus, it is important to identify the knowledge, rules, and skills that workers should have to fulfill particular roles in the organization effectively. For example, the competencies that the crew of workers would need to function effectively in a supervisory control mode should be identified. Worker competency analysis is described in detail in chapter 11.

The five concepts can be roughly categorized into two groups. The first group comprises the work domain, control tasks, and strategies. It describes the characteristics of the problem demands that must be satisfied. The second group of concepts comprises social organization and cooperation and worker competencies. It describes the characteristics of the organization and actors who will be responsible for satisfying those problem demands. An ideal sociotechnical system is one that provides a seamless fit between these two groups of concepts.

At this point, we do not expect you to have an intimate understanding of the five conceptual distinctions comprising CWA—that is what the rest of the book is for. Our goal for now is merely to introduce the distinctions and to get you to think about how the five concepts are related to each other. With this objective in mind, we now turn to addressing two questions: (a) Why are the concepts listed in this order? (b) How do the concepts relate to classes of design interventions?

Why This Order?

In chapter 2, we argued that work analysis for complex sociotechnical systems should be based on an ecological approach. That is, work analysis should begin with the environment because those constraints must be dealt with if there is to be any hope of achieving reliable and effective performance. Only once ecological compatibility has been obtained is it then useful to ensure that cognitive compatibility is also achieved. The latter objective can be accomplished by analyzing cognitive constraints to close the remaining degrees of freedom for action.

This ecological orientation explains the order of the five conceptual distinctions listed earlier. At a coarse level, this rationale can be observed in Fig. 5.2 by noting the gradual transition from ecological to cognitive considerations. The first layer of constraint to be analyzed is the work domain, which is a map of the environment to be acted on. The second layer of constraint is the set of control tasks, which represents what needs to done to the work domain. The third layer of constraint is the set of strategies, which represents the various processes by which action can be effectively carried out. The fourth layer of constraint is the social-organizational structure, which represents how the preceding set of demands are allocated among actors, as well as how those actors can productively organize and coordinate themselves. Finally, the fifth layer of constraint is the set of worker competencies, representing the capabilities that are required for success. Figure 5.2 shows a successive narrowing in with each phase because degrees of freedom are reduced with each layer of constraint (see later discussion).

Each transition in Fig. 5.2 also represents one step away from the ecological and toward the cognitive. This rationale is well captured for engineering systems by the following quote from Rasmussen (1986b): "The concepts used in the dimensions should be able to bridge the gap between an engineering analysis of a technical system and a psychological description of human abilities and preferences in a consistent way, useful for systematic design" (p. 195). CWA allows us to span ecological and cognitive considerations in one integrated framework for work analysis, thereby cashing in the claim made in chapter 2—it is possible to respect the primacy of ecological constraints while also considering all-important cognitive constraints.

The ecological orientation behind CWA can also be observed at a more fine-grained level in Fig. 5.3 by noting how the five layers of constraint are logically nested from a formative perspective. The size of each set in this diagram represents the productive degrees of freedom for actors, so large sets represent many relevant possibilities for action whereas small sets represent fewer relevant possibilities for action. The first phase of CWA, work domain analysis, shows what the controlled system is capable of doing. This level is a fundamental bedrock of constraint on the actions of *any* actor. No matter what control task is being pursued, what strategy has been adopted, what social-organizational structure is in place, or what the

FIG. 5.2. The transition from ecological to cognitive considerations in CWA.

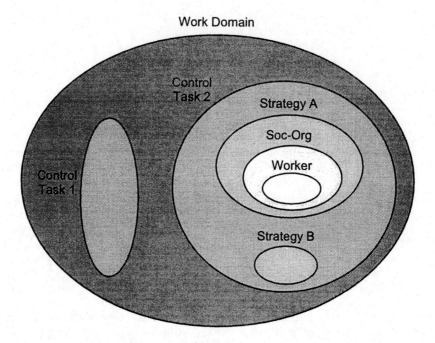

FIG. 5.3. The dynamic reduction in degrees of freedom associated with the nested relation between layers of constraint identified in Fig. 5.2. See text for detailed description.

competencies of the workers are, there are certain constraints on action that are imposed by the functional structure of the system being acted on. For example, pilots cannot use engines for functions that they are not capable of achieving, and library patrons cannot retrieve books that do not exist. Thus, the work domain delimits the productive degrees of freedom that are available for action.

The second phase, control task analysis, inherits the constraints of the first phase but adds additional constraints as well. For the sake of clarity, only two hypothetical control tasks are illustrated in Fig. 5.3. Although the work domain provides a large number of total degrees of freedom, when actors are pursuing a particular control task, only a subset of those degrees of freedom are usually relevant. For example, when pilots are navigating at cruising altitude, the constraints associated with the landing gear and brakes are usually not relevant. Furthermore, there are new constraints that must be respected above and beyond those imposed by the work domain. For example, for some control tasks, it is important that some actions be performed before others. This constraint is a property of the control task, not the work domain. As shown in Fig. 5.3, the net result is a reduction in the relevant degrees of freedom. When workers are pursuing a particular control task, only certain actions are meaningful and require consideration. It is for this reason that the sets depicting control tasks 1 and 2 in Fig. 5.3 are nested within the set for the entire work domain.

The third phase of CWA is depicted in Fig. 5.3 by two hypothetical strategies, A and B, for control task 2 (strategies for control task 1 are not shown for the sake of clarity). The strategy phase also inherits the constraints associated with previous phases of analysis. After all, a strategy cannot make a work domain do something that it is not capable of doing. This is why the two sets for strategies A and B in Fig. 5.3 are subsets of the work domain set. In addition, a strategy must also work within the constraints associated with its corresponding control task, otherwise it will not reliably achieve the required task goals. It is for this reason that strategies A and B in Fig. 5.3 are nested inside the set for control task 2. The strategies phase also introduces new constraints of its own, however. The control task level merely identifies the degrees of freedom associated with achieving a particular goal. There are conceivably many different ways in which control task 2, for instance, can be performed. All of these processes are encompassed by the control task 2 set in Fig. 5.3. When a particular strategy for performing the control task is identified, some degrees of freedom are usually not relevant because they are only required for other strategies. A specific strategy imposes a certain flow or process that adds constraints on top of those that are imposed by merely achieving the desired outcome. It is for this reason that strategies A and B are nested within the set for control task 2 in Fig. 5.3.

The fourth phase of CWA, social-organizational analysis, follows a similar pattern. It too inherits the constraints imposed by previous phases of work analysis, and it too adds a new layer of constraint. There are multiple organizational structures that could conceivably be adopted for any one strategy. To take a very simple example, a strategy may be performed by one worker alone, by two workers in a collaborative manner, by one worker supervising another, or by a worker supervising automation. In each of these cases, the same strategy is being adopted, the same control task is being pursued, and the same work domain is being acted on. Nevertheless, these different organizational architectures have different constraints associated with them. A worker executing the strategy alone will likely draw on a different, although probably overlapping, set of relevant actions than two workers executing the strategy in a cooperative fashion. Thus, a particular social-organizational structure represents a further narrowing of degrees of freedom. This logic explains why the social-organizational set in Fig. 5.3 is nested within the set for strategy A (analogous constraints for strategy B are not shown for the sake of clarity).

Finally, the fifth phase of worker competencies reduces the degrees of freedom even further. There are certain things that people are simply not capable of doing. Consequently, particular ways of working are not feasible. For example, some activities require too much working memory load, too much time, too much knowledge, or too much computational effort for people to perform. These constraints are specifically associated with workers' competencies, not with any of the other preceding phases of analysis alone. This final narrowing down of degrees of freedom is illustrated in Fig. 5.3 by the worker competency set, which is nested within the social-organizational set for strategy A.

It is important to point out that the relevant set of constraints will change as a function of the context. Although Fig. 5.3 does not show this explicitly, the available degrees of freedom are dynamic. They will change with the task, strategy, and so on. Thus, the relationship between the five layers is much more fluid than the figure would indicate.

Another way to illustrate the logical nesting between the layers in Figs. 5.2 and 5.3 is to show how changes to one layer of constraint propagate logically to others during a work analysis. Because the layers are nested from a formative perspective, propagation only logically occurs in a downward direction. For example, if the work domain itself is changed (e.g., one piece of equipment in a process plant is replaced by a new piece), then all subsequent levels may be affected. New control tasks associated with the new equipment may be required, or some control tasks may no longer be possible or necessary because they were only feasible or required with the old equipment, respectively. In addition, the strategies level may also be affected because the same control tasks may have different constraints associated with them because of the new equipment. Also, some new organizational configurations may be possible or desirable because of the change in equipment (e.g., if it becomes easier to colocate workers). Finally, the change to the work domain can also have implications for worker competencies as well (e.g., new skills or knowledge may be required to operate the new equipment). Note, however, that a change in lower layers of constraint will not logically affect preceding layers. For example, adding a new control task may affect what strategies are possible or desirable but it will not, by itself, change the work domain. In this case, what is being modified is the set of goals to be achieved by actors, not the functional structure of the underlying system being acted upon. Similar relationships exist between other layers of constraint.[4]

A final way of briefly illustrating the nested relationship between the phases of CWA is to consider the logical precedence of formative decisions that must be made by analysts or designers. For example, before we can decide how something can be done (strategies analysis), we should know what goals are to be achieved (control tasks analysis) and what the system being controlled is capable of doing (work domain analysis). That is, before we can choose a process, we should know what we should be trying to achieve and what means we have available for achieving it. Similarly, before we can decide how work should be allocated among human and machine actors (social-organizational analysis), we should establish what needs to be done to satisfy the relevant goals (control tasks), how those goals may be satisfied (strategies analysis), and what the controlled system is capable of doing (work domain analysis). In short, before we can make decisions about how to allocate work, we should find out what there is to be allocated in the first place. An analogous precedence relationship exists in identifying the knowledge, rules, and skills that workers should exhibit to function effectively. Before we can make decisions about these issues (worker competencies analysis), we should know what or who workers will be cooperating with (social-organizational analysis), what processes they may be following (strategies analysis), what goals need to be achieved (control tasks analysis), and the means that are available for action (work domain analysis). In other words, before we can make decisions about what somebody needs to know, we should find out what there is to know about first.

It is because of these formative precedence relationships that earlier phases of work analysis try to be independent of later phases. For example, a work domain analysis describes the capabilities of the system being controlled independently of any particular task, strategy, actor, or device. From a formative perspective, decisions about what needs to be done, how it can be done, who can do it, and using what tools are better made after the capabilities of the system being controlled have been identified. Similarly, a control task analysis describes what should be done independently of how or by whom. The reason for this is that decisions about how

---

[4]The logical nesting and the unidirectional propagation between levels only hold from a formative perspective. An example can help to make this point. From a descriptive perspective, we could argue that a change in worker competencies might affect the strategies that could be adopted, for instance (e.g., if less skilled workers are hired, then they may not be able to adopt certain strategies). However, from a formative perspective, this propagation would not occur because the workers should be selected so that they can effectively cope with all of the requirements imposed by earlier layers of analysis.

something can be done and who can do it are better made after that something has been identified and described. In general terms, earlier phases of analysis try to avoid making a priori commitments about issues that naturally emerge only later during the work analysis. In this way, we can try to discover new possibilities for design by determining what options are feasible and useful, given the insights gained from the work analysis.[5]

These interdependencies between phases of CWA are illustrated by detailed, concrete examples in Part III. For now, the main point we want to convey is that the five phases are nested in an integrated, coherent, and logical fashion that is consistent with an ecological approach.

Relation to Design Interventions?

We mentioned earlier that if a work analysis framework is to be formative, then each of the conceptual distinctions on which that framework is based (the first column in Fig. 5.1) must be closely linked with a particular class of systems design interventions (the fourth column in Fig. 5.1). Only this coupling will ensure that the results of the work analysis have direct, strong, and obvious implications for design. So far, we have described the five conceptual distinctions in CWA and the nested relationships between them, but we have not yet discussed how each of these layers of constraint map onto particular types of design interventions.

Table 5.1 shows this mapping. The list is merely intended to be illustrative, not definitive or exhaustive. Beginning with the work domain, analyzing the system being controlled provides a great deal of insight into what information is required to understand its state. This analysis, in turn, has important implications for the design of sensors and models (Reising & Sanderson, 1996). The work domain analysis also reveals the functional structure of the system being controlled. These insights can then be used to design a database that keeps track of the relationships between variables, providing a coherent, integrated, and global representation of the information contained therein.

The control task analysis deals not with data structures, but with control structures. The goals that need to be satisfied for certain classes of situations, and the constraints on the achievement of those goals (see chap. 3) are identified here. This knowledge can then be used to design either constraint-based procedures that guide workers in achieving those goals, or automation that achieves those goals autonomously or semiautonomously. In addition, this analysis will also identify what variables and relations in the work domain may be relevant for certain classes of situations (see Fig. 5.3). Those insights can be used to design context-sensitive interface mechanisms that present workers with the right information at the right time (Mitchell & Saisi, 1987; Woods, 1991).

The strategies analysis deals, not just with what needs to be done, but also with how it is to be done. Each strategy can be thought of as a different frame of reference for

---

[5]As we pointed out in chapter 4 (see Footnote 2), this approach is only useful for revolutionary design conditions (Alexander, 1964). If designers must work strongly within the context of previous designs, then some of the design decisions that could have emerged as an output of CWA may have to be specified up front. These situations are addressed toward the end of this chapter. In the meantime, note that the resulting design will probably not be nearly as effective as it could be in these cases, because it is inheriting whatever deficiencies or limitations existed in the previous design.

TABLE 5.1 Relationships Between the Five Phases of CWA and Various Classes of Systems Design Interventions

1. *Work Domain*

    - What information should be measured? (sensors)

    - What information should be derived? (models)

    - How should information be organized? (database)

2. *Control Tasks*

    - What goals must be pursued and what are the constraints on those goals? (procedures or automation)

    - What information and relations are relevant for particular classes of situations? (context-sensitive interface)

3. *Strategies*

    - What frames of reference are useful? (dialogue modes)

    - What control mechanisms are useful? (process flow)

4. *Social-Organizational*

    - What are the responsibilities of all of the actors? (role allocation)

    - How should actors communicate with each other? (organizational structure)

5. *Worker Competencies*

    - What knowledge, rules, and skills do workers need to have? (selection, training, and interface form)

pursuing control task goals, each with its unique flow and process requirements. Thus, identifying what strategies can be used for each control task provides some insight into what type of human-computer dialogue modes should be designed. Ideally, each mode should be tailored to the unique requirements of each strategy. The strategies analysis also reveals the generative mechanisms (i.e., rules or algorithms) constituting each strategy, which, in turn, helps specify the process flow for each dialogue mode.

The social-organizational analysis deals with two very important and challenging classes of design interventions. Given the knowledge uncovered in previous phases, analysts can decide what the responsibilities of the various actors are, including workers, designers, and automation. These role allocation decisions define the job content of the various actors. In addition, analysts should also determine how the various actors can effectively communicate with each other. That analysis will help identify the authority and coordination patterns that constitute a viable organizational structure.

The final phase shown in Table 5.1 is the analysis of worker competencies. Because the work demands have been thoroughly analyzed by this point, the knowledge, rules, and skills that workers must have to function effectively can be determined. This analysis will help develop a set of specifications for selection (if relevant) and training. In addition, this analysis will also provide some insight into how information should be presented to workers because some competencies may not be triggered unless information is presented in particular forms (Vicente & Rasmussen, 1992).

Although the list in Table 5.1 is not definitive, we hope that it will show you how—right from the very start—CWA is deliberately geared toward uncovering implications for systems design. The detailed examples presented in Part III will help you appreciate the nature of this pragmatic orientation in more intimate detail.

DELIBERATELY SUPPORTING ADAPTATION: WORKERS FINISH THE DESIGN

So far in this chapter, we have described the generic characteristics of formative approaches to work analysis and the basic structure of one such framework, CWA. In Part III, we describe the five phases of this particular framework in much more detail. First, however, we want to give you a better intuitive feel for the overall philosophy behind CWA. Accordingly, in this section, we address a number of related questions, including: What is the intended role of workers in CWA? How does CWA try to support that role? What is the envisioned impact of CWA on safety, productivity, and health? These are very important questions to address because the philosophy behind CWA is easily subject to misinterpretation.

What Is the Intended Role of Workers?

As we discussed in chapter 3, complex sociotechnical systems are open systems. They are subject to disturbances that are not, and frequently cannot be, anticipated by designers. For example, in a field study of flexible manufacturing systems, Norros (1996) found that workers had to cope with an average of *three disturbances per hour* (i.e., events during which the system was functioning in ways that were not anticipated). To deal with these disturbances effectively, workers must exhibit context-conditioned variability—they must use their expertise and ingenuity to create a solution to counteract the disturbance in question. In complex sociotechnical systems, the primary value of having people in the system is precisely to play this adaptive role. Workers must adapt online in real time to disturbances that have not been, or cannot be, foreseen by designers. As more and more routine tasks become automated in various application domains, this requirement for worker adaptation will only increase (see chap. 13).

As we see later, workers have typically tried to play this role in an informal fashion. But given the importance of adaptation in complex sociotechnical systems and the potential hazards involved (see chap. 1), these ad hoc adaptive activities are unsatisfactory. We cannot expect workers to play the role of adaptive problem solvers in a consistent and reliable fashion unless we provide them with information support that is tailored to the demands of this challenging role. Rather than merely expecting workers to be adaptive actors, we should instead deliberately design computer-based information systems to help workers be effective and reliable adaptive actors. This is precisely the philosophy behind CWA—uncover the requirements that will help workers be flexible, adaptive problem solvers. *CWA is all about designing for adaptation.*

Reconciling Modeling & Flexibility: Formative ≠ Normative

You might be thinking: How can you reconcile the analytical goal of formative modeling with the design goal of supporting flexible, adaptive action? After all, previous models of work (e.g., traditional task analysis) have usually been associated with normative approaches. And as we discussed in chapter 3, such approaches are deeply limited for our purposes. A key insight, however, is that the limitations of normative approaches lie in the assumptions that they make, not in the fact that they frequently rely on work modeling (Sheridan, in press). Proponents of descriptive approaches have justifiably criticized the unrealistic prescriptive nature of normative approaches (see chap. 4). However, in the process, the value of models seems to have been overlooked and, to some extent, lost. It is important to keep in mind that not every model of work is "rationalistic," and thus subject to the limitations of normative approaches that we reviewed earlier. CWA shows that the dichotomy between modeling work and providing support for flexible, adaptive behavior is a false one. In this section, we argue that it is possible for work modeling and adaptive flexibility to coexist (cf. Sheridan, in press). Even more strongly, we believe that the former may even be a prerequisite for effectively realizing the latter.

Formative modeling and designing for flexibility seem to be conflicting goals because the former involves foresight on the part of analysts, whereas the latter involves giving some responsibility and discretion to workers. If we rely on analysts' insights, aren't we telling workers what to do? How can workers have the flexibility to adapt to local contingencies if analysts engage in formative modeling? Conversely, if we give discretion to workers to adapt to disturbances online in real time, how can we plan anything ahead of time? How can you possibly do formative modeling for the unanticipated?!

The key to reconciling these apparent contradictions lies in the concept of constraints. As we discussed in detail in chapter 3, constraints specify what should not be done, rather than what should be done (cf. Kugler et al., 1980).[6] Thus, constraints are what should remain constant, regardless of how workers decide to act in any one particular situation. Flexibility does not mean that anything goes, a point recognized by Suchman (1987) in her analysis of situated action:

> The emergent properties of action means that it is not predetermined, but neither is it random. A basic research goal for studies of situated action, therefore, is to explicate the relationship between structures of action and the resources and constraints afforded by physical and social circumstances. (p. 179)

> The organization of face-to-face interaction is the paradigm case of a system that has evolved in the service of orderly, concerted action over an indefinite range of essentially unpredictable circumstances. What is notable about that system is the extent to which mastery of its constraints localizes and thereby leaves open questions of control and direction, while providing built-in mechanisms for recovery from trouble and error. The constraints on interaction in this sense are not determinants of, but are rather "production resources" . . . for, shared understanding. (p. 95)

In other words, constraints remain invariant in the presence of context-conditioned variability. As a result, they provide a basis for reconciling formative modeling (by specifying boundaries on action) and worker adaptation (by giving workers the flexibility to adapt within those boundaries). *This one idea is, without a doubt, the single most important prerequisite to understanding CWA.* Unless you understand the value of a constraint-based approach, you will not be able to appreciate the

---

[6]Chapter 3 focused primarily on the constraints that emerge from analyzing task goals (e.g., action A must be done before action B). We have seen in this chapter that there are several other sources of constraints as well (see Fig. 5.2). In this subsection, we are interested in the generic value of the concept of constraints, regardless of their type.

remainder of this book. Accordingly, we devote the remainder of this section to explaining the power of this perspective.

Behavior-Shaping Constraints

CWA is based on the notion of intrinsic work constraints. Each of the five layers in the framework identifies a category of constraint that needs to be respected. To use the terminology introduced in chapter 1 (see Fig. 1.14), intrinsic work constraints are *behavior shaping* (Rasmussen & Pejtersen, 1995) because they define the boundaries on action (cf. T. J. Smith et al., 1994). This idea is shown in highly simplified form in Fig. 5.4. The constraint types that were illustrated in Fig. 5.3 can be integrated, thereby collectively creating a dynamic constraint boundary defining possibilities for action. Following the logic shown in Fig. 5.3, this constraint boundary will of course change as a function of the context (e.g., the control task being performed, or the strategy being adopted).

If analysts identify the set of behavior-shaping constraints, then these can be embedded into an information system. As shown in Fig. 5.4, workers would then have the flexibility to adapt within the remaining space of possibilities. In one circumstance, they may select one trajectory (i.e., sequence of actions), whereas in another circumstance, they may have to select a different trajectory to achieve the same task goals (see chap. 3). Thus, the constraint space provides the flexibility that is required to support context-conditioned variability. In addition, the same worker can also choose different trajectories, even when the circumstances remain the same. As a result, the constraint space is also flexible enough to support the intrinsic variability that is frequently observed in human action. Finally, different workers can choose different trajectories to achieve the same outcome in different ways. Therefore, the constraint space is also flexible enough to support individual differences between workers. In summary, the constraint-based approach illustrated in Fig. 5.4 allows

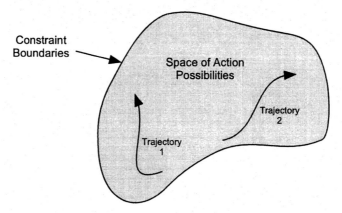

FIG. 5.4. The constraint-based approach. Analysts identify constraints on action that can be embedded in an information system. Workers then have the flexibility to adapt within the remaining space of possibilities. Although not shown in the figure, the constraint boundaries, and thus the available degrees of freedom, are situation-dependent and therefore dynamic.

workers to respond to unanticipated contingencies and to follow their subjective preferences (by choosing a different trajectory through the space), while at the same time, satisfying the demands of the job (by staying within the constraint boundaries). As odd as it may sound, the constraint-based approach actually supports workers in "finishing the design" (Rasmussen & Goodstein, 1987) of the sociotechnical system.

Supporting Workers in "Finishing the Design"

It is well known that workers frequently informally tailor the design of their device and work practices to better meet domain demands. In some cases, workers make *permanent* changes that could have been, but were not, originally introduced by designers. For example, Seminara, Gonzalez, and Parsons (1977) described many examples where workers added labels to similar-looking displays to make them easier to identify, and even changed the knobs on some controls to make them easier to distinguish. In other cases, workers make *temporary* changes to their device or work practices to make it easier to deal with the transient demands of the moment. For example, Vicente, Burns et al. (1996) found that workers in a nuclear power plant control room changed the set points on alarms to make them more sensitive to the current context. In that control room, some analogue meters have two adjustable alarm set points, one representing an upper limit and the other representing a lower limit for that variable. If the meter value exceeds these limits, then an alarm message and an auditory signal are presented to workers. For each meter, the alarm set points are always supposed to be set at prespecified values determined by designers.

Vicente, Burns et al. (1996) found, however, that workers sometimes change the alarm set points on a particular meter to values that are different from the predefined set points. Typically, workers select a temporary set point value that will lead to an auditory alarm when it is time to perform a certain action. For example, the "official" upper and lower alarm set points for a particular tank level may be specified at ± X units of the nominal tank level. So if the nominal value of a tank is 70% full, then the alarm limits may be set at 65% and 75%, for instance. This range defines the desired tank level for normal operations, but there are situations where it is necessary to have the tank level go outside of this range—if the tank needs to be emptied, for instance. In those situations, workers will move the lower alarm set point so that it will trigger an auditory alarm when the tank level is low, thereby indicating that it is time to turn off the pump that is being used to empty the tank. This informal practice actually involves using the alarm system in ways for which it was not intended. However, it makes it easier for workers to perform their job. Instead of having to remember to check the meter periodically until its value reaches the point at which the pump should be turned off, workers are free to perform other activities until the alarm signal "tells" them that it is time to turn off the pump. Thus, manipulating the adjustable alarm set point obviates the need for periodic monitoring. The downside is that these ad hoc activities may lead to problems. For example, the workers may forget to reset the alarm set points to their normal values. The implications of such dangers are discussed later.

There is a difference between these two types of informal tailoring activities that is very relevant for our purposes. The permanent changes observed by Seminara et al. (1977) could have been part of the original design because designers had the information required to introduce those features. For example, there is no reason why designers should not have been able to provide easy-to read labels to allow workers to distinguish between displays. The fact that such labels were not introduced can be considered a design deficiency. In contrast, the temporary changes observed by Vicente, Burns et al. (1996) would have been far more difficult, if not impossible, for designers to anticipate completely and reliably. These tailoring activities are responding to local circumstances and thus require information that is only fully available online to workers, not beforehand to designers. In the case of the alarm set point manipulations just described, it would be very difficult for designers to predict when an alarm set point would need to be changed, which setpoint should be changed, what value it should be set at, and when it should be returned to its normal value. As a result, it is very difficult for designers to specify a detailed procedure, automation, or work flow that explicitly supports these activities. Workers, on the other hand, do have access to the local information that is required to make these decisions online. There are numerous situations analogous to this one in complex sociotechnical systems, many being much more involved and thus even more difficult to anticipate.

This difference between anticipated and unanticipated factors is represented graphically in Fig. 5.5. There are some factors that can be anticipated by designers because the information to make decisions is available a priori during the design process. In these cases, designers can create the device to take into account these factors explicitly. However, there are other factors that cannot be reliably anticipated because the relevant information is only available during operation, not during design (see chap. 3). In these cases, designers will not be able to provide detailed, "pre-packaged" information support. Instead, workers must deal with the contingency online in real time because the relevant information is only available locally. As we pointed out at length in chapter 3, the more open a system is, the more these unanticipated factors are likely to occur.

Figure 5.5 allows us to view the temporary tailoring activities described previously in a new light. Following Rasmussen and Goodstein (1987), these informal worker activities can be interpreted as attempts at "finishing the design" of the sociotechnical

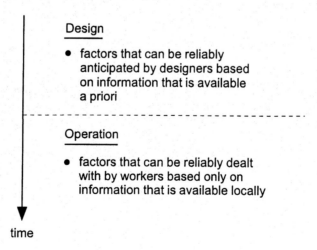

FIG. 5.5. Some factors can be anticipated by designers, and thus, explicitly designed for. Other factors are unanticipated, and thus, must be dealt with on-line in real-time by workers.

system.[7] Designers try to anticipate as much as possible beforehand, but some situations can be dealt with only in real time. In these cases, workers have to finish the design by making decisions that could not be reliably made by designers a priori because the relevant information was not available. Worker attempts at finishing the design such as those observed by Vicente, Burns et al. (1996) are commonplace (e.g., R. I. Cook & Woods, 1996; Henderson & Kyng, 1991; Xiao, Milgram, & Doyle, 1997), particularly in complex sociotechnical systems because they are substantially open systems.

Despite the frequency and importance of such adaptation activities, workers are usually required to finish the design in a largely informal manner. Our story in the Preface provides an example. The nuclear power plant operators did not have any aids that helped them determine when the procedures they were supposed to be following were no longer appropriate, nor did they have any aids that would help them determine how the procedure should be modified in such circumstances. As a result, they had to make such decisions unaided and "on the fly" by relying on their conceptual knowledge of the plant. There is a danger that such ad hoc activities may lead to unforeseen errors in unusual situations (R. I. Cook & Woods, 1996). It would be preferable if we could design a device so that workers could finish the design in a more systematic fashion, without creating such potentially dangerous side-effects.

Centralized Versus Distributed Control

This apparently straightforward insight actually has profound implications for work analysis and systems design (Rasmussen & Goodstein, 1987). As shown in Fig. 5.6, these implications can be appreciated by contrasting the traditional *centralized* control approach to analysis and design with the *distributed* control approach being advocated here (cf. Resnick, 1994; Wegner, 1997). The centralized/distributed distinction actually lies on a continuum, but to simplify the discussion, we describe it in terms of the two extremes.

The traditional approach to work analysis and systems design is to adopt a centralized control approach. As shown in Fig. 5.6a, this approach begins with a detailed work analysis. The designer analyzes the work and, in the extreme, tries to identify the "optimal" way of doing the job. The insights derived from this analysis are then used to develop a plan that is implemented by one or more means, including: (a) automation that performs the job according to the optimal plan, (b) a computerized work flow that forces workers to follow the optimal plan, and (c) paper- or computer-based procedures that guide workers in following the optimal plan. With this approach, there is comparatively little intellectual work left for workers. As shown in Fig. 5.6a, they are supposed to either rigidly execute the designers' optimal plan or sit back and let the automation do the work.[8] This approach is centralized

---

[7]The phrase "finishing the design" does not refer to relatively permanent modifications made by workers to tailor the interface to their own personal preferences. Instead, it refers to continual adaptations of the structure and state of the sociotechnical system to particular local contingencies. Thus, the design is never really finished in the sense that work practices are dynamic and evolving. As Norros (1996) put it, "unpredictability is a source of disturbances throughout the lifetime of the system. . . . attempts to fight against them are also continuous" (p. 175). Thus, workers are continually finishing the design.

[8]We use the phrase "supposed to" because, in practice, workers are known to deviate from the official, centralized work procedures in order to cope with local contingencies (e.g., Vicente, Burns et al., 1996).

a) Centralized Control:

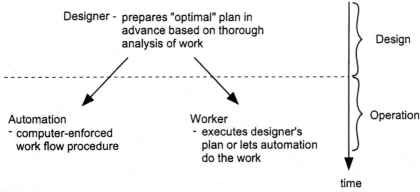

b) Distributed Control:

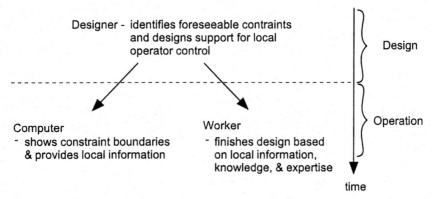

FIG. 5.6. A comparison of centralized and distributed approaches. See text for description.

because decisions about what should be done and how it is to be done are made beforehand by designers, who are the central authority in this scheme.

As mentioned in chapter 3, Taylor's (1911) approach to work analysis was the ultimate in centralized control. This extreme is illustrated in the following characterization of Taylorism taken from a recent biography of Taylor:

> The control of work must be taken from the men who did it and placed in the hands of a new breed of planners and thinkers. These men would think everything through beforehand. The workmen—elements of production to be studied, manipulated, and controlled—were to do as they were told. (Kanigel, 1997, p. 371)

The antithesis of centralized control is distributed control. As illustrated in Fig. 5.6b, designers play a more modest role in this case. Rather than trying to plan everything up front and then try to enforce rote procedural compliance, designers explicitly acknowledge the fact that some factors cannot be anticipated (see chap. 3). Accordingly, designers instead try to identify the constraints that are reliably foreseeable and then design an information system that deliberately tries to help workers finish the design in real time. In this case, the role of the computer system is to show workers the constraints that must be respected for safe and reliable performance (see Fig. 5.4), and to present local information about the current context. Workers can then finish the design (i.e., make local decisions) based on: their knowledge of the current structure of the work domain (e.g., what components are currently being maintained or repaired), the constraint boundaries and local state information provided by the computer system, and their expertise. This approach is distributed because decisions about what should be done and how it is to be done are made in a cooperative fashion between designers, computers, and workers. There is no central authority dictating how work should be carried out.

Rasmussen and Goodstein (1987) described the rationale behind this distributed control scheme for the particular case of process control:

- Instead of continuing their efforts to make their preplanning of responses and countermeasures more and more complete and thus restrict the operator's own initiative, designers should take advantage of modern information technology to make available to the operators their conceptual models and their processing resources so as to allow the operators to function as their extended arm in coping with the plant. (p. 663)

- Since the designer will not be able to foresee the necessary control responses for all possible disturbances, he needs a representative on site—the operator(s)—who have to be able to take over in a competent way. The operator's supervisory control task is indeed in many respects a completion of the system design for the particular, perhaps infrequent, situation being dealt with. As a consequence, the operator will need information about the problem space underlying the design, and the designer will have to communicate this kind of information to the operator. (pp. 663–664)

The contrast with centralized Tayloristic approaches could hardly be stronger.

The discussion so far has been simplistic in that it has contrasted two diametrically opposed approaches. But as mentioned earlier, the centralized/distributed distinction lies on a continuum. In between the two extremes we have been discussing, many different modes of control are possible, some more centralized and others more distributed (Sheridan, 1987). Which mode does CWA advocate? In other words, just how distributed should the design be?

The answer to this question is perhaps subtle, but quite important. Because of its constraint-based nature, CWA lets the control architecture emerge as an output of the work analysis rather than advocating any one particular mode as being the most appropriate. For example, if there really is only one correct way to perform a control task (i.e., if there are no degrees of freedom), then this fact should be revealed in the analysis of behavior-shaping constraints. In this case, it would be appropriate to have a very centralized form of control, either in the form of autonomous automation or a rigid procedure (see Footnote 7 in chap. 3). Usually, in complex sociotechnical systems, there will be many ways of performing a control task because of the open nature of the system. In those situations, it would be more appropriate to have a more distributed form of control, letting the worker(s) finish the design. Note that it is highly likely that different forms of control may be required under different circumstances in the very same domain. For example, in a nuclear power plant, some control tasks may be completely automated (e.g., safety system activation) whereas others may require workers to be adaptive problem solvers (e.g., diagnosing an unanticipated fault). The results of the CWA should help analysts determine that these different forms of control are required, and should be supported, in these different situations.

The general point is that we are not suggesting that workers should always be solely responsible for all decisions (i.e., that it is always appropriate to implement a highly distributed form of control). The extent to which a particular situation should be dealt with in a centralized or distributed fashion depends on the predictability of the circumstances (see Fig. 5.5). As we pointed out in chapter 3, such circumstances can include, not just situational factors (e.g., initial conditions, instigating events), but psychological factors as well (e.g., what strategies workers will choose to adopt, when they will switch from one strategy to another). In some cases, all of the information that is needed to make a reliable decision about these circumstances is available beforehand, in which case control can be more centralized. In other cases, the information that is needed to make a reliable decision is only available on the spot, in which case control should be more distributed. In the words of Rasmussen (1986c):

> How much of the decision sequence . . . the designer can prepare in advance depends on the invariance across occurrences of the class of situations considered, and on the designer's ability to foresee the relevant conditions for decisions. What is not included in the designer's decisions . . . must be left to the operator's decision during the actual situation. (pp. 45–50)

As a result, there may very well be situations where a particular task should be fully automated or proceduralized. Examples of such situations include those where there are no degrees of freedom available for action, the time to react is beyond human capabilities, and the consequences of failure are extremely severe and the designers are confident that their preplanned solution will work in every conceivable situation (Rasmussen & Goodstein, 1987). Not only is CWA capable of accommodating these situations, it can be effectively used to help identify and support them as well. In general, however, the more open a system is, the greater the need for worker discretion, and thus, the greater the need for distributed control (cf. Wegner, 1997). Thus, in complex sociotechnical systems, distributed control will be the rule, rather than the exception.

With Freedom Comes Responsibility

The CWA philosophy is to, wherever appropriate, give workers the freedom to finish the design. Instead of having to rotely follow designers' procedures, or having to let automation do all of the work, workers are more intimately involved in making decisions about how work is to be done (see Fig. 5.6). With this freedom comes responsibility, however. The demands on workers in a distributed control architecture are greater than in a centralized architecture. Does this mean that the philosophy of finishing the design is flawed?

There are at least two reasons to answer this question in the negative. First, recall that the motivation for going to a distributed architecture and having workers finish the design is based on necessity. Given an open system with unanticipated events requiring context-conditioned variability, centralized control architectures will simply not be able to deal with the entire range of task demands effectively. Designers cannot possibly anticipate, and design for, all situations. Thus, returning to a centralized form of control is not a viable option for complex sociotechnical systems if we want to improve safety, productivity, and health. Second, the demands on workers can be made manageable by providing workers with the type of information support that is required to finish the design effectively. The role of adaptive problem solver is indeed a very challenging one, which means that it is essential that workers be provided with the appropriate knowledge and tools. CWA attempts to achieve this goal by adopting an ecological perspective, factoring in both the demands of the ecology and the capabilities and limitations of human information processing (see Fig. 5.2). The results should be a device and an organizational structure that allow workers to finish the design in a reliable and efficient manner.

Two Examples of the Paradigm

In this subsection, we provide two examples of devices that embody this philosophy. Although the design of these devices was not based on CWA, they are nevertheless good illustrations of distributed control to support worker adaptation. The first device is a novel alarm system for process control plants that was developed in industry based, in part, on the tailoring activities observed by Vicente, Burns et al. (1996). The second device is a novel decision support system that was developed in academia for a medical application.

Example #1: User-Initiated Notification. Traditionally, alarms for process control systems have been designed according to a centralized scheme (see Fig. 5.6a). Based on a thorough analysis, designers determine what the normal ranges of values are for many variables. Then, they identify upper and lower alarm limits that will notify workers when a variable is outside of its normal range. Typically, these set point limits are static in that they do not change over time. As Lees (1983) pointed out, this practice results in many false alarms, thereby greatly reducing the efficacy of the alarm system. These false alarms occur because the definition of what is a normal range for any particular variable frequently depends on the context. For example, if a maintenance test is being performed, a particular value may be perfectly normal, whereas if a test is not being performed, that same value may be indicative of a catastrophic failure. But because the alarm set points are supposed to be fixed, false alarms are generated in situations not considered by designers. As mentioned earlier, Vicente, Burns et al. (1996) found a case where workers informally manipulate alarm set points to take into account these contextual dependencies, thereby "creating" a more informative alarm system that is sensitive to context.

Guerlain and Bullemer (1996) created a user-initiated notification (UIN) device that overcomes some of the limitations of traditional alarm systems by adopting a distributed control approach. Rather than having designers decide once and for all what the best alarm set point values are for all variables, the UIN device provides a systematic means for workers to define temporary, context-specific notifications. For example, with UIN, workers can program the device to issue a message and auditory signal when a particular event occurs (e.g., when tank level is almost empty, signifying the time to close an output valve). Note also that deliberate safety measures have been built into the design of the UIN device (the normal alarms signifying a dangerous plant state are never overridden by the UIN). Because of this bounded flexibility (see Fig. 5.4), worker adaptation can be realized without the errors that can occur when tailoring activities are performed in the typical informal manner. The UIN device thereby allows workers to perform the local adaptation activities observed by Vicente, Burns et al. (1996) in a controlled, rather than an ad hoc, fashion.

UIN results in a distributed control scheme (see Fig. 5.6b) because it supports a coordinated cooperation between designers, computer, and workers. Performance benefits from the safety constraints introduced by the designer, the information acquisition and condition-checking ability of the computer, and the workers' inge- nuity, knowledge, and access to local information. This distributed architecture thereby accommodates adaptive flexibility while still respecting goal-relevant con- straints.[9]

Example #2: Critiquing Decision Support System. Traditionally, decision support systems have been designed according to an expert system philosophy that embodies a centralized control approach (see Fig. 5.6a). The designer consults one or more experts to figure out the best way to perform a task, and then embodies the knowledge that has been obtained into a set of rules. The resulting rule-based expert system provides decision guidance to workers who then follow this advice (or not). As Guerlain (1995) pointed out, the problem with these expert systems is that they are brittle. The advice they provide in some (undetermined set of) situ- ations is inappropriate, in part because it is not possible for designers to anticipate and explicitly design for all of the situations that can occur in practice (see chaps. 1 and 3).

Guerlain (1995) developed a decision support system based on a critiquing philosophy that embodies the distributed control approach. Rather than try to create a perfect expert system—an unrealistic goal—Guerlain instead tried to identify a set of constraints that should be respected by workers if task goals are to be reliably achieved. These constraints were then built into a critiquing system that allows workers to be in charge of making decisions, but that offers advice when a constraint has been violated. Thus, rather than telling workers what to do, the critiquing system merely advises workers when it believes that they have done something wrong (i.e., that they have crossed a boundary like that in Fig. 5.4). For example, in the device developed by Guerlain to support medical technologists in a blood bank, the critiquing system would warn workers when they: (a) ruled out hypotheses incor- rectly, (b) failed to rule out a hypothesis when it was appropriate to do so, (c) failed to collect converging evidence, (d) gave an answer that was implausible given the available data, and, (e) gave an answer that was implausible given prior probabilities. These are some of the behavior-shaping constraints that defined the space of action possibilities for this application domain.

Like the UIN device, this critiquing system implements a distributed control scheme (see Fig. 5.6b) because it results in cooperative problem solving between designer, computer, and worker. Performance benefits from the analysis conducted by the designer, the rule-checking ability of the computer, and the workers' ingenuity, knowledge, and access to local information. This distributed architecture thereby

---

[9]UIN has undergone a formal usability evaluation, and prototypes that were introduced into process plants have been very favorably received by workers. The device is now available from Honeywell as a commercial product.

achieves adaptive flexibility while still respecting goal-relevant constraints. The empirical results obtained by Guerlain (1995) illustrate the advantages of her critiquing system over a traditional decision support system.

Taking Stock: Implications for Safety, Productivity, and Health

Now that the CWA philosophy has been laid out, we can consider how this framework fulfills the requirements for success that we identified in chapter 1. There, we determined that at least four objectives must be satisfied for computer-based work in complex sociotechnical systems to be safe, productive, and healthy:

1. Support workers in adapting to, and coping with, events that are unfamiliar to them and that have not been anticipated by system designers.
2. Identify the functionality that is required to accomplish intellectual tasks requiring discretionary decision making (i.e., usefulness issues).
3. Be based on an understanding of human capabilities and limitations (i.e., usability issues).
4. Improve decision latitude by providing workers with the autonomy to make decisions and the opportunity to exercise and develop skill.

Is CWA capable of meeting these needs?

Beginning with the first requirement, support for worker adaptation and flexibility plays a central role in CWA, as evidenced by the concept of behavior-shaping constraints. Constraints accommodate many ways of accomplishing a task. This flexibility provides the freedom to cope effectively with context-conditioned variability, thereby improving safety. In addition, accommodating greater variability in action means that workers have more opportunities to learn, providing a means for the development of expertise and thus health as well. Second, CWA makes few assumptions about the properties of the device available to workers. Consequently, it is more likely that the new design will result in novel functional possibilities, rather than being constrained by designers' current assumptions about functionality. Moreover, because CWA is deliberately geared to supporting worker adaptation, it is very well suited to identify the information needs associated with intellectual tasks requiring discretionary decision making. Given Landauer's (1995) arguments and evidence (see chap. 1), we would thereby expect improved productivity. Third, CWA is based on an ecological perspective which, though giving primacy to environment considerations, also integrates human capabilities and limitations as well (see chap. 2). By taking usability issues into account, we also have a better chance of improving productivity (Landauer, 1995). Fourth, adopting a constraint-based approach also gives workers the discretion to make decisions online about how to perform the task, leading to greater worker autonomy. By creating a distributed control architecture wherever appropriate, workers are given the responsibility and the resources to "finish the design" of the sociotechnical system. Based on Karasek and Theorell's (1990) demand-control model (see chap. 1), we would therefore expect better worker health as a result.

We hope this brief discussion illustrates how the characteristics of CWA are closely coupled with the criteria we outlined in chapter 1. This connection should become stronger as we elaborate on the various phases of CWA in Part III.

TWO LOOSE ENDS

Before wrapping up this chapter, there are a few additional issues that must be dealt with. First, we explain the complementary relationship between descriptive and formative approaches to work analysis. Second, we discuss the applicability of CWA to evolutionary, rather than revolutionary, design problems.

Data Describe, Models Generalize

Now that we have described formative approaches in general and CWA in particular, we are in a position to revisit the descriptive approaches discussed in chapter 4. Figure 5.7 is a simplified representation of the relationship between descriptive and formative work analysis. Descriptive approaches focus on the process of data collection. Analysts study current work practices and their development over time to better understand the demands on workers. The result is a detailed description of the way things are and, perhaps, have been. As the four case studies in chapter 4 illustrate, many valuable insights can be gained from such analyses.

As shown in Fig. 5.7, formative approaches focus on the modeling of intrinsic constraints. Analysts develop various models of work, each directly linked to a certain class of systems design interventions (see Fig. 5.1 and Table 5.1). As a result, it becomes possible to identify requirements that a sociotechnical system must have if it is to support work effectively in the application domain of interest. As we have tried to show in this chapter, the power of formative approaches lies in their strong link to systems design.

To be sure, the characterization in Fig. 5.7 is simplified, merely representing the emphasis in each approach. After all, attention has been paid to some form of modeling in a few descriptive approaches (e.g., Klein's, 1989, RPD), and attention has been paid to data collection in formative approaches (e.g., Beyer & Holtzblatt, 1998; Rasmussen et al., 1994). Thus, the differences captured in Fig. 5.7 are a matter of degree.

Having said that, Fig. 5.7 shows that descriptive and formative approaches to work analysis are largely complementary. Data describe, whereas models generalize. But useful models cannot be developed in the absence of data, and data cannot be strongly linked to design without being organized into models. These complementary

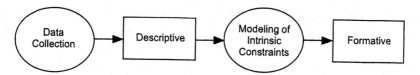

FIG. 5.7. Simplified representation of the relation between descriptive and formative approaches to work analysis.

relationships are well captured by the quote in chapter 4 from Ahl and Allen (1996), which is worth repeating here:

> Although they are a critical part of science, data are not the purpose of science. Science is about predictability, and predictability derives from models. Data are limited to the special case of what happened when the measurements were made. Models, on the other hand, subsume data. Only through models can data be used to say what will happen again, before subsequent measurements are made. Data alone predict nothing. (p. 45)

Therefore, data, and descriptive approaches, are necessary but not sufficient. This was one of the main conclusions of chapter 4. Similarly, models are impotent without a foundation of data to support them.

We have focused more on the modeling of intrinsic work constraints than on methods for data collection because the latter have received far more attention in cognitive engineering than the former. Discussions of various techniques that can be used to gather data about work abound (e.g., see the review in Schraagen et al., 1997). However, we regularly encounter colleagues who have collected a great deal of data but have then found it very difficult to derive design implications from those data. They become overwhelmed in details, and find it hard to see the forest for the trees. Consequently, they also find it difficult to use their findings to envisage new ways of doing work. The process of moving from analysis to design thereby becomes largely dependent on intuition and creative insight. Although creativity will always be a part of design (Ferguson, 1977), if we can reduce the gap between data and requirements, then we can make the transition from analysis to design much more reliable and defensible. The promise of CWA lies in providing a systematic framework for collecting and then organizing data about work practices, so that the development of truly novel ways of working becomes more effective and efficient. That is why we continue to emphasize modeling of intrinsic work constraints in the remainder of this book.

What About Evolutionary Design?

In chapter 1, we discussed the fact that it is important that designers of complex sociotechnical systems "get it right the first time around" and design a system that is capable of functioning effectively, especially under rare, unanticipated events. To paraphrase Alexander (1964), what is required is more of a revolutionary rather than an evolutionary design: "[The designer] has to make clearly conceived forms without the possibility of trial and error over time" (p. 4). Having said this, we also realize that, in many cases, designers face evolutionary design problems where there is a requirement to work strongly within the constraints imposed by previous generations of designs (e.g., Burns & Vicente, 1995). For instance, in commercial aviation, new levels of cockpit automation are usually added as an extra layer on top of the existing automation, rather than "throwing away" the previous design and designing from scratch. As a result, the new automation will not be able to provide functionality that the previous layers are incapable of supporting, even if that functionality is highly desirable from a pilot's point of view. Similarly, in the banking industry, new interfaces (e.g., to allow customers to do their banking on the WWW) are built on top of existing "legacy" databases that were designed for other purposes, rather than "throwing away" the previous information systems and designing from scratch. As a result, the new interface will not be able to provide information that the content and structure of the existing databases are incapable of supporting, even if that information is highly desirable from a customer's point of view. In such strongly evolutionary design projects, the capabilities that can be built into the new design are limited because of constraints inherited from previous designs. Can CWA be applied to design projects that are more evolutionary in nature-or only to those that are more revolutionary?

So far, we have been writing from the perspective of revolutionary design, but CWA can in fact be used in evolutionary design projects as well. The same five behavior-shaping constraints apply (see Figs. 5.2 and 5.3). The primary difference lies with the classes of design interventions listed in Table 5.1. As we have described it for the case of revolutionary design, CWA allows these design decisions to emerge as outputs of the work analysis. For example, rather than assuming up front as an input what the interface should look like, the CWA specifies as an output the requirements that a good interface should have. This relationship is reversed in evolutionary design projects. In these cases, there are some design decisions that are frozen and therefore must be considered as inputs into the work analysis.

The five layers of behavior-shaping constraints comprising CWA can be used to represent these frozen decisions, as well as the implications they have for other design decisions. Consider the example where a new financial information system is being introduced into a university. Let's assume that, in such a case, the organizational structure of the university cannot be altered substantially. However, the training of the workers can be changed to accommodate the new information system. Under these conditions, the social-organizational layer of CWA can be used to represent what the organizational structure is, rather than what it could be (as in the revolutionary design case). Typically, these will not be the same. Having represented that layer, the implications for training can then be addressed by determining how decisions at the social-organizational layer propagate to the worker competencies layer (see Fig. 5.3). By studying the implications represented by this propagation, the analysis of worker competencies can reveal what type of training program would need to be put in place for workers to function effectively with the organizational structure that was specified a priori. Note that the results of such an analysis could reveal that no training program could realistically achieve this objective! This conclusion would indicate that the existing organizational structure is inappropriate, given the structure of the work domain, the tasks that need to be performed, the strategies that can be used, and what we know about human capabilities and limitations.

This last point raises a more general issue. The practice of inheriting design decisions before starting a work analysis (as opposed to letting those decisions emerge as an output of the analysis) is certainly commonplace but is fraught with problems. A priori design decisions are frequently vestiges of idiosyncratic historical factors that may no longer be relevant (e.g., in the banking case described earlier, the old databases were not originally designed to support Web-based banking applications). In contrast, the design decisions made as an output of CWA are instead shaped by the constraints that are inherent in the problem. Consequently, although CWA can be used in this evolutionary fashion, the resulting design will probably not be as effective as it could otherwise be. The full value of CWA is obtained by letting the characteristics of the problem, represented by the five layers of behavior-shaping constraints, inform the design of the sociotechnical system.

SUMMARY

The diagram in Fig. 5.8 is a partially filled-in version of Fig. 5.1 that graphically summarizes this chapter, as well as outlining the next six chapters. We began this chapter by describing the generic characteristics of formative approaches to work analysis. As shown in Fig. 5.8, such frameworks focus on modeling classes of intrinsic work constraints that have a tight and direct connection to classes of design interventions. One particular formative framework, CWA, was introduced. This framework is based on an ecological approach and is deliberately geared to the unique characteristics of complex sociotechnical systems. Because of the open nature of such systems, the philosophy behind CWA is that the primary role of workers is to act as flexible, adaptive problem solvers. Thus, uncovering the kind of information support that workers need to deal with decisions that cannot be anticipated by designers is a primary concern. As shown in Fig. 5.8, CWA is based on five conceptual distinctions. Frequently, the end result of this analysis will be a distributed control architecture that gives workers the responsibility for finishing the design. Two innovative examples of this philosophy were illustrated. Finally, the connections between the characteristics of CWA and the all-important criteria of safety, productivity, and health were briefly described. The evidence and arguments presented in these first five chapters all suggest that CWA has a good chance of meeting the demands of complex sociotechnical systems.

Next, in chapter 6, we introduce a case study that is used in Part III to illustrate the phases of the CWA framework. Chapters 7–11 each describe a layer of behavior-shaping constraint (the first column in Fig. 5.8) in more detail, introduce a modeling tool for that layer of constraint (the second column in Fig. 5.8), and then show how that modeling tool can be applied in the context of a case study (the third column in Fig. 5.8). By describing these layers and case studies in detail, we hope to show you how CWA can directly and productively inform systems design (the final column in Fig. 5.8).

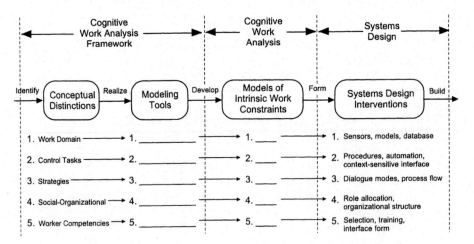

FIG. 5.8. A summary of the CWA framework described in the remainder of the book.

COGNITIVE WORK ANALYSIS IN ACTION **III**

Case Study: Process Control 6

Example is not the main thing in influencing others. It is the only thing.
—Albert Schweitzer

PURPOSE

In the next five chapters, we describe the five phases of CWA corresponding to the five layers of behavior-shaping constraints identified in chapter 5. In most of those chapters, we illustrate CWA by using it to analyze one sample application. The purposes of this brief chapter are to explain the criteria by which this case study was selected, and to describe the case according to the characteristics of complexity identified in chapter 1.

SELECTING THE CASE STUDIES

Our Criteria

In choosing application domains to be used as case studies, we considered a number of different criteria:

1. We originally wanted to select several application domains so that we could better illustrate our concepts. By seeing more than one example, you would have been able to see how the same concepts could be consistently applied to different application domains.
2. We also wanted to select application domains that were quite diverse in nature, so that we could better illustrate the wide breadth of applicability of our concepts. We wanted to try to reflect the broad set of demands imposed by complex sociotechnical systems.
3. We thought it would be best to include application domains with which we were already intimately familiar, and that we had already used extensively in our own research activities. In addition to making our job easier, this choice would also allow us to draw on an existing set of broad and rich insights, thereby enhancing the informativeness of the book.
4. We also felt we had to include application domains that were rich enough to exercise fully the framework we wanted to present. Given the multifaceted demands imposed by complex sociotechnical systems, the case studies had to be sufficiently complex if we were going to illustrate the value of our framework.
5. We also wanted to include application domains that had already been used for design purposes, so that we could indirectly show the practical value

of CWA. Although this book is primarily about analysis, the goal of analysis is to impact design, so it is important to show the connection between the two.

6. We also decided that the number and complexity of the application domains had to be manageable enough to be described in a book of reasonable length. Case studies are a means of illustrating CWA, not the primary focus of the book. Thus, they should not swamp the other important material that we wanted to include, nor should they greatly extend the length of the book.

7. We also had to choose application domains that did not require too much domain expertise. Modeling work, especially with CWA, is an activity that requires intimate familiarity with the domain of interest. Choosing esoteric application domains (e.g., nuclear power plants) would make the examples accessible to only a very limited audience.

8. We also wanted to choose application domains that already had existing software that you could use on your own to explore and verify the results of our CWA.

Our Choice

In the end, we decided to focus on only one application domain, process control. More specifically, we chose a thermal-hydraulic process control microworld as our case study. There are two significant disadvantages associated with this decision. First, it would have been better to have at least two examples to show the diversity of the CWA framework. However, we decided to take advantage of the fact that Rasmussen et al. (1994) had already presented a detailed CWA for a "humanistic" application, namely library information retrieval (that example is summarized briefly in chap. 12). The case study we chose is complementary in that it is representative of a "technical" application. Although this choice has the disadvantage of cross-referencing across books, it does have the advantage of making this book more concise and hopefully easier to read as a result. Second, although the case study we chose is relatively rich and complex, it does a poor job of representing the social demands of work. We could not find a case study that was broad enough to amply demonstrate organizational factors, while still manageable enough to be explained concisely in detail.

Having pointed out the major limitations associated with our choice, it is only fair to point out the main advantages as well. First, the case study we chose benefits from drawing on previous research and design work. This process control microworld case continues to be the subject of an extended research program (see Vicente, 1996, for a review) that has prompted the design of novel interfaces in both industrial and laboratory contexts (see Itoh, Sakuma, & Monta, 1995; Vicente & Rasmussen, 1990, respectively).

Second, a simulation of this process control microworld has been put on the WWW. You can download it by accessing the following URL: www.mie.utoronto.ca/labs/cel/duressII.html. Currently, the simulation runs on only a Silicon Graphics workstation. You are encouraged to explore the simulation as it will allow you to independently verify the results of our CWA. In addition, you are free to use the simulation as a test-bed for your own experiments.

Finally, and perhaps most important, the case study we chose should be accessible and readily comprehensible. The process control application domain is one that will be familiar to those of you with a science or engineering background, and that should be accessible with a little effort to those of you who have some rudimentary mathematics and physics knowledge.

To avoid confusion, note that the object of the case study is not a particular device. That is, we describe the behavior-shaping constraints associated with the case, not a particular interface design. Following the rationale in chapter 5, the attributes of the device should emerge as outputs from the CWA (see the examples in chap. 12). Thus, the case study illustrates the process of conducting a work analysis to discover how a device should be designed, rather than assuming the details of a particular device up front.

THERMAL-HYDRAULIC PROCESS CONTROL MICROWORLD

In this section, we introduce our case study and then describe its application domain according to the 11 factors of complexity outlined in chapter 1 (see Table 6.1). Reviewing the different ways in which this case is complex should be an effective means of introducing you to the case study.

Description

DURESS (DUal REservoir System Simulation) II is a thermal-hydraulic process control microworld that was designed to be *representative* (Brunswik, 1956) of industrial process control systems, thereby promoting generalizability of research results to operational settings (Vicente, 1991). The physical structure of DURESS II, illustrated in Fig. 6.1, consists of two redundant feedwater streams (FWSs) that can be configured to supply water to either, both, or neither of two reservoirs. Each reservoir has

TABLE 6.1 The Aspects of Complexity Outlined in Chapter 1, as Represented in This Case Study

Characteristic	DURESS II
1. large problem spaces	many variables
2. social	operators, engineers, technicians
3. heterogeneous perspectives	different goals and backgrounds
4. distributed	spatially distributed
5. dynamic	first-order lags
6. potentially high hazards	equipment failures
7. many coupled subsystems	many-to-many mapping, intergoal constraint
8. automated	negative feedback control loops
9. uncertain data	missing, failed, or noisy sensors
10. mediated interaction via computers	indirect information presentation
11. disturbances	faults

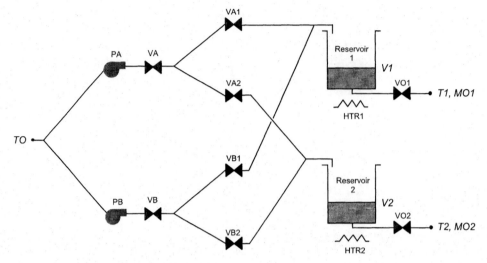

FIG. 6.1. A schematic diagram of the DURESS II process control microworld, showing the various components and their topological connections.

associated with it an externally determined demand for water that can change over time. The work domain purposes are twofold: to keep each of the reservoir temperatures (T1 and T2) at a prescribed temperature (40° C and 20° C, respectively), and to satisfy the current mass (water) output demand rates (MO1 and MO2). To accomplish these goals, workers have control over eight valves (VA, VA1, VA2, VO1, VB, VB1, VB2, and VO2), two pumps (PA and PB), and two heaters (HTR1 and HTR2). All of these components are governed by first-order lag dynamics, with a time constant of 15 s for the heaters and 5 s for the remaining components. Other important variables include the system input temperature (T0), and the volumes for both reservoirs (V1 and V2).

Figure 6.1 depicts a schematic diagram of DURESS II. Beginning on the extreme left, the normal inlet water temperature is 10° C. The input water stream splits and flows to two pumps (PA and PB) that operate as discrete switches (on or off). The maximum flow rate through each pump is 10 units/s.

The next set of components are the primary valves (VA and VB), which have a continuous range from 0 to 10. From these primary valves, each FWS splits into two secondary valves connecting each stream to both reservoirs. The secondary valves (VA1, VA2, VB1, and VB2) operate in the same manner as the primary valves. The water then flows to each of the two reservoirs, where it is heated and removed, through the use of the heaters (HTR1 and HTR2) and the output valves (VO1 and VO2), in order to meet the temperature and demand goals, respectively. The reservoirs have a maximum capacity of 100 units.

The heaters (HTR1 and HTR2) also have a continuous range of 0 to 10. The goals for the water temperatures in the reservoirs (T1 and T2) have a tolerance of ± 2°C.

Finally, the outlet valves (VO1 and VO2) are used to meet the demand goals. These valves operate in the same manner as the other valves, except that their maximum setting is 20. The goal area is ±1 unit around the desired level, and moves as a function of changes in the demand.

Sources of Complexity

In this subsection, the nature of the demands imposed by DURESS II are revealed by describing it in terms of the sources of complexity identified in chapter 1 (see Table 6.1).[1] Some sources of complexity are not adequately represented by DURESS II because it is only a microworld. For those situations, we have also described how the respective dimension would manifest itself on a larger scale in the generic application domain of industrial process control:

1. *Large problem spaces*—There are 37 variables that can be used to meaningfully describe DURESS II. Only a subset of these are shown in Fig. 6.1. Table 6.2 provides a complete list of the variables, with definitions for each. In subsequent chapters, we refer to these variables during our CWA of DURESS II. In an industrial-scale process control plant, the number of variables is typically several thousand.

2. *Social*—DURESS II is only a microworld, so it is of small enough scale that it can be operated by a single worker. However, industrial-scale process control plants require cooperation between many individuals, including: plant manager, plant engineers, control room operators, field operators, and maintenance technicians.

3. *Heterogeneous perspectives*—The individuals just described have different interests and thus hold different perspectives on the very same plant. For example, a plant manager will be interested in production and safety goals, whereas a field operator will be interested in finding the location of a particular piece of equipment. These different perspectives must somehow be resolved to achieve effective communication and coordination.

4. *Distributed*—The aforementioned individuals are also physically distributed. For example, field operators spend much of their time out in the plant itself, whereas plant managers and engineers are frequently located in a separate office area. Despite their different locations, these individuals must work together if a plant is to work effectively.

5. *Dynamic*—As shown in Table 6.3, which provides a partial list of the equations governing the behavior of the process, DURESS II is governed by first-order dynamics. As a result, there is a lag between worker actions and their full effects on the work domain. Because the effects of their actions are delayed, workers have to anticipate the future state of the work domain and act well before the point when a response is desired.

6. *Potentially high hazards*—There are a number of actions that can lead to equipment failures in DURESS II: (a) If either pump is turned on without any of the downstream valves being opened, the pump will fail after approximately 5 s; (b) it is possible to overflow either of the reservoirs, if the input flow rate is consistently greater than the output flow rate; (c) heating an empty reservoir for an extended period will damage the reservoir; (d) causing the water in the reservoir to boil also damages the work domain. Although these failures do not threaten the environment or human life, as can faults in industrial systems, they nevertheless represent situations to be avoided by workers.

---

[1]Most, but not all, of the characteristics described here have been implemented in the simulation of DURESS II described by Vicente (1996).

TABLE 6.2 Complete List of Process Variables in DURESS II

Variable	Description
Temperature	
T0	Inlet water temperature
T1	Temperature of res 1
T2	Temperature of res 2
Mass	
MO1	Mass output flow rate for res 1
MO2	Demand output flow rate for res 2
MI1	Mass input flow rate for res 1
MI2	Mass input flow rate for res 2
V1	Volume of res 1
V2	Volume of res 2
Energy	
E1	Total energy stored in res 1
E2	Total energy stored in res 2
EI1	Energy input flow rate for res 1
EI2	Energy input flow rate for res 2
EO1	Energy output flow rate for res 1
EO2	Energy output flow rate for res 2
Heat transfer	
FH1	Flow from heater HTR1
FH2	Flow from heater HTR2
Flow rates	
FA1	Flow rate from valve VA1
FB1	Flow rate from valve VB1
FA2	Flow rate from valve VA2
FB2	Flow rate from valve VB2
FPA	Flow rate from pump PA
FPB	Flow rate from pump PB
FVA	Flow rate from valve VA
FVB	Flow rate from valve VB
Heaters	
HTR1	Setting for heater of res 1
HTR2	Setting for heater of res 2

(Continued)

TABLE 6.2 (Continued)	
Variable	**Description**
Pumps	
PA	Setting of pump in fws A
PB	Setting of pump in fws B
Valves	
VA	Setting of initial valve in fws A
VB	Setting of initial valve in fws B
VA1	Setting of valve 1 in fws A
VB1	Setting of valve 1 in fws B
VA2	Setting of valve 2 in fws A
VB2	Setting of valve 2 in fws B
VO1	Setting output valve in res 1
VO2	Setting of output valve 2 in res 2

Note. fws = feedwater stream, res = reservoir. From Vicente (1991). Copyright 1991 by Kim J. Vicente. Adapted by permission.

7. *Many coupled subsystems*—The equations in Table 6.3 show that there are strong interactions governing the behavior of DURESS II. For example, the process consists of two FWSs, each of which can be coupled to either reservoir. This many-to-many mapping can create interactions between subsystems. In addition, the mass and energy topologies for DURESS II are also coupled. Specifically, the state of the temperature goal is affected, not only by the heater setting, but also by the inflow of cold water into the reservoir that is used to satisfy the demand goal. This intergoal constraint creates an interaction that must be respected by workers.

8. *Automated*—Process control systems such as DURESS II are almost always automated, being controlled by negative feedback control systems. Aside from relieving workers from the tedious job of having to manually control the process on a regular basis, automation can also frequently result in economic gains due to more efficient control. However, automation is fallible, thereby requiring worker involvement. Not only can automation fail, but it is not capable of controlling the process during all fault situations (see later discussion). Consequently, it is important that workers supervise the automation to ensure that it is operating as designed, and take over control during faults, if necessary. Note, however, that one of the purposes of CWA is to determine if, and how, the process should be automated. Following the philosophy outlined in chapter 5, automation design is an output from, rather than an input to, CWA.

9. *Uncertain data*—The data about the state of DURESS II would be obtained from sensors placed at particular locations in the plant. These data would be subject to several types of uncertainty. First, the sensors are not perfectly accurate, and thus the resulting readings always have at least some noise. Second, sensors can fail and thereby provide no information or misleading information, depending on their failure mode. Third, the data available are also limited by sensor technology. There are

Algebraic Equations	Purposes
1. $EI(t) = TI(t)\, VI(t)\, c_p\, \rho$	1. Temperature set point
• relationship between energy, volume, and temperature	2. Demand set point
2. $EII(t) = FHI(t) + c_p\, T_I\, MII(t)$	
• conservation of energy from heater and inflow	
3. $MII(t) = FAI(t) + FBI(t)$	
• conservation of mass from two feedwater streams	
4. $EOI(t) = MOI(t)\, c_p\, TI(t)$	
• energy leaving reservoir	
5. $FVA(t) = FAI(t) + FA2(t)$	
• conservation of mass in feedwater stream	
6. $FHI(t) = HTRI(t)$	
• conservation of energy from heater	
7. $FAI(t) = \dfrac{FVA(t)\, VAI(t)}{VAI(t) + VA2(t)}$	
• flow split relation	
8. If pump is OFF then $FPA(t) = 0$,	
otherwise:	
IF $[VAI(t) + VA2(t)] > VA(t)$ THEN	
$FPA(t) = FVA(t)$	
ELSE	
$FPA(t) = FVAI(t) + FVA2(t)$	
• flow through pump	
9. $FVA(t) = FPA(t)$	
• conservation of mass in pipe	

State Equations

1. $\dfrac{dTI(t)}{dt} = \dfrac{FHI(t) - [MII(t)][TI(t) - T_I]\, c_p}{VI(t) c_p\, \rho}$

   • conservation of energy in reservoir

2. $\dfrac{dVI(t)}{dt} = \dfrac{MII(t) - MOI(t)}{\rho}$

   • conservation of mass in reservoir

Note. *Constants:* ρ—density of water; c_p—specific heat capacity; T_I—inlet water temperature. Not shown are the first-order state equations with a time constant of 5 seconds for the pumps and valves, and 15 seconds for the heaters. The other half of DURESS II is governed by a symmetrical set of equations. From Vicente (1991). Copyright 1991 by Kim J. Vicente. Adapted by permission.

some variables (e.g., energy) that we might like to measure but that cannot be sensed with existing technology. It may be possible to derive these variables analytically from other sensed data using models or rules. However, the resulting data are still limited by the reliability of the original sensor readings and the fidelity of the models or rules used for the derivation. Fourth, the data are also limited by the number and placement of sensors. If sensors are not available in certain places, then the true state of the work domain in that location may be uncertain.

10. *Mediated interaction via computers*—As with most other process control systems, many of the variables that describe the state of DURESS II (e.g., energy in the reservoir, heat transfer from the heater) cannot be directly and reliably observed by unaided human perceptual systems. In the past, these variables were displayed to workers using traditional analogue instruments, which thereby served as a mediating representation between workers and work domain. More recently, computer interfaces are being increasingly used to serve this essential function.

11. *Disturbances*—Like all process control systems, DURESS II is an open system that is subject to different types of disturbances. For example, although the temperature of the incoming water is supposed to be 10°C, it is possible for the water to be warmer or cooler because of some unforeseen influence. Furthermore, the components themselves are subject to failures. For instance, the valves can fail open or shut, the reservoirs can leak, and the heaters can transfer heat at a greater or lesser rate than they are supposed to for a given heater setting. In addition, it is possible for multiple, independent failures to occur simultaneously. During many of these situations, the automation would not be able to compensate effectively, so it would be up to the workers to intervene and satisfy system purposes in the face of disturbances.

SUMMARY

This chapter has introduced the process control microworld that is used as a case study in this book. In the five chapters that follow, this case study is used to illustrate most of the five layers of behavior-shaping constraints defined by CWA (see Fig. 5.8).[2] By complementing the information retrieval example presented in detail by Rasmussen et al. (1994) and summarized in chapter 12, we thereby show that the very same set of concepts can be consistently used to model two diverse application domains in a manner that reveals important insights for systems design.

---

[2]DURESS II is only a microworld simulation rather than an actual industrial process. As a result, the CWA representations we develop for DURESS II will be different from those that would be developed for an industrial process. This qualifier is particularly true of the first phase of CWA, work domain analysis. Despite this reduced level of richness, however, the CWA models we present for DURESS II should serve our primary goal of enhancing your understanding of the CWA framework.

Phase I: Work Domain Analysis 7

When we look closely at behavior, we see lots of local variability. No two acts are ever exactly alike.
—Thelen and L. B. Smith (1994, pp. 215–216)

The generic we can know, but the specific eludes us.
—Bateson (1979, p. 41)

PURPOSE

Our goal in this chapter is to explain in full detail the first of the five phases of CWA. The concept of work domain analysis, introduced in chapter 3, is probably the most important and unique phase of CWA. We begin by discussing the nature of field descriptions to give you an appreciation for why work domain analysis is so important. Then, we describe Rasmussen's (1979b, 1985) *abstraction-decomposition space,* a two-dimensional modeling tool that can be used to build work domain models (cf. Fig. 5.8). The characteristics and advantages of one dimension in this space, the abstraction hierarchy (AH), are then described in more detail because the AH is more difficult to understand than the decomposition hierarchy. Next, examples of how the abstraction-decomposition space can be used to conduct a work domain analysis are presented for the case study introduced in chapter 6. By the end of this chapter, you should understand why work domain analysis identifies a fundamental set of constraints on the actions of any actor, thereby providing a solid foundation for subsequent phases of CWA.

> At several points in this chapter, the discussion proceeds at what might seem to be an unnecessarily detailed and rudimentary level. Our reason for adopting such a didactic style is to make sure that you can clearly distinguish between different types of hierarchical relations. Our experience in teaching this material during the past 6 years has repeatedly shown that this level of detail is essential to learning these concepts. People who are not familiar with CWA are not used to distinguishing between different types of hierarchical relations in an explicit and consistent fashion. We hope the simple examples we present in this chapter will help you develop the skills that are required to conduct a proper work domain analysis.

THE IMPORTANCE OF FIELD DESCRIPTIONS

In chapter 1, we reviewed empirical evidence showing that the greatest threat to system safety is posed by events that are not familiar to workers and that have not been anticipated by designers. In chapter 3, we argued that task analysis is not capable of identifying the information workers need to deal with such unanticipated events. Instead, a work domain analysis is required to develop an event-independent representation that can be used to cope with novelty. Recall that the relationship between work domain and task descriptions is similar to that between maps and directions for spatial navigation. A map describes the "lay of the land" independent of any actor or action on that land, whereas directions describe activities that should be performed to reach certain locations. Similarly, a work domain representation describes the structure of the controlled system independent of any particular worker, automation, event, task, goal, or interface, whereas a task representation describes what goals should be achieved and perhaps how as well. Whereas a task is what actors do, a work domain is what actors do it on (i.e., the object of action).[1]

This analogy allows us to understand the unique value added by work domain analysis. Just as maps allow people to deal with novelty (see Table 3.2), a work domain analysis allows workers to cope with the unanticipated (see Table 3.3). This capability is shared by what are called *field descriptions*, a concept originally developed in physics and mathematics (e.g., Abraham & Shaw, 1982, 1983, 1984, 1988; Einstein & Infeld, 1938) but that is being increasingly used in cognitive science (e.g., Kelso, 1995; Port & van Gelder, 1995; Thelen & L. B. Smith, 1994). In this section, we try to give you a better intuition for the unique and valuable properties of field descriptions (and thus, work domain analysis) by coming at the concept from several different directions. First, we describe Simon's (1981) well-known parable about an ant on a beach. Second, we describe the difference between the two areas of physics known as kinematics and dynamics. Third, we review an old, but still insightful, example showing how field descriptions are relevant to understanding human behavior. By seeing what these diverse examples have in common, we are able to appreciate better the value of field descriptions. And because work domain analysis is a form of field description, these examples should also illustrate the unique and important role that work domain analysis plays in CWA.

Simon Says: The Parable of the Ant on the Beach

Nobel laureate Herbert Simon's (1981) parable about an ant on a beach is a useful place to start in building intuitions about field descriptions (Vicente, 1995b). Figure 7.1 provides a graphical illustration of Simon's description of an ant traversing the terrain of a beach: "Viewed as a geometric figure, the ant's path is irregular, complex, hard to describe. But its complexity is really a complexity in the surface of the beach, not a complexity in the ant" (Simon, 1981, p. 64).

This simple parable has tremendous implications for work analysis. The first important lesson is that the ant's trajectory is an emergent property that is shaped, not just by the constraints of ant "psychology," but also by the constraints imposed by the beach itself. Stated another way, the question "Why did the ant follow this path as opposed to some other?" actually has (at least) two answers: because the ant is the way that it is, and because the beach is the way that it is. To ignore the

---

[1]Admittedly, the work domain representation is purpose-dependent. Just as different types of maps are required for different purposes (e.g., landscape elevation vs. average annual rainfall), different representations of the same work domain can be developed for different purposes. For example, the work domain representation developed by an analyst who is interested in demolishing a power plant would not be the same as that developed by an analyst who is interested in using that same power plant to supply power to the local electrical grid.

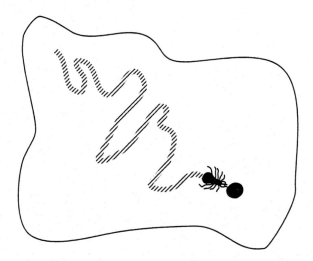

FIG. 7.1. Simon's (1981) parable about an ant on the beach. "Viewed as a geometric figure, the ant's path is irregular, complex, hard to describe. But its complexity is really a complexity in the surface of the beach, not a complexity in the ant" (Simon, 1981, p. 64).

latter factor is to make an enormous mistake because it is not possible to understand the ant's path without an understanding of the contributions made by the beach. Note that this is just a restatement of the ecological approach described in chapter 2. To understand behavior, we must have a description of the environmental constraints on action. In short, we must have a description of the beach.

There is a second, deeper lesson to be gathered from Simon's (1981) parable. If we were to study the paths that the ant takes in different circumstances, or that different ants take under the same circumstances, we would probably find a great deal of variability. This variability makes analysis very difficult because the object of analysis—action—is constantly changing (see chap. 3). However, there is something that remains relatively constant, and thus can be analyzed, in the presence of all of this variability. No matter what point the ant starts at, what point it stops at, and what path it takes in between, the constraints imposed by the beach remain relatively constant. Of course, only the constraints in the area navigated by the ant are relevant on any one occasion, but the constraints imposed by other areas of the beach are always present. Thus, a description of the beach is a very useful thing to have because it is *relatively invariant over particular initial conditions, task goals, and trajectories.* It provides a relatively stable object that can be analyzed. No matter where the ant is going or how it is getting there, it is subject to the constraints imposed by the structure of the beach.

What implications does this parable have for work analysis? The first point is that we must have a description of the system being acted on. Just as we need a description of the beach because it imposes constraints on the ant's behavior, we need a representation of the work domain because it imposes constraints on the action of actors. Note that this statement is true regardless of whether the actor in question is a worker or automation. The second point is that such a representation is very useful and powerful because it is valid for many different situations. No matter what goals actors are pursuing or how they are pursuing those goals, actors should take into account the fact that the work domain is the way that it is. Just as the structure of the beach imposes constraints on the behavior of the ant in any situation, the functional structure of the work domain imposes constraints on the actions of actors in any situation. Trajectories can vary widely from one situation to the next, but the structure of the work domain is relatively constant.[2]

Kinematics Versus Dynamics or Trajectories Versus Fields

The beach in Simon's (1981) parable is analogous to what physicists call a *field* (Einstein & Infeld, 1938). To appreciate this point, consider the particular case of a magnetic field illustrated in Fig. 7.2 (cf. Kugler, Shaw, Vicente, & Kinsella-Shaw, 1990).[3] The upper part of the figure shows a top-down view of a hypothetical magnetic field created by two fixed, positively charged particles. Each particle is of the same charge, so their respective fields are of identical strength. Because the two particles are close together, the two fields overlap. The bottom of the figure shows a cross-sectional view of the combined magnetic field. The x axis represents spatial position, and the y axis represents the strength or magnitude of the field at any one position.

Say we place a negatively charged object near these two positively charged particles. What trajectory will this object take? The answer, of course, depends on where we put the object. If we put it outside of the range of the magnetic field in Fig. 7.2, then the object will not move. If we place it within the field and just to the left of the left-most particle, it will be attracted to that particle and move toward it to the right. If we place the object within the field and just to the right of the right-most particle, it will be attracted to that particle and move toward it to the left. The region in between the two positively charged particles is particularly interesting because the direction of movement of the negatively charged object can be very sensitive to initial conditions. If we place the object ever so slightly to the right of the valley in the middle of the magnetic field in Fig. 7.2, then the object will move to the right. If we place it ever so slightly to the left of that valley, then the object will move to the left. Thus, a tiny change in initial position will completely reverse the direction of movement.

What does any of this have to do with work domain analysis? Before we answer this question, consider a few thought experiments. First, imagine that we observed the movement of the negatively charged object under three or four of the scenarios just described. Imagine further that we had no knowledge of the underlying magnetic field or the position of the two positively charged particles shown in Fig. 7.2. All we could see were the specific trajectories followed in each instance. What could we infer about the behavior of this system? It would be very difficult to come to any conclusions. In one case, the object might move to the left, whereas in the next it might move to the right. Sometimes it may move a very small distance; other times it may move a larger distance. In some situations, it does not move at all. Depending on where the object was placed, you may even notice that, in some cases, if the

---

[2]It is certainly true that the structure of some work domains can change considerably over time (e.g., flexible manufacturing systems). Nevertheless, even in these cases, the rate of change in work domain structure will be much slower than the rate of change in actor trajectories.

[3]The diagram is meant for illustration purposes only, so the shape of the magnetic field was not designed to reflect the laws of physics accurately.

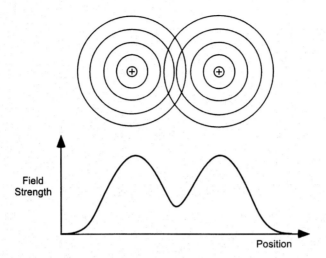

FIG. 7.2. A hypothetical magnetic field created by two positively charged particles.

initial position of the object is changed by just a little bit, it reverses its direction of movement. In other cases, if the initial position of the object is changed by the same distance, it still goes in the same direction.

As a second thought experiment, say the same conditions held but that we were allowed to observe many trajectories, not just three or four. After a while, we would see that the negatively charged object usually comes to rest at one of two positions that, unbeknownst to us, correspond to the locations of the two positively charged particles. Note, however, that our inferences would be limited by the number of trajectories that we observed and the extent to which those trajectories comprehensively sampled the space around the particles. As a result, it would be difficult to be sure what would happen if we placed the negatively charged object in a new location. For example, the first time we put the object far from the two particles, we would probably be very surprised to find that it would not move at all.

The third thought experiment adds a dynamic twist. Say that, after each trajectory, the positions of the two positively charged particles are changed within some bounded area. Then, the negatively charged object is placed in some random initial condition, and a new trajectory is generated. What could we infer without knowledge of the position of the two positively charged particles and the field? Under these conditions, it would be very difficult to come to any conclusions at all because the trajectories would not seem to follow any pattern. The starting points, ending points, and trajectories taken would differ from situation to situation.

Finally, consider a fourth thought experiment. What if the same conditions held as in the third experiment with the exception that we also had knowledge of the underlying field? We would notice that the field would change each time the two positively charged particles were moved. Moreover, knowledge of the field would allow us to see that, for each trajectory, the negatively charged particle is merely following the gradient set up by the field. All of a sudden, there is no mystery anymore. Different behaviors arise in different circumstances, simply because of the constraints imposed by the field. What a big difference it makes to know the field.

These thought experiments illustrate the difference between kinematics and dynamics (Abraham & Shaw, 1982, 1983, 1984, 1988; Port & van Gelder, 1995). *Kinematics* is the study of motion. It is concerned with trajectories. Objects are described in terms of *state variables* such as position, velocity, acceleration, and so forth. *Dynamics*, on the other hand, is the study of the forces that shape motion. It is concerned with fields. Objects are described in terms of *structural parameters*, such as damping constants and spring constants. Whereas kinematics *describes motion*, dynamics *predicts possibilities for motion*. Each representation has its advantages, of course, but dynamics can be considered to be more fundamental. If we know the field and the initial conditions, we can begin to understand why different trajectories are followed in different situations. On the other hand, if we only know the kinematics, it will not generally be possible to understand why a different path is taken when the initial conditions or final state change.

If we have done our job to this point, these differences between kinematics and dynamics should sound very familiar. They are reminiscent of the distinction between task analysis and work domain analysis, respectively. Although action is not governed solely by work domain constraints (see chap. 5), having a work domain representation helps us understand why actors may have to adopt different actions under different circumstances. In contrast, having a task description is not generally sufficient to understand why different actions may be required when the initial conditions or the goal to be achieved change in unforeseen ways.

This difference illustrates the value added by work domain analysis. Being a field description, a work domain representation shows the possibilities for action, and is thus generally invariant over any particular situation.[4] Thus, if we can develop a work domain representation, we have identified the information and relationships that actors can use to reason under virtually any situation, including those that have not been anticipated by designers.[5] Based on the evidence and arguments reviewed in chapter 1, we would thereby expect a commensurate improvement in safety as a result.

A Psychological Field Description: Automobile Driving

So far in this section, we have been talking about an ant on a beach and an object moving in a magnetic field. Hopefully, these examples have provided you with an intuition for the unique importance of field descriptions. Now, it is time to show that field descriptions can also be applied to psychological problems that are closer to the ones we are concerned with in this book.

The example we have chosen is old but still remarkably fresh—J. J. Gibson and Crooks' (1938) field description of automobile driving (for a more recent example of a psychological field description for a different application domain, see Kirlik, R. A. Miller, & Jagacinski, 1993). This analysis shows the importance of describing psychological demands in terms of fields (i.e., possibilities for actions) rather than

---

[4]As we discuss later in this chapter, faults result in a very localized change in the structure of the work domain. Metaphorically, faults cause a "dent" in the field. However, the redundancy in the field provided by the areas around the dent allow workers to cope with the fault (see the example later in this chapter).

[5]The qualifier is necessary because analysts and designers can make mistakes and because the state of knowledge in a particular application may not be sufficiently advanced to develop work domain models that are always useful (Vicente & Rasmussen, 1992). Therefore, although work domain analysis provides much broader coverage than task analysis, it does not guarantee perfection. Note that these limitations are not specific to work domain analysis; they are true of any modeling technique and any activity involving (fallible) people.

trajectories (i.e., actions), thereby generalizing the lessons learned from the previous two examples. Gibson and Crooks' key insight was that the constraints on behavior can be identified by developing a functional description of the work domain. Rather than adopting a context-free language to describe objects in the work domain, Gibson and Crooks used an action-relevant (i.e., functional) language to describe those objects. For automobile driving, terms such as *obstacle, collision, path,* and *destination* play a key role.

When these functional possibilities are integrated, the result is a *field of safe travel.* Several examples are given in Figs. 7.3 to 7.5. This field description of the work domain represents the possible paths that the car may safely follow. Note that the field of safe travel is a description of the work domain, not of the task. The object of description is the environment, not the actions taken on that environment. The field merely describes what is possible from a functional point of view. It is still up to the driver to choose a particular trajectory on any one occasion. In short, the field describes constraints on action, not action itself. As a result, field descriptions inherit the advantages of the constraint-based approach described in chapters 3 and 5.

Summary

Table 7.1 provides a summary of the relationship between the examples discussed in this section. We hope we have convinced you of the important and unique value of work domain analysis. Of the several examples that we reviewed, J. J. Gibson and Crooks' (1938) field of safe travel was the closest to our needs, but it was developed for the particular case of automobile driving. What we require is a modeling tool that has the same general characteristics (and thus, advantages), but that can be applied to

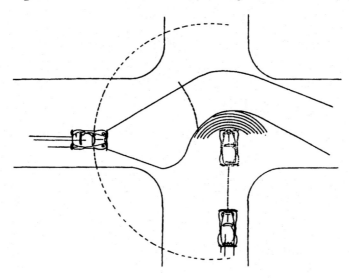

FIG. 7.3. An example of a field of safe travel (J. J. Gibson & Crooks, 1938). The environment is described in terms of its functional degrees of freedom. The resulting field description represents the possibilities for action (i.e., the possible paths that the car may safely follow). From *American Journal of Psychology.* Copyright © 1938 by the Board of Trustees of the University of Illinois. Used with permission of the University of Illinois Press.

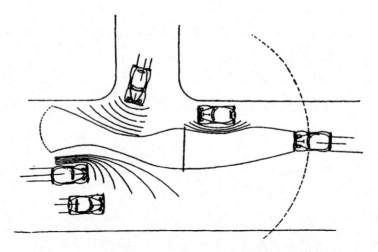

FIG. 7.4. A second example of a field of safe travel (J. J. Gibson & Crooks, 1938). From *American Journal of Psychology*. Copyright © 1938 by the Board of Trustees of the University of Illinois. Used with permission of the University of Illinois Press.

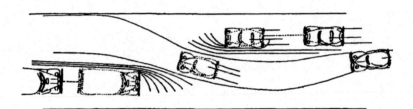

FIG. 7.5. A third example of a field of safe travel (J. J. Gibson & Crooks, 1938). From *American Journal of Psychology*. Copyright © 1938 by the Board of Trustees of the University of Illinois. Used with permission of the University of Illinois Press.

TABLE 7.1 Comparison of Various Analogues Used to Show the Difference Between Work Domain Analysis and Task Analysis

Navigation	map	directions
Simon's Parable	beach	ant's actions
Mathematics & Physics	field	trajectory
	dynamics	kinematics
	predict possibilities	describe instances
	possibilities for motion	motion
	structure	behavior
Automobile Driving	field of safe travel	driver's actions
Work Analysis	work domain	task

a diverse range of complex sociotechnical systems. We turn now to the description of such a tool.

THE ABSTRACTION-DECOMPOSITION SPACE

In this section, we describe Rasmussen's (1979b, 1985) abstraction-decomposition space, a two-dimensional modeling tool that can be used to conduct a work domain analysis in complex sociotechnical systems. First, we provide an example of how the abstraction-decomposition space can be used as a field description in which workers' problem-solving activities can be mapped as trajectories. Second, we show how the abstraction and decomposition hierarchies that comprise this space can be differentiated from other types of hierarchies.

The Abstraction-Decomposition Space as a Field: An Example

Figure 7.6 shows an example of how the abstraction-decomposition space can serve as a work domain representation. The nodes in the diagram are based on verbal reports from a professional electronic technician engaged in troubleshooting computer equipment (Rasmussen, 1986c; Rasmussen & Jensen, 1974). Each node corresponds to one verbal statement, and the statements are numbered in the order in which they were

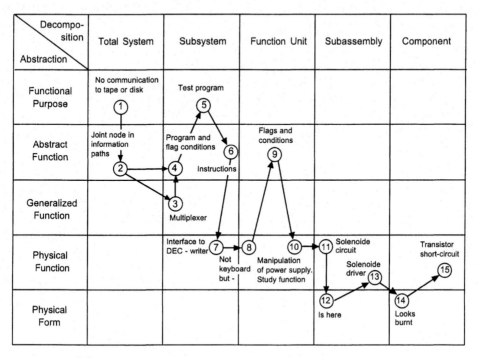

FIG. 7.6. A problem solving trajectory showing how the troubleshooting activities of a professional electronics technician can be mapped onto a field description of the equipment being repaired. (Adapted from Rasmussen, 1986a). Reprinted with permission of the author.

made. Thus, the sequence of nodes, from 1 to 15, represents the entire verbal protocol for one technician trying to find a particular fault in a particular piece of computer equipment. Using the terms in the previous section, this sequence represents a trajectory of the verbalized cognitive activities of one technician during one situation.

As shown in Fig. 7.6, the underlying space in which the trajectory is plotted consists of two dimensions. Along the top, there is a *decomposition (or part–whole) hierarchy* with five levels of resolution, ranging from the coarsest level of total system to the finest level of component. Each of these levels represents a different level of granularity with which to represent the work domain. Moving from left to right is equivalent to "zooming in" because each successive level provides a more detailed representation of the same work domain. Conversely, moving from right to left is equivalent to "zooming out" because each successive level provides a less detailed representation of the same work domain. Along the side of Fig. 7.6, there is an *abstraction (or means–ends) hierarchy* with five levels of abstraction, ranging from the most abstract level of purposes to the most concrete level of form. Each of these levels represents a different language with which to represent the work domain. In general, higher levels of the AH represent the work domain in terms of its functional properties, whereas lower levels represent it in terms of its physical properties. Thus, the AH spans the gap between purpose and material form.

Each of these hierarchy types is described in more detail later. For now, we merely want to use Fig. 7.6 to make a few points. First, each cell in the two-dimensional space is a place holder for a different but complete representation of the very same work domain. For example, the top, left cell in the space represents the purposes of the entire system—a very coarse, functional representation of the work domain. The bottom, right cell in the space represents the material form of each individual component—a very fine-grained, physical representation of the work domain. Because each cell in Fig. 7.6 is merely a place holder, a more comprehensive depiction would show the details of each of these representations—one for each cell—as well as the connections between them (see the examples at the end of this chapter).[6]

Second, each node in the trajectory in Fig. 7.6 was placed in a particular cell because the verbal report of the technician indicated that he was thinking about the work domain at that level of abstraction and decomposition (see Vicente, Christoffersen, & Pereklita, 1995, for a detailed example from another verbal protocol study). For example, the first node represents a statement about the current state of the overall purposes for which the entire system was designed. The 12th node represents a statement about the material form of a particular subassembly. The last node, number 15, represents a statement about the physical state of a particular component. Because the trajectory in Fig. 7.6 spans many cells in the space, we can infer that the technician spontaneously adopted, and switched between, different representations of the very same work domain (i.e., the computer equipment being repaired).

Third, the trajectory shown in Fig. 7.6 was idiosyncratic in the sense that it was unique. Figures 7.7 and 7.8 provide two additional examples from the same study of electronic troubleshooting (Rasmussen & Jensen, 1974). Despite the fact that the same task was being performed in each case—find the faulty component—Figs. 7.6 to 7.8 show a great deal of variability in the trajectories that were taken across particular cases.

---

[6]As the case study presented later in this chapter shows, it is sometimes useful to connect objects in the same representation (i.e., in the same cell in the two-dimensional space) using topological links.

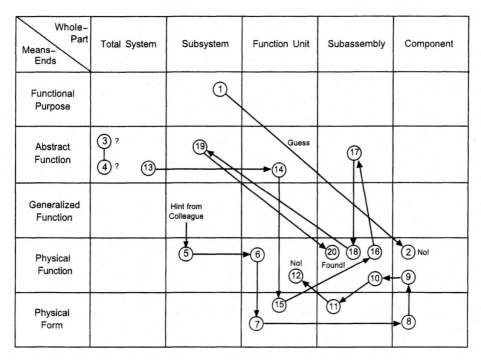

FIG. 7.7. A second problem solving trajectory showing how the troubleshooting activities of a professional electronics technician can be mapped onto a field description of the equipment being repaired.

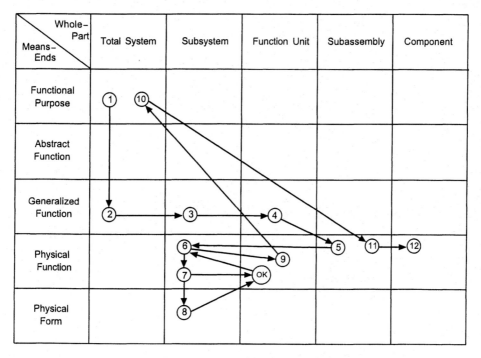

FIG. 7.8. A third problem solving trajectory showing how the troubleshooting activities of a professional electronics technician can be mapped onto a field description of the equipment being repaired.

Fourth, despite this context-conditioned variability in trajectories, the underlying representations of the computer equipment being repaired remained relatively constant. For example, the purposes for which the equipment was designed did not change, nor did the functions that were built into the equipment, nor did the number, location, or type of components. All of these work domain properties remained relatively stable across different trajectories. Although the statements about the work domain changed, the object of those statements was essentially the same across instances.

Based on these observations, it should be clear that the two-dimensional abstraction-decomposition space shown in Figs. 7.6 to 7.8 is like a field description on which many different problem-solving trajectories can be mapped. As a result, this space inherits the important advantages of field descriptions described in the previous section. Note that the field does not determine the trajectory in this case because factors other than the work domain shape behavior (see chap. 5). Nevertheless, knowing the structure of the field gives us some insight into why trajectories may differ across instances. Moreover, the field itself is capable of accommodating a bounded but infinite number of trajectories, including those that are required to deal with events that have not been anticipated by designers. This is a critical prerequisite for improved safety (see chap. 1). Moreover, it is a prerequisite to which task descriptions cannot lay claim (see chap. 3). No matter what goal actors are currently pursuing, or how they are pursuing it, knowledge about the functional structure of the work domain, represented in the form of an abstraction-decomposition space, shows the degrees of freedom actors have available for action.

Distinguishing Abstraction and Decomposition From Other Types of Hierarchies

Different types of hierarchical structures have frequently been used to model complex systems in a variety of disciplines (e.g., Ahl & Allen, 1996; Allen & Starr, 1982; Booch, 1994; Korf, 1987; Mesarovic, Macko, & Takahara, 1970; Pattee, 1972; Simon, 1981). However, not all hierarchies are created equal. On the contrary, hierarchies actually come in many different flavors (cf. Mesarovic et al., 1970). The key characteristic that distinguishes various types of hierarchies is the nature of the relationship between levels. As we see shortly, many different relations are possible (e.g., spatial scale, temporal scale, authority, flow of information, etc.).

How do the means–ends and part–whole links used in work domain analysis differ from other types of links? A few examples can help us answer this question. Figure 7.9 shows four, very simple hierarchies, each with one parent node and two leaf nodes.

> Take a moment to examine each example and identify in your mind the differences between them. Then, see if you can determine what type of link connects the two levels in each hierarchy. Rather than telling you straightaway what type of hierarchy each example in Fig. 7.9 is, we instead use an indirect technique that you will find valuable when conducting a work domain analysis. We have found that, while constructing or evaluating hierarchies, it is useful to ask questions about the relationships between nodes to ensure that the links are of the proper type.

For example, one type of hierarchy commonly found in corporate organizational charts is based on an authority link. To identify, or make sure that you are properly

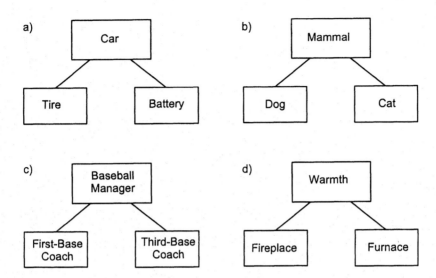

FIG. 7.9. Examples of different types of hierarchies, each defined by a different type of link between levels.

constructing, an authority hierarchy, it is useful to ask two questions of every node, one top-down and the other bottom-up. First, does the node in question have authority over all of the nodes that it is linked to at the level below? Second, is the node in question subordinate to all of the nodes that it is linked to at the level above? Take a moment to ask these two questions of each of the examples in Fig. 7.9. You should be able to determine that three of the four examples clearly do not fit this structure. After all, a car does not have authority over a battery, a mammal does not have authority over a cat, and warmth does not have authority over a furnace. The only example that fits is c). The manager of a baseball team has authority over the first- and third-base coaches. Or conversely, the first- and third-base coaches are subordinates of the team manager.

A second type of hierarchy that is frequently found in the fields of computer science, library science, and biology is based on a classification or "is-a" link. To identify, or make sure that you are properly constructing, a classification hierarchy, it is useful to ask two questions of every node. First, is the node in question the superordinate category for all of the nodes it is linked to at the level below? Second, is the node in question an exemplar of all of the nodes it is linked to at the level above? Take a moment now to ask these two questions of each of the examples in Fig. 7.9. You should be able to determine that two of the remaining three examples do not fit this structure. After all, a battery is not an exemplar of a car, and a fireplace is not an exemplar of warmth. The only remaining example that fits is b). Mammal is the category to which both dog and cat belong. Or conversely, dogs and cats are exemplars of the category mammal.

A third type of category that is very common in many disciplines is a decomposition hierarchy defined by a part–whole link. This hierarchy type is one of the two that is used in work domain analysis (see Figs. 7.6 to 7.8). To identify, or make sure that you are properly constructing, a decomposition hierarchy, it is useful to ask two questions of every node. First, is the node in question made up of all of the nodes that it is linked to at the level below? Second, is the node in question a part of all of the nodes that it is linked to at the level above? Take a moment now to ask these two questions of each of the remaining examples in Fig. 7.9. You should be able to determine that one of the two remaining examples does not fit this structure. After all, a fireplace is not a part of a warmth, and warmth is made not up of furnace and fireplace. The only example that fits is a). Tires and batteries are both parts of cars. Or conversely, a car can be decomposed into tires and a battery (as well as other parts, of course).

We have left the hierarchy type that is the most important for work domain analysis until the end. The AH is defined by a structural means–ends link between levels. To identify, or make sure that you are properly constructing, an AH, it is useful to ask two questions of every node. First, is the node in question the end that can be achieved by all of the nodes that it is linked to at the level below? Second, is the node in question a structural means that can be used to achieve all of the nodes that it is linked to at the level above? Even though you already know which is a means–ends hierarchy, take a moment now to ask these two questions of each of the examples in Fig. 7.9. You should verify that three of the four examples do not fit this structure. After all, tire cannot be used to achieve car, dog cannot be used to achieve mammal, and the first-base coach cannot be used to achieve manager. The only example that fits is d). Warmth is the end that can be achieved by both fireplace and furnace. Or conversely, fireplace and furnace are both structural means that can be used by an actor to achieve warmth.

You may have noticed that we used the term *structural means* rather than just *means* in the preceding paragraph. The reason for using this wording is to distinguish between two different kinds of means–ends relationships. The more familiar type— which, unfortunately, happens to be the one that we are not concerned with here—is an action means–ends link (cf. A. Newell & Simon, 1972) rather than a structural one. The difference between the two is best illustrated by example. In Fig. 7.9d, we gave an example of a structural means–ends relationship. Furnace and fireplace are both *objects* that can be used to achieve warmth. Thus, each one is a *noun*. A related example of an action means–ends link would be: Going down to the basement and then lighting the fireplace are both means for achieving warmth. Note the difference. In this case, the two means are *actions*, not objects. Thus, they are both *verb* phrases. This difference is very, very easy to overlook because it is quite subtle. Yet, it is of profound importance. Why?

The answer can be found in Table 7.1. A work domain analysis represents the thing being acted on. That is why nouns are almost always used; work domains are objects of action. In contrast, a task analysis represents the goals to be achieved by an actor's actions. That is why verbs are always used; tasks are activities that need to be performed. Therefore, a work domain analysis simply cannot be conducted with an action means–ends relation. Although such relations are very valuable for conducting a control task analysis (see chap. 8), they are not compatible with the orientation of a work domain analysis. After all, to describe the thing being controlled, we need a relation that represents the structure of the object of action, not the structure of actions themselves.[7]

---

[7]There are exceptions where it is appropriate to use verbs when describing a work domain (e.g., see the analysis for baseball conducted by Vicente & Wang, 1998). Nevertheless, even in these cases, the work domain still describes the object of the actor's (e.g., the baseball manager's) actions rather than the actions themselves.

It is absolutely essential to keep this distinction between structural means–ends and actions means–ends firmly in mind. Most people are much more used to thinking in terms of the latter than the former. It takes a great deal of practice—and perhaps a certain cognitive style (cf. Torenvliet, Jamieson, & Vicente, 1998)—to think fluently in terms of structural means–ends relations. Yet, this skill is essential in conducting a CWA. Along with distinguishing between different kinds of hierarchies (see earlier discussion), this is one of the largest obstacles that needs to be overcome to carry out a work domain analysis effectively.

Summary

There are many different kinds of hierarchies. The two that are of value in conducting a work domain analysis are a decomposition (or part–whole) hierarchy and an abstraction (or means–ends) hierarchy. Together, these two hierarchies define a two-dimensional space (e.g., Figs. 7.6 to 7.8). Each cell in this space consists of a different representation of the very same work domain. Part–whole hierarchies are relatively easy to understand because they involve aggregating parts into larger wholes, or conversely, decomposing wholes into their constituent parts. Means–ends hierarchies are more difficult to understand, so we describe them in more detail next.

MORE ABOUT THE ABSTRACTION HIERARCHY

In this section, we begin by describing the generic properties of the AH. Then, we review some of the evidence showing that the AH is a psychologically relevant way of describing complex sociotechnical systems. Next, we review the arguments showing that the AH can be used to identify the information support that workers need to cope with disturbances that have not been anticipated by designers. Some hints for conducting work domain analyses are also provided. By the end of this section, you should have a deep appreciation for why the AH plays a critical role in work domain analysis.

Stratified Hierarchies

The AH belongs to the class of *stratified hierarchies* described by Mesarovic et al. (1970), the properties of which are as follows:

1. Each stratum, or level, deals with the very same system, the only difference being that different strata provide different descriptions, or different models, for observing the system.
2. Each stratum has its own unique set of terms, concepts, and principles.
3. The selection of strata for describing a particular system depends on the observers, and their knowledge and interest in the control of the system. For many systems, however, there may be some strata that appear to be natural or inherent.
4. The requirements for proper system functioning at any level appear as constraints on the meaningful operation of lower levels, whereas the evo-

lution of the state of the system is specified by the effect of the lower levels on the higher levels.

5. Understanding of the system increases by crossing levels: By moving up the hierarchy, we obtain a deeper understanding of system significance with regard to the purposes that are to be achieved, whereas in moving down the hierarchy, we obtain a more detailed explanation of the system's functioning in terms of how those purposes can be carried out.

As we already mentioned, the structure of the AH is further specified by a means–end relationship between levels (Rasmussen, 1979b, 1985).

The properties just described define a family of representations for an analyst. Thus, the AH is not a specific representation but rather a framework that an analyst can use to develop representations for various problem domains. One point that is frequently overlooked is that the number of levels, and especially their content, may vary as a function of the types of constraints in each domain. For example, five levels of constraint have been found to be useful for describing process control systems (Rasmussen, 1985): the purposes for which the system was designed (functional purpose); the intended causal structure of the process in terms of mass, energy, information, or value flows (abstract function); the basic functions that the plant is designed to achieve (generalized function); the characteristics of the components and the connections between them (physical function); and finally, the appearance and spatial location of those components (physical form). Although these levels have been found to be useful for a variety of application domains (Rasmussen et al., 1994), there is no reason to believe that the same levels, or even the same number of levels, would always be appropriate for representing any work domain. For instance, medical applications require a different set of abstraction levels (Hajdukiewicz, 1998; Sharp, 1996). Nevertheless, an AH representation will have the properties just described, regardless of the domain.

In the next two subsections, we describe two important features of the AH. First, we show how it provides a psychologically useful way of representing complex work domains. Second, we demonstrate that it also provides a basis for identifying the information workers need to deal with unanticipated events.

Psychological Relevance

There are many different ways to represent a work domain. In the specific case of DURESS II, for instance, one possible representation is the set of algebraic and state equations depicted in Table 6.3. As we see in the case study discussed later, other representations of the same work domain are possible. In choosing how to represent a work domain, one important consideration is how easy it is for workers to understand and utilize alternative representations. For example, it seems likely that the equations in Table 6.3 are not the most intuitive or useful representation from the perspective of workers. Most people do not naturally think in terms of equations. In this subsection, we describe how the AH provides a psychologically relevant way of representing work domains.

From a psychological perspective, one important property of the AH is that higher levels are less detailed than lower levels. Shifting our representation from a low, detailed level to a higher level of abstraction with less resolution makes complex work domains look simpler. In effect, this provides a mechanism for coping with complexity. Metaphorically, moving up one or more levels allows workers to "see the forest for the trees." Thus, part of the psychological relevance of the AH is that it allows resource-bounded actors, as people are, to deal with work domains that would be unmanageable if they had to be observed by workers in full detail all at once.

This advantage is not unique to the AH, as the examples in Fig. 7.9 show. Virtually all hierarchies allow one to observe systems at a less detailed level. From a psychological point of view, the unique and important characteristic of the AH is that it is explicitly purpose oriented. The various levels in the hierarchy are linked by a structural means–end relation. This relationship provides a very important source of constraint that can be exploited by actors. Problem solving can be constrained by starting at a high level of abstraction, deciding which lower level function is relevant to the current situation, and then concentrating on the subtree of the hierarchy that is connected to the function of interest. This type of problem solving is efficient (cf. Korf, 1987) because all work domain objects not pertinent to the function of current interest can be ignored.

As a result, an AH representation supports goal-directed problem solving in a directed and computationally economic manner. It should allow actors to: (a) structure their overall problem-solving process, (b) frequently start their problem-solving activities at a higher level of abstraction to avoid detail, and (c) iteratively "zoom in" on lower levels of abstraction to selectively examine only those parts of the domain that are relevant to the goal or function of current interest. These predictions have been empirically confirmed by verbal protocol data obtained from experts engaged in representative problem-solving activities in a number of complex problem domains, including electronic troubleshooting (Rasmussen, 1986c), nuclear power plants (Itoh, Yoshimura, Ohtsuka, & Masuda, 1990), and process control simulations (Vicente et al., 1995). These predictions are also generally consistent with many studies of expert problem solving not explicitly based on the AH. Such studies have repeatedly shown that experts spend a great deal of their time analyzing the functional structure of a problem at a high level of abstraction before narrowing in on more concrete details (Glaser & Chi, 1988). Some notable examples include Selz (1922; see Frijda & de Groot, 1981, for an English account), Duncker (1945), and de Groot (1946/1965).

The psychological advantages listed previously are not enjoyed by other types of hierarchies, such as the ones shown in Figs. 7.9a–c. The links between levels in these other representation formats are not explicitly related to work domain purposes or functions. Although it is possible to examine the work domain at a high level of the hierarchy, the subtree of the hierarchy that is connected to that parent node may not necessarily contain objects that are relevant to the selected node. In other words, other hierarchies constrain search but not in a way that is explicitly related to purpose. It is this latter type of constraint that is needed to support goal-directed behavior.

Three Goal-Oriented Questions: Why? What? How? The rationale behind the psychological advantages of the AH can be made more concrete with an example. Figure 7.10 shows a subset of an AH representation for a hypothetical scientific research program in human factors. Although we may not usually think of it this way, science is a complex work domain with scientists playing the role that is equivalent to that of workers in a sociotechnical system. To support their problem-solving processes, scientists also need a representation of their work domain. The AH framework is one candidate for developing such a representation. We have

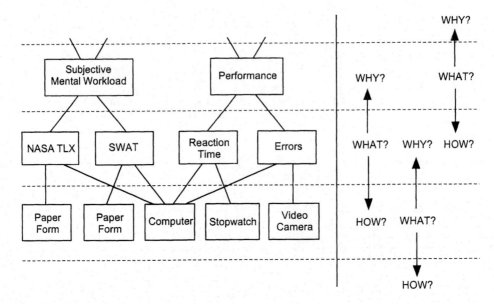

FIG. 7.10. A portion of an abstraction hierarchy for a hypothetical scientific research program in human factors.

chosen this particular example because it will be familiar to researchers and because it shows how the AH can be generalized outside of complex sociotechnical systems. Explicitly showing this generalization should help us get a good grasp of the underlying concepts, independent of any one particular example.

Although the purposes of this hypothetical research program are not depicted in Fig. 7.10, assume that subjective mental work load and performance are relevant constructs. They are both illustrated at the top level in the figure. The second level shows how each construct can be operationalized. For subjective mental work load, two well-known options are the NASA TLX scale (Hart & Staveland, 1988) and the SWAT scale (O'Donnell & Eggemeier, 1986). For performance, two typical options are reaction time and number of errors. The third level in Fig. 7.10 illustrates the apparatus that can be used to record data using each measure. For example, the NASA TLX measure can be administered using a paper form or a computer. SWAT can also be administered using a computer or a (different) paper form. Reaction time can be determined using a computer or a stopwatch. The number of errors can be determined using a computer or a video camera. Of course, many other degrees of freedom are available to a researcher at the second and third levels, but these should suffice for our purposes.

This example can be used to illustrate how an AH representation naturally supports goal-directed problem solving. There are three questions that are important to any goal-oriented actor: Why? What? How? An AH representation provides information that is relevant to answering each of these questions. As illustrated to the right in Fig. 7.10, the linkages between the three questions and the levels of the hierarchy are not absolute in any sense, but rather can "slide" up or down (this point is described in more detail later). The level at which workers observe the work domain at any one point in time defines the *what* level. Given this choice, the level above specifies *why* and the level below specifies *how*.

The pragmatic value of these relationships can be made clear by showing how the particular example in Fig. 7.10 could be used by scientists to make research decisions. For example, if the researcher entered the work domain at the reaction time node, then that level would be the *what* level and the levels above and below would be those of *why* and *how*, respectively. Moving up in the hierarchy, the answer to the question "Why reaction time?" is "Because it serves the *end* of performance evaluation." Moving down the hierarchy, the answer to the question "How reaction time?" is "By *means* of either a computer or a stopwatch." Analogous upward and downward relations exist for each of the other nodes in the second level.

As we already mentioned, the Why? What? How? relation is a sliding one. As shown on the right in Fig. 7.10, if the researcher entered the work domain at a node in the third level, then that would be the *what* level and the level above would now be the *why* level. Similarly, if the researcher entered the work domain at a node in the first level, then that would be the *what* level and the level below would now be the *how* level. In either case, the means–ends structure helps researchers in answering questions like the ones discussed earlier.

From a psychological perspective, one of the most important features of an AH representation is that it identifies the structural work domain constraints on achieving goals. By following the linkages identified by the analysis, people can focus their attention on the aspects of the work domain that are relevant to the current context. Other aspects of the work domain need not be consulted. Referring to Fig. 7.10, a stopwatch is not a means for administering the NASA TLX scale, and the SWAT scale is not a means of evaluating performance. This constraint provides a tremendous amount of leverage because, not only can people focus solely on the nodes that are relevant to the current context, but they also do not have to remember what aspects of the work domain are relevant to the functions or purposes of interest. This information is provided by the structural means–ends links between levels, which can be built into an information support system.

Another point worthwhile mentioning is that in complex work domains, there can be many-to-many mappings between nodes at various levels of abstraction (cf. Dörner, 1989/1996). Thus, a particular function (e.g., reaction time) can be fulfilled by a number of different structural alternatives (e.g., computer and stopwatch). Also, a single node (e.g., computer) can be related to more than one higher order function (e.g., NASA TLX, SWAT, reaction time, errors). Revealing this complexity by explicitly showing the links between levels of abstraction provides people with a way to cope with that complexity.

Note also that the nodes at each level in the hierarchy in Fig. 7.10 share a common language that differs from that used to describe other levels. The nodes in the bottom level are all examples of physical equipment, the nodes in the middle level are all examples of empirical measurements, and the two nodes at the top level are both examples of theoretical constructs. These examples emphasize the observation made earlier—each level of the AH constitutes a different model, with its own unique language, for describing the very same work domain.

Finally, we can use the example in Fig. 7.10 to reemphasize, yet again, the difference between a work domain analysis and a task analysis. Note that the hierarchy in Fig. 7.10 does not describe the actions of the actor (i.e., the researcher). In fact, no actions are represented at all. Instead, the hierarchy describes the constraints that the work domain imposes on the actions of any actor. The representation is of the constraints imposed by the objects being acted on, not of the actions themselves.

Summary. Actors in complex sociotechnical systems must have some way of coping with the complexity of the work domain with which they are interacting. AH representations provide a way of representing complex work domains in a psychologically useful manner, directly supporting goal-directed problem solving. In the next subsection, we describe the second unique property of the AH, namely its capability to support workers in dealing with events that have not been anticipated by designers.

Coping With the Unanticipated

As we mentioned earlier, the greatest threat to safety in complex sociotechnical systems is events that are unfamiliar to workers and that have not been anticipated by designers. Some researchers believe that it is not possible to identify the information needs associated with such events (e.g., Mitchell, 1996; Shepherd, 1993). Therefore, it is important to explain the rationale behind the claim that the AH can be used to identify the information that workers need to cope with the unanticipated.

This rationale can be made clear by examining the control requirements posed by such events. Rasmussen (1974) analyzed this problem at a fundamental level and uncovered several important insights. First, when a work domain is functioning correctly, it can be described by a set of constraints that are imposed on the observed data set by the functioning and anatomy of that work domain. Because it was designed for a certain purpose, and in a certain way, there will be relationships between variables. These relationships can be described as constraints. These constraints represent "rules of rightness" (Polanyi, 1958); to say that the work domain is behaving normally is equivalent to saying that the constraints in question hold. When a disturbance of some sort occurs (e.g., a fault), the work domain state or structure will change. Consequently, a disturbance results in the breaking of one or more constraints that govern the work domain under normal circumstances (see Footnote 4). From this perspective, the task of disturbance detection is equivalent to detecting the breaking of constraint(s). However, to be able to detect such a change, the states of all of the variables entering into the violated constraint must be represented, otherwise it will not be possible for workers to uniquely determine if a constraint has indeed been broken. The problem is compounded by the fact that it is not possible to know beforehand which constraint will be violated. Thus, the complete set of goal-relevant constraints governing the work domain must be represented to permit workers to determine when a constraint has been broken, and thereby allow them to diagnose the disturbance.

That is precisely what the AH tries to do. It provides a framework for identifying and integrating the set of goal-relevant constraints that are operating in a given work domain. Each level in the hierarchy represents a different class of constraint, or in the terms of Mesarovic et al. (1970), a different stratum (see the properties of stratified hierarchies listed earlier). One way to think of the AH, then, is as a set of models of the work domain, each defining a level of the hierarchy. Higher levels represent relational information about purpose, whereas the lower levels represent more elemental data about physical implementation.

With respect to information system design, the important implication is that because higher order, functional relations are explicitly represented, it should be possible for workers to determine when work domain constraints have been

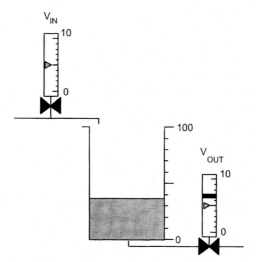

FIG. 7.11. A simple process that will be used to illustrate how analytical redundancy provides a way of coping with the unanticipated. There are three components: an input valve (V_{IN}), output valve (V_{OUT}), and a reservoir holding water (level 0 to 100).

broken.[8] Consequently, developing an AH representation for a work domain allows analysts to identify the information that workers need to cope with the entire range of operating demands, including unanticipated events.

An Example: The Logic of Analytical Redundancy. The technique of using multiple constraints to detect and diagnose unanticipated disturbances is not original to the AH (although the particular form of the latter is relatively unique). In control theory, this technique is known as *analytical redundancy* (Frank, 1990). The basic idea is to feed the current inputs to a system into a model of that system, use the model to derive what the system outputs should be, and then compare the expected outputs with the actual outputs to determine if there is a residual. If the model is a useful one, then any discrepancy is an indication that the system is not behaving as it should be. In principle, this logic can be applied to any system (e.g., automation, worker, work domain). In this subsection, we apply it to a work domain. More specifically, we use analytical redundancy in a very simple, concrete example to illustrate how the AH can be used to help workers cope with the unanticipated.

The context for this example is illustrated in Fig. 7.11, and consists of an input valve (V_{IN}), output valve (V_{OUT}), and a reservoir holding water (level 0 to 100). We do not construct a full AH for this very simple work domain. Instead, we merely illustrate how the analytical redundancy in work domain constraints can be exploited to help workers cope with unanticipated events.

One of the constraints that governs the process in Fig. 7.11 is the law of conservation of mass. If the process is operating normally, we would thereby expect that the rate of change of volume in the reservoir is determined by the difference between

---

[8]The tentative wording is required because making the necessary information available does not guarantee that it will be attended to or interpreted correctly.

the flowrate of mass going through V_{IN} and that going through V_{OUT}. This constraint can be stated more formally as follows:

$$\frac{dVol(t)}{dt} = \frac{MI(t) - MO(t)}{\rho} \qquad (1)$$

where Vol(t) = reservoir volume or level, MI(t) = mass input flow rate, MO(t) = mass output flow rate, ρ = density. Other constraints that govern the work domain could be identified (e.g., the relationship between each valve setting and the respective flow through that valve), and represented in an analogous fashion.

Now consider what kinds of disturbances could affect this work domain. There is literally an infinite number. There could be a leak in the reservoir. If the building housing the process develops a hole right above the reservoir, then water can leak into the reservoir on a rainy day. If there is a fire in the building housing the process, then someone may decide to bail water out of the reservoir to put out the fire. A disgruntled employee may decide to pour quantities of some other liquid into the reservoir on a regular basis. We could go on and on with this list, adding more and more implausible events. If we begin to consider multiple faults occurring simultaneously, then the possibilities increase exponentially. The main point to take away is that it is not possible to enumerate a priori all of the things that can go wrong. There are always events that can, and do, happen that designers did not anticipate (see chap. 1). What can we do to help workers deal with those situations?

The logic of analytical redundancy, on which the AH is based, provides a viable answer to this question. The solution is to turn the problem on its head. Instead of trying to figure out all of the things that can go wrong—a hopeless task—we can instead define how the work domain should be functioning. That is precisely what work domain constraints do—they provide an operational definition of what it means for the work domain to be operating normally. Then, when constraints are violated, there is an indication that something is wrong. The technique is equivalent to definition by exclusion—no matter how bizarre the fault, it should affect one or more of the constraints governing the work domain.

For example, each of the faults described earlier will violate the mass balance constraint in Equation 1. Therefore, if workers have information about each of the variables that enter into the constraint (i.e., dVol(t)/dt, MI(t), MO(t)) and the way in which those variables are normally interrelated, then workers should be able to detect the disturbance by noticing the violated constraint. Moreover, they can use information about exactly how the constraint is violated to diagnose the nature of the disturbance in sufficient detail to take appropriate compensatory actions. For instance, in the case of a leaky roof, workers could detect that the rate of change of volume is greater than it should be, given the measured inflow and outflow. This information alone is not sufficient to identify the leaky roof. However, it is enough to determine that too much water is entering the reservoir, and thus, that it may be appropriate to decrease the input valve setting or increase the output valve setting. This example should show you how the technique of analytical redundancy, and thus the AH, provide support for coping with unanticipated events.

In practice, the situation is much more complicated, of course. As we mentioned earlier, we never know beforehand which constraints will be violated by a particular fault. That is why it is important to identify all of the goal-relevant constraints. The AH provides a structured and systematic framework for conducting this analysis.

Summary. To improve safety, we must provide workers with the support required to deal with the unanticipated. The AH achieves this goal by exploiting the logic of analytical redundancy. Each level in an AH representation represents a layer of goal-relevant constraint. Disturbances result in a violation of constraint(s), so if the information in the AH is presented to workers, then they have a much better chance of detecting, diagnosing, and compensating for unanticipated events.

Hints for Conducting a Work Domain Analysis

In this subsection, we provide a few hints that should be kept in mind when conducting a work domain analysis using the abstraction-decomposition space as a modeling tool:

1. The first step is defining the scope of the analysis by drawing a boundary to delimit the work domain of interest. There is no one correct way to draw this boundary. Different definitions are useful for different kinds of analyses. Analysts must decide which boundary definition is most appropriate for the problem with which they are faced. In general, it is useful to define a boundary in such a way that the interactions across the boundary (i.e., between the work domain and its environment) are relatively weak.

2. Use a matrix diagram like that in Fig. 7.6 as an overview of the various representations that are going to be developed in the two-dimensional abstraction-decomposition space.

3. It may be easier to begin by constructing the part–whole hierarchy. Identify all of the smallest parts that are worthwhile modeling, and then determine how they can be aggregated into functionally meaningful wholes (see Jørgensen, 1993). Aggregates of parts should be tightly coupled to each other, but weakly coupled to other aggregates (cf. Simon, 1981). Use as many levels of aggregation as are required for the problem. Do not decompose beyond the lowest level at which action is possible.

4. For each node in the decomposition hierarchy, ask whether the node connected at the level below is a part of the node in question, and whether the node connected at the level above is the whole to which the node in question belongs.

5. In developing the AH representation, it is usually easier to begin with the contents of the top level and the bottom two levels. These are usually the easiest to identify. Once you have analyzed these, try to identify the bridging intermediate levels. Use as many levels of abstraction as necessary, but do not confuse part–whole and means–ends links. For work domains with multiple purposes, it may be useful to develop an AH representation for each purpose individually, and only then combine them into one overall representation.

6. Always keep in mind the distinction between action means–ends and structural means–ends. Your AH representation should be based solely on the latter.

7. Make sure that all of the nodes at a particular level of the AH are in the same modeling language. Make sure that nodes at different levels use different modeling languages.

8. Recursively use the why, what, how questions to verify that all of the nodes in the AH have the proper structural means–ends linkages between levels.

9. Make sure that connections between cells in the abstraction-decomposition space are only structural means–ends and part–whole links and that these are not mixed. Other types of links should not be present (cf. Fig. 7.9).

10. Keep in mind that movement along the decomposition hierarchy results in a description of a *different* object (e.g., car vs. battery). In contrast, movement along the AH results in a new (functional) description of the *same* object (e.g., warmth vs. furnace).

11. Be prepared to revise and correct your representation many times. Work domain analysis is a highly iterative process.

12. Refer to the small examples provided earlier in this chapter, and the larger case study that follows when you are conducting your work domain analysis. Additional detailed examples can be found in: Rasmussen et al. (1994); Dinadis and Vicente (1996, in press); Reising and Sanderson (1996); Sharp (1996; Sharp & Helmicki, 1996); Xu (1996; Xu, Dainoff, & Mark, 1996); Burns and Vicente (1997); Hajdukiewicz (1998); Vicente and Wang (1998); and, Jamieson and Vicente (1998). These examples can be used as analogues that can help you conduct your analysis.

WORK DOMAIN ANALYSIS FOR PROCESS CONTROL MICROWORLD

This section uses the abstraction-decomposition modeling tool described earlier in the chapter to conduct a work domain analysis of the process control microworld case introduced in chapter 6. The goal is to take an understanding of the structure of DURESS II, as represented by the constraints listed in Table 6.3, and develop a psychologically relevant work domain representation that will allow workers to cope with the unanticipated. The description is based on the analyses conducted by Vicente (1991) and Bisantz and Vicente (1994).

Overview

Figure 7.12 provides an overview of the abstraction-decomposition space for DURESS II. The part–whole dimension consists of three levels of resolution: whole system, subsystems, and components. The AH dimension consists of the five levels of abstraction described earlier: functional purpose, abstract function, generalized function, physical function, and physical form. Note that not all of the cells in Fig. 7.12 are being used to describe the work domain. Results from empirical studies have shown that workers tend to think about a work domain at a coarser level of description when using higher levels of abstraction (Vicente, 1992b). Conversely, at lower levels of abstraction, workers tend to think about the work domain at finer levels of decomposition. Therefore, only the representations roughly along the diagonal in Fig. 7.12 have been developed for DURESS II.

Figures 7.13, 7.14, and 7.15 show the means–end, part–whole, and topological links between objects in the representation, respectively. The topological links have not been shown in previous examples. They provide a way of showing relationships

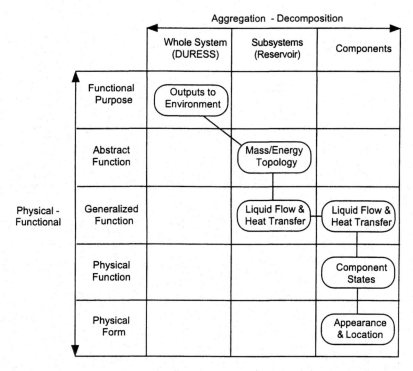

FIG. 7.12. Abstraction-decomposition space for DURESS II (adapted from Vicente, 1991).

between objects in the same cell of the abstraction-decomposition space. The value and meaning of topological connections should become clear as the example unfolds.

Decomposition Hierarchy

The part–whole dimension of the space consists of three levels of resolution. The objects at the *component* level of decomposition are the pumps, valves, heaters, and reservoirs. At the next level, these components are aggregated into meaningful *subsystems*. Thus, the objects are now transport subsystems, storage subsystems, and heating subsystems. Finally, at the *system* level, the entire work domain is described as a single whole. Part–whole links are shown in Fig. 7.13.

Abstraction Hierarchy

The AH, which is orthogonal to the part–whole dimension, consists of the five previously defined levels of description. These are shown in Fig. 7.14, and are described next.

Functional Purpose. Objects at this level of abstraction correspond to work domain purposes, and therefore are appropriately described at the system level of the part–whole decomposition. There are four purposes in DURESS II: Keep the water at the set point temperature for each reservoir (two purposes), and keep enough water in each reservoir to keep up with the current demand flow rate (two purposes).

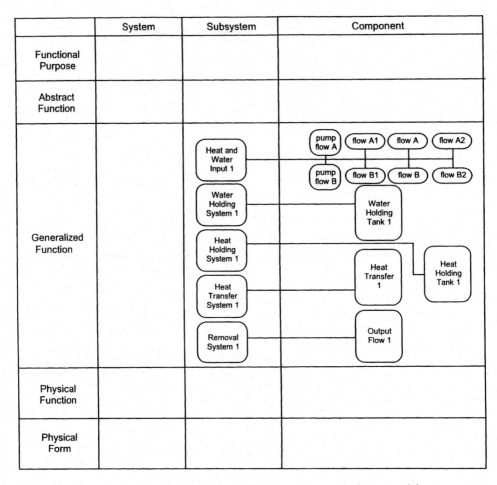

	System	Subsystem	Component
Functional Purpose			
Abstract Function			
Generalized Function			
Physical Function			
Physical Form			

FIG. 7.13. Part-whole links in DURESS II (Bisantz & Vicente, 1994). Reprinted from *International Journal of Human-Computer Studies, 40,* A. M. Bisantz & K. J. Vicente, "Making the abstraction hierarchy concrete," pp. 83–117, Copyright © 1994. Reprinted by permission of the publisher, Academic Press.

Abstract Function. This level can be described in terms of the flow of mass and energy that balances the conservation of mass and energy for each storage subsystem. In addition to shifting downward in abstraction from the functional purpose level, this corresponds to a decomposition from the system to subsystem level (see Fig. 7.12). As shown in Figs. 7.14 and 7.15, each subsystem has one mass and one energy store (the reservoir), one source of mass (input water), two sources of energy (input water and the heater), and one sink of mass and energy (the output valve). Topological links at this level, shown in Fig. 7.15, indicate the flows of mass and energy through the subsystems.

Generalized Function. Flows and storage of heat and water are described at this level of abstraction. At the subsystem level of decomposition (see Figs. 7.14 and 7.15), the rate of flow of water and heat transfer from the input stream, rate of heat transfer from the heating system, storage of heat and storage of water in the reservoirs,

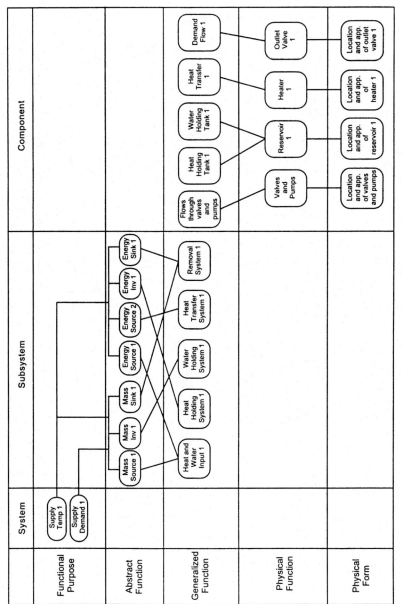

FIG. 7.14. Means-ends links in DURESS II (Bisantz & Vicente, 1994). Reprinted from *International Journal of Human-Computer Studies, 40*, A. M. Bisantz & K. J. Vicente, "Making the abstraction hierarchy concrete," pp. 83–117, Copyright © 1994. Reprinted by permission of the publisher, Academic Press.

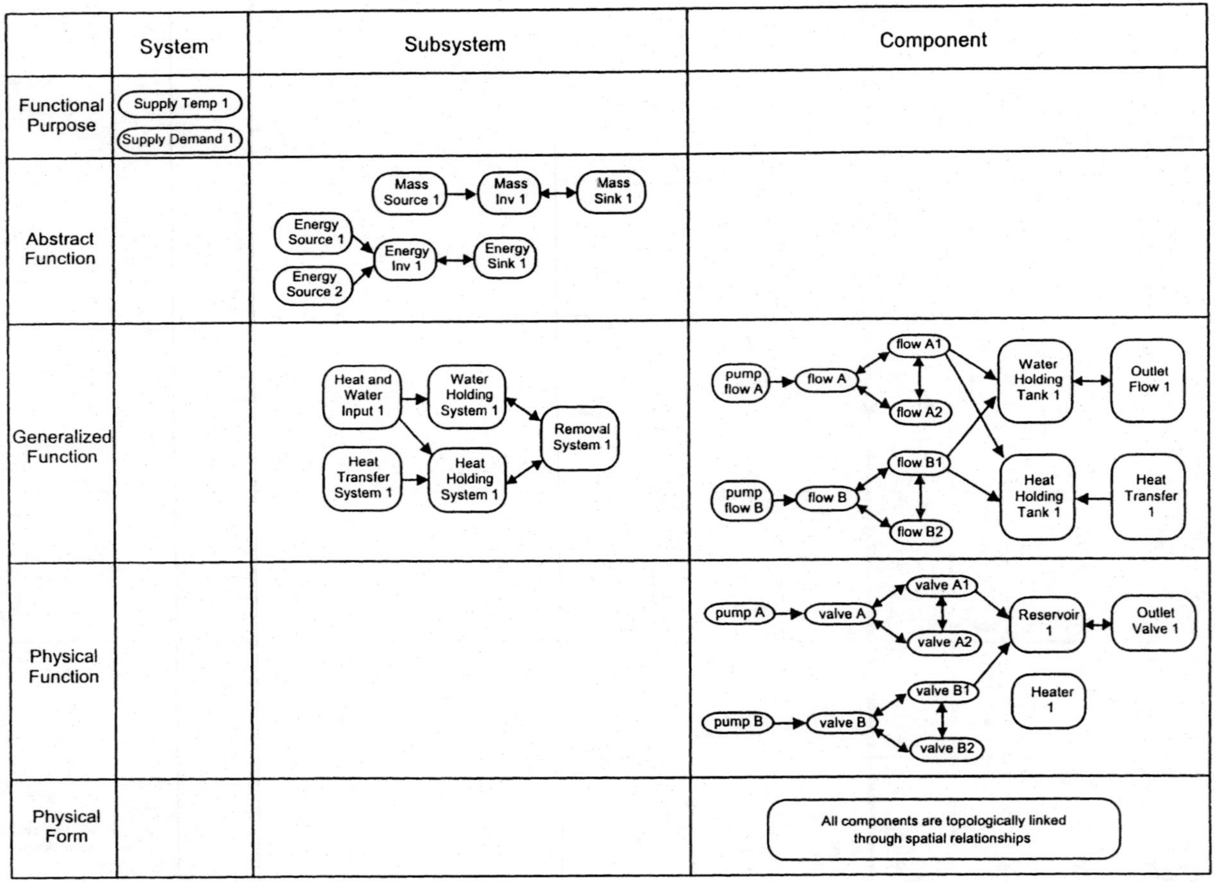

FIG. 7.15. Topological links in DURESS II (Bisantz & Vicente, 1994). Reprinted from *International Journal of Human-Computer Studies, 40,* A. M. Bisantz & K. J. Vicente, "Making the abstraction hierarchy concrete," pp. 83–117, Copyright © 1994. Reprinted by permission of the publisher, Academic Press.

and rate of removal of heat and water due to the output valve are described for both subsystems. A further decomposition to the component level, shown in Fig. 7.13, allows the description of the rate of heat transfer and water flow through each valve and pump, as well as the rate of heat transfer from the heater, storage of heat and water in the reservoir, and rate of removal of heat and water due to demand. For both the subsystem and component descriptions, the topological links, shown in Fig. 7.15, indicate the flows of water and heat through the components.

Physical Function. The states of the components are described at this level of abstraction. Because only individual components have measurable states in this work domain, the descriptions are at only the component level of decomposition. The settings of valves, pumps, and heaters are described, along with the volume and temperature in the reservoir. Topological links at this level indicate physical connections between components (see Fig. 7.15).

Physical Form. At this level, the appearance, condition, location, and anatomical configuration of each component are described. The topological links reflect spatial relationships between components.

Exercising the Work Domain Analysis

We can obtain greater insight into the value of this AH representation by examining the information provided by the various links. These relations reveal some important properties that actors need to take into account when controlling the work domain. First, the means–ends mappings between functional purpose and abstract function shown in Fig. 7.14 indicate that the temperature goal is connected to both the mass and energy balances. The reason for this is that temperature is defined as energy per unit mass. In contrast, the demand goal is only connected to the mass balance, which means that changing the mass flow to affect demand may have the unintended side-effect of changing temperature, whereas the reverse need not be true. A second property of the work domain revealed by the means–ends links shown in Fig. 7.14 is the many-to-many mapping between abstract function and generalized function. There are several instances wherein a single subsystem is a means for controlling both mass and energy. For example, each feedwater stream serves both as a mass input and an energy input. Therefore, there is a structural coupling between the mass and energy configurations. A third critical property of the work domain is the many-to-one mapping between physical function and generalized function. This property results from the fact that either feedwater stream can be used to supply water to either reservoir. As a result, changing the state of one of the components in one feedwater stream to control the flow to one reservoir could unintentionally change the flow to the other reservoir. All of these relationships should be taken into account by actors. Specific examples can be found in Vicente et al. (1995), who showed how this representation of DURESS II can be used to interpret verbal protocol trajectories, and in Bisantz and Vicente (1994), who showed how this representation can be used to simulate problem-solving trajectories.

Although it may be difficult to believe, the work domain representation just described is actually a qualitative, function-oriented representation of the same relations that were identified in a quantitative, function-independent form by the equations in Table 6.3 (reproduced here as Table 7.2). The structural means–ends

Algebraic Equations

Purposes

1. $EI(t) = TI(t) \, VI(t) \, c_p \, \rho$

    • relationship between energy, volume, and temperature

2. $EII(t) = FHI(t) + c_p \, T_I \, MII(t)$

    • conservation of energy from heater and inflow

3. $MII(t) = FAI(t) + FBI(t)$

    • conservation of mass from two feedwater streams

4. $EOI(t) = MOI(t) \, c_p \, TI(t)$

    • energy leaving reservoir

5. $FVA(t) = FAI(t) + FA2(t)$

    • conservation of mass in feedwater stream

6. $FHI(t) = HTRI(t)$

    • conservation of energy from heater

7. $FAI(t) = \dfrac{FVA(t) \, VAI(t)}{VAI(t) + VA2(t)}$

    • flow split relation

8. If pump is OFF then $FPA(t) = 0$,

    otherwise:

    IF $[VAI(t) + VA2(t)] > VA(t)$ THEN

        $FPA(t) = FVA(t)$

    ELSE

        $FPA(t) = FVAI(t) + FVA2(t)$

    • flow through pump

9. $FVA(t) = FPA(t)$

    • conservation of mass in pipe

1. Temperature set point

2. Demand set point

State Equations

1. $\dfrac{dTI(t)}{dt} = \dfrac{FHI(t) - [MII(t)][TI(t) - T_I]c_p}{VI(t) \, c_p \, \rho}$

    • conservation of energy in reservoir

2. $\dfrac{dVI(t)}{dt} = \dfrac{MII(t) - MOI(t)}{\rho}$

    • conservation of mass in reservoir

Note. Constants: ρ—density of water; c_p—specific heat capacity; T_I—inlet water temperature. Not shown are the first-order with a time constant of 5 seconds for the pumps and valves, and 15 seconds for the heaters. The other half of DURESS II is governed by a symmetrical set of equations. From Vicente (1991). Copyright 1991 by Kim J. Vicente. Adapted by permission.

and part–whole relationships identified by the work domain analysis are actually implicitly represented in these equations governing DURESS II. The problem is that those relationships are difficult for most people to extract because the equations tacitly mix together different kinds of links. For example, Equation 5 in Table 7.2 showing how the flows from the two FWSs sum up to the combined flow into a reservoir represents a part–whole relationship. In contrast, Equation 7 in Table 7.2 showing how the valve settings affect the flow rates represents a structural means–ends relation. Furthermore, the relation between variables is not explicitly identified in the functional manner that is required to support goal-directed problem solving. Therefore, although the equations in Table 7.2 provide a useful quantitative representation of DURESS II, the work domain representation described previously provides a complementary and psychologically relevant qualitative representation that workers can use productively to cope with the unanticipated.

SUMMARY AND IMPLICATIONS

To improve safety, we must provide workers with a means of coping with unanticipated events. The first phase of CWA, work domain analysis, is intended to identify the information requirements associated with this challenging objective. Field descriptions are particularly useful for dealing with the unanticipated because they generalize beyond particular trajectories. The abstraction-decomposition space is a field description that is particularly useful to represent the work domain constraints imposed by complex sociotechnical systems. In addition to providing a way to cope with the unanticipated, this two-dimensional space also has a number of psychological advantages as well. It allows us to represent complex work domains in a way that makes them manageable, and understandable, to workers.

What are the implications of work domain analysis for design? The answer to this question can be obtained by examining Fig. 7.16, which is an updated version of Fig. 5.8. This chapter has dealt with the first row in this figure. The concept of a work domain can be analyzed using the abstraction-decomposition space as a modeling tool. We have used this tool to develop a work domain representation for our case study. This representation can be used to address several design issues. The variables identified in the work domain analysis help determine what information needs to be measured, and thus, sensed (Reising & Sanderson, 1996). These variables

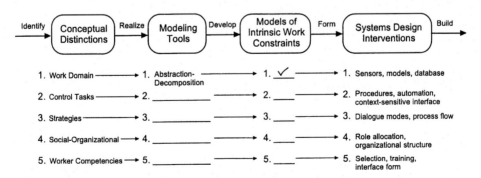

FIG. 7.16. An outline of the CWA framework with the first phase (i.e., row) filled in.

also help determine what information needs to be derived by the use of models. Finally, the means–ends and part–whole links help determine how all of this information can be organized into a database that supports goal-directed problem solving (Vicente & Rasmussen, 1990).

This first phase of CWA is particularly important because it provides support for coping with the unanticipated. Nevertheless, as we have mentioned several times in this chapter, the work domain is not the only source of constraint in a complex sociotechnical system (see chap. 5). Furthermore, although a work domain representation has certain advantages, it also has some disadvantages as well (see chap. 3). To overcome these disadvantages, we need to conduct a complementary constraint-based analysis of the control tasks that need to be performed by actors. Now that we have described the structure of the object of control, we can turn to analyzing the structure of control itself.

Phase 2: Control Task Analysis 8

The efficiency of skilled performance is due to the ability to compose the process needed for a specific task as a sequence of familiar subroutines which are useful in different contexts. This implies the existence of links in the sequence at standard key nodes or "states of knowledge" which are characteristic of the specific skill. The data process stops at such links, the mode of processing and frequently the level of abstraction changes, and to study and identify the processes, the activity must be structured according to such key nodes.
—Rasmussen (1976, p. 374)

PURPOSE

Our goal in this chapter is to explain the second phase of CWA, control task analysis, in full detail.[1] As we explained in chapter 3, control task analysis and work domain analysis are complementary. Although an analysis of control tasks does not identify the support required to deal with unanticipated events, it does allow us to identify the requirements associated with known, recurring classes of situations. Such an analysis can identify the constraints on *what* needs to be done, independently of how or by whom. We describe the *decision ladder* (Rasmussen, 1974, 1976), a modeling tool that can be used to develop control task models. An example of how the decision ladder can be used to conduct a control task analysis is presented for the DURESS II process control case study introduced in chapter 6. By the end of this chapter, you should understand how control task constraints both inherit, and build on, the work domain constraints discussed in the previous chapter.

DIFFERENT APPROACHES TO ANALYZING CONTROL TASKS

The need for some kind of task analysis was described in detail in chapter 3. We outlined three kinds of approaches: input–output, sequential flow, and timeline. The first is an example of constraint-based analysis, whereas the latter two are examples of instruction-based analysis. We went on to show that a constraint-based approach is required to deal with open systems. Instruction-based approaches do not have the flexibility required to deal with substantial disturbances. Thus, the input–output analysis is the most appropriate for complex sociotechnical systems.

---

[1]Note that, here and elsewhere, the term *control* is being used in a broad sense to cover all phases of information processing activity (see Fig. 8.2), not just motor output. This usage is consistent with that adopted in control theory.

In this section, we review some additional considerations that should be taken into account in choosing a control task analysis method. In particular, we focus on the need to choose a method that will allow us to identify the requirements that are associated with expert action. We do not want to build an information system that forces workers to have to engage in all of the activities exhibited by a novice. Instead, we would like to develop a design that both induces, and supports, efficient expert action. To accomplish this goal, the method of control task analysis we choose must be able to identify the requirements associated with skilled activity.

What Is Control Task Analysis?

There are at least three important points that need to be kept in mind in this chapter. The first is the objective of control task analysis, and how it differs from the objective of work domain analysis. Figure 8.1 illustrates the basic relationship between control tasks and a work domain. On the input side, control tasks require information about the state of the work domain. On the output side, they require action on the work domain. Accordingly, a work domain analysis describes the functional structure of the system being acted on. In terms of Simon's (1981) parable, it is equivalent to describing the beach. CWA begins by analyzing the work domain because the field in which action takes place imposes constraints independent of any particular task, event, or goal.

But once we understand the layout of the beach, we need to know something about the goals that the ant must satisfy in certain situations. For example, the constraints that are relevant to the ant when it is going home will be different than when it is foraging for food. For one, it is likely that different parts of the beach will be traversed in these two situations, so the parts of the work domain that are relevant will differ. In addition, the goals associated with going home and looking for food are different. If we wanted to support the ant in each of these classes of situations, we would have to identify the requirements associated with each class (in addition to the constraints imposed by the beach). In more general terms, a control task analysis should seek to identify the requirements associated with known classes of events in a particular application domain.

Note in Fig. 8.1 that the object of analysis moves from the thing being controlled (i.e., the work domain) to the requirements associated with effective control (i.e., the control tasks). That is, now that we have completed the analysis of the work domain, we are no longer interested in describing the beach. Instead, we are

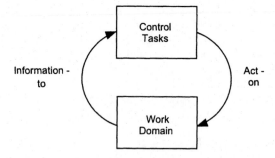

FIG. 8.1. Simplified diagram of the relation between control tasks and work domain.

interested in describing the requirements associated with going home, foraging for food, and other classes of recurring situations. Linguistically, we move from an emphasis on nouns (i.e., the object of action) to an emphasis on verbs (i.e., action itself; cf. chap. 7).

The second characteristic that needs to be kept in mind is that control task analysis describes only what needs to be done, not how or who. Recall from chapters 4 and 5 that one of the aims of CWA is to develop device-independent models so that design decisions emerge as outputs of the analysis rather than being assumed up front as inputs. One implication of this objective is that we should first analyze what needs to be done, before we consider how it can be done or by whom. In other words, it is easier to identify how tasks can be done (the strategies phase) and how they should be allocated to different actors (the social-organizational phase) if we explicitly identify what needs to be achieved in the first place (the control task phase). After all, before we focus on strategies, we should be clear about what goals those strategies should be achieving. Similarly, before we focus on role allocation, we should be clear on what there is to allocate.

A third characteristic that needs to be considered during control task analysis is that complex sociotechnical systems are open systems (see chap. 3). Consequently, they require context-conditioned variability to adapt to disturbances. The same goals may have to be accomplished in different ways on different occasions. To accommodate this requisite variability, an input–output analysis is required (see chap. 3). Such an analysis identifies the inputs that are required to perform the task, the outputs that are achieved after it is completed, and the constraints that must be taken into account in selecting the actions that are required to achieve the task (e.g., Figs. 3.1 and 3.4). Input–output analysis usually specifies what needs to be done without dictating precisely how it should be done, thereby giving workers the responsibility of finishing the design based on their knowledge of the current context—knowledge that designers frequently do not have access to a priori (see chap. 5).

In summary, a control task analysis should identify what needs to be done, independently of how or by whom, using a constraint-based approach.[2] In the next subsection, we begin to address the issue of how to identify the requirements associated with expert action by reviewing a traditional approach to describing control tasks.

The Traditional Information-Processing Approach

Perhaps the best known form of task analysis is that based on a human information-processing approach (e.g., Card et al., 1983; Kirwan & Ainsworth, 1992; A. Newell & Simon, 1972; Norman, 1986; J. W. Payne, Bettman, & Johnson, 1993; Wickens, 1992). This approach is reductionistic in the sense that it tries to break down molar tasks into their constituent elemental information-processing steps. Typically, these elemental steps are ordered in a linear sequence that progresses from perception to decision making to action.

---

[2]You may be wondering how any of this differs from traditional task analysis (e.g., Kirwan & Ainsworth, 1992). To be sure, the objectives of control task analysis are very similar to those of traditional task analysis. The primary difference is that the vast majority of traditional task analysis techniques are instruction based, not constraint based (see chap. 3). In addition, some traditional task analysis techniques do not separate the question of what needs to be done from those of how it is to be done or by whom (again, see chap. 3).

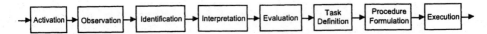

FIG. 8.2. A linear sequence of information processing tasks, typical of the human information processing approach to human factors, HCI, and psychology.

A prototypical example is provided in Fig. 8.2. In this particular case, there are eight elementary information-processing steps. The sequence begins with a signal from the work domain that causes a need for activation or detection. This step is followed by observation of the work domain. Based on the data collected, an identification of the current state can be made. Next, this state identification can be interpreted to determine the consequences for the criteria of interest. Then, an evaluation can be made to determine which criteria are more important, and should therefore take precedence, for the present situation. Once this evaluation is made, the task to be pursued can be defined. Given this definition, a procedure for accomplishing the task can be formulated. Finally, the actions on the work domain can be executed.

For our purposes, the particular steps in Fig. 8.2 are not important. Instead, the key is that the information-processing approach breaks down a task into a linear sequence of elementary information-processing activities. To be sure, different models use different activities. For example, Fig. 8.3 illustrates another exemplar of the same

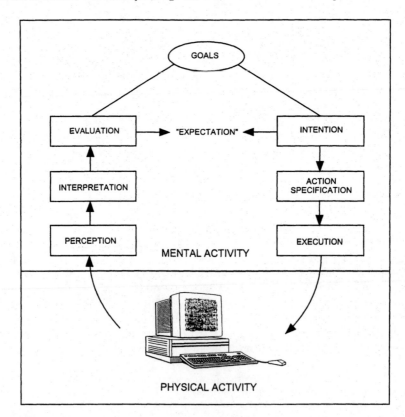

FIG. 8.3. Norman's (1986) seven-stage model of user activities. Reprinted with permission. Copyright © by Donald A. Norman, 1986.

general approach, Norman's (1986) seven stages of user activities. Norman's framework is well known in the HCI community. Although the number and labels for the steps differ from those in Fig. 8.2, the basic idea is the same. There is a progression from perception to decision making to action in a generally linear sequence of elementary information-processing steps.

What Descriptive Studies Show: Expert Performance

Traditional information-processing approaches to task analysis are quite useful for many purposes. Nevertheless, they have important limitations, namely that it is not always necessary to follow all of the steps in the linear sequence and that the steps need not be followed in the order that is usually shown. These limitations were pointed out by Rasmussen (1974) in a field study of workers starting up a fossil fuel power plant.

Ironically, Rasmussen (1974) began by adopting the information-processing approach to analyze his verbal protocol data. In fact, he used the linear sequence illustrated in Fig. 8.2 to code the control tasks being performed by the workers he was observing. As he analyzed the data, however, he realized that the steps were rarely followed in sequence. Instead, these expert workers were frequently able to recognize situations, and as a result, take efficient shortcuts. For example, in very familiar situations, workers may know exactly what procedure they should be using merely by conducting a few observations. In those cases, the intervening information-processing steps between observation and execution could all be bypassed. In addition, Rasmussen also noted that the flow of activity need not follow the left-to-right sequence specified by the sequence depicted in Fig. 8.2. Sometimes, it was possible for workers to move from right to left. For example, once the task definition step was completed, workers sometimes moved to the observation step to determine if the plant was in the proper state to proceed with the required task (see the example discussed later).

Given these deviations from the normative linear sequence specified by the information-processing approach, Rasmussen (1974) reached the following conclusion:

> The characteristic steps of a mental task—i.e. the sequence of steps between the initiating cue and the final manipulation of the system—can be identified as the steps a novice must necessarily take to a carry out the sub-task. Study of actual, trained performance may then result in a description of his performance [in] terms of shunting leaps within this basic sequence. (p. 26)

Thus, the information-processing approach identifies the steps that a novice would have to take to accomplish the task. Because Rasmussen observed experts (i.e., professional process control operators) in his study, he rarely observed this linear sequence. Instead, shortcuts and shunts were the norm rather than the exception.

These insights are similar to some of the findings obtained by Klein (1989) and colleagues in their descriptive field studies (see chap. 4). In familiar types of situations, experts rely on their previous experiences and domain knowledge to bypass certain cognitive activities (H. L. Dreyfus & S. E. Dreyfus, 1988). As a result, they are able to accomplish task demands in a more economic and efficient manner. Novices, on the other hand, do not possess the same experience or knowledge, so they must laboriously work through each step in the sequence.

An Alternative Approach: Expertise as an Active, Constructive Process

Rasmussen's (1974, 1976) field observations led to the following view: Expertise is the ability to compose a process needed for a specific task as a sequence of familiar subroutines that are useful in different contexts (Rasmussen, 1976). We need to unpack this description so that we can better appreciate its implications for control task analysis.

The first point is that experts have developed a set of "subroutines" that can be used in different situations. Metaphorically, each of these routines is like a "bag of tricks," specific enough to be applicable to a particular task, but general enough to be relevant to various situations. As workers become more experienced, they increase the number of bags of tricks in their repertoire.

The second point is that expert performance is a constructive process rather than an attribute of an actor. It is a set of actions performed by someone, not a thing that someone possesses (cf. Bartlett, 1932; Kolers & Roediger, 1984; Morris, Bransford, & Franks, 1977; Thelen & L. B. Smith, 1994). In this view, expert workers do not *retrieve preplanned solutions* from memory (Dreyfus & Dreyfus, 1988). Instead, they actively *generate a contextually tailored sequence of cognitive activities* that is appropriate for the present situation.

The final, related point is that a particular action sequence is composed through the aforementioned constructive process by sequencing together a string of routines that, collectively, meet task demands. Thus, expert behavior can be metaphorically likened to the concatenation of a string of skills that are tailored to task demands. If this is the case, then it makes sense to carve the global task according to the "joints" defined by these feasible expert routines. At the same time, we should retain the flexibility to string together those routines in different ways, not just in the ideal sequence identified by the information-processing approach.

Implications for Control Task Analysis

Viewing expertise as an active, constructive process gives us some insights into how to conduct a control task analysis that will support expert performance. Given that one of the benefits of control task analysis is cognitive efficiency (see chap. 3), it is desirable for the analyst to identify, and subsequently provide information support for, any reliable opportunities for shortcuts. Thus, rather than trying to determine what it would take for a novice to do the job (as the information-processing framework does), we can instead try to determine what preconditions have to be satisfied for expert performance. By adopting such an approach, we should be able to design computer-based information systems that deliberately induce and support expert action, thereby leading to gains in cognitive efficiency.

The discussion in this section suggests that one way to accomplish that objective is to parse tasks according to the routines that experts could use, and to provide for the flexibility to combine those routines in various ways, as a function of the domain demands. Whatever modeling tool we adopt for conducting a control task analysis should take these basic insights into account. In the next section, we describe a tool that has been developed to achieve this objective.

THE DECISION LADDER

Based on his field study of power plant operators, Rasmussen (1974, 1976) developed the decision ladder as a modeling tool that can be used to meet the objectives described in the previous section. Figure 8.4 illustrates the basic structure of the ladder. In this section, we discuss the characteristics of this modeling tool, its relationship to the abstraction-decomposition space, and two descriptive examples of its use.

Basic Structure

By comparing Figs. 8.2 and 8.4, we see that the decision ladder was created roughly by bending a linear sequence of information-processing steps in half and adding

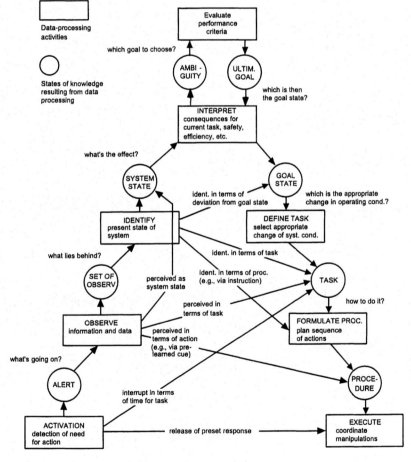

FIG. 8.4. The decision ladder. Adapted from J. Rasmussen, "Outlines of a hybrid model of the process plant operator," in T. B. Sheridan & G. Johannsen (Eds.), *Monitoring behavior and supervisory control.* Plenum Publishing, 1976, with permission.

shortcuts that connect the two sides of the ladder. In this subsection, we describe the features associated with this structure.

States of Knowledge Versus Information-Processing Activities. . Some nodes in Fig. 8.4 are boxes whereas others are circles. These graphical symbols discriminate between two kinds of constructs. The boxes correspond to *information-processing activities,* whereas the circles correspond to *states of knowledge.* The information-processing activities correspond to the expert routines that we described earlier. They are activities in which actors need to engage. In contrast, states of knowledge are the results of information-processing activities. They are not activities in which actors need to engage to accomplish task goals. They are rather the outputs or products of those processing activities. For example, after engaging in the information-processing activity labeled as the *identify* box in Fig. 8.4, an actor may know what the current system state is.

Given this distinction between knowledge states and processing activities, it is not feasible to go directly from one box to another. Each box in the ladder is an information-processing activity, so once an activity is completed, the result must logically be a state of knowledge (i.e., a circle). Only after a state of knowledge has been attained is it possible to engage in a new information-processing activity. This syntax has implications for the types of shortcuts that can be made in the ladder (see later discussion).

Opportunistic Movement. The most obvious distinctive characteristic of the decision ladder is the set of shortcuts that connect the two sides. These shunting paths consist of stereotypical processes frequently adopted by experts. Note that these *shunts* connect a box (i.e., an information-processing routine) to a circle (i.e., a state of knowledge). The reason for this is that the state of knowledge is the output of the information-processing activity. For example, it is possible to perceive the work domain in terms of the task to be performed, a shunt that is represented by a direct link between the *observe* box and the *task* circle in the ladder (see Fig. 8.4).

Figure 8.5 shows that there is another type of shortcut that can be mapped onto the decision ladder, namely associative leaps. *Leaps* are different from shunts because they link together two circles (rather than a box and a circle), because one state of knowledge can be directly associated with another. In other words, leaps do not require any information-processing activity; they are direct associations. For example, in some cases, knowledge of system state can be directly associated with knowledge of what task to perform (see Fig. 8.5).

Another important feature of the decision ladder that is not graphically represented in Figs. 8.4 and 8.5 is that it accommodates various starting points. Contrary to what is frequently assumed, cognitive activity need not start in the bottom left box (*activation*). In fact, it need not even start with the left leg (see Fig. 8.7 for an example). Similarly, cognitive activity need not end in the bottom right box (*execution*). Different types of situations will usually require different starting points in the ladder. Furthermore, the shunts and leaps labeled in Figs. 8.4 and 8.5 are not the only ones that are possible. These are just the ones that were frequently identified in the field study that led to the development of the ladder (Rasmussen, 1974, 1976). Different situations in other application domains may require different shunts or leaps. Thus, the decision ladder is a very flexible modeling tool.

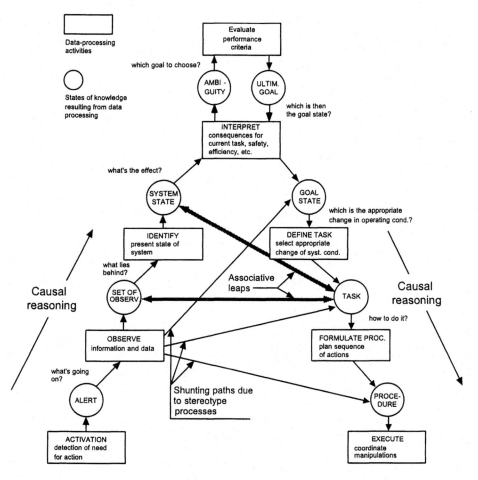

FIG. 8.5. The associative leaps and shunting paths in the decision ladder (Rasmussen, 1980). Reprinted from H. T. Smith & T. R. G. Green (Eds.), *Human interaction with computers.* J. Rasmussen, "The human as a systems component," pp. 67–96, Copyright © 1980, by permission of the publisher, Academic Press.

Having said that, a novice worker may have to engage strictly in the analytical causal reasoning path depicted in Fig. 8.5. Such a process is equivalent to following the normative linear sequence shown in Fig. 8.2. However, the shunts and leaps allow for a much more opportunistic form of expert cognitive activity that deviates from this rigid sequence. As Rasmussen (1976) observed:

> Frequently, a skilled operator does not enter the sequence at its entry; his process feel can initiate consideration first of a step later in the sequence, he may change the order of the steps, and he only occasionally has to move through all the steps in the basic sequence. (p. 376)

The decision ladder is flexible enough to accommodate such expert behavior. From the viewpoint of CWA, this is an important feature because it allows analysts to identify, and therefore support, shortcuts that can induce and support more skilled, and thus more efficient, performance.

Template, Not a Model. Another important feature of the decision ladder is that it is a template rather than a model. It is not difficult to find authors criticizing the decision ladder because it is an impoverished model of human cognitive activity. However, such critiques are unwarranted because the ladder was never intended to be a model. This point was made, albeit cryptically, in the very first technical report describing the decision ladder: "This is clearly a very idealized description" (Rasmussen, 1974, p. 28). The point was later made more clearly in Rasmussen's (1986c) well-known monograph: "The decision ladder is not a model of the decision process itself, but rather a map useful to represent the structure of such a model" (p. 70; see also p. 12). Metaphorically, the ladder is a skeleton without any flesh. The flesh (i.e., the content) has to be added for each application domain. In fact, this is one way to think about a control task analysis; it is the act of using the decision ladder as a scratchpad to identify the control task requirements for a particular work domain.

This clarification should make you realize that the ladder itself is quite modest. After all, it is not a model and a great deal of work must be undertaken by an analyst to use the ladder to uncover information requirements. There is a great deal of domain knowledge that is required that the ladder simply does not help with. Nevertheless, as we hope the case study presented later shows, the decision ladder is still a very useful tool.

"What," Not "How" or "Who". Not only is the decision ladder not a model of human information-processing activity, it is not even a template of human information-processing activity. Instead, it is a template of *generic* information-processing activity. A closer look at the labels in Fig. 8.4 reveals why. No specific reference is made to people, or to any psychological mechanisms. In fact, any of the constructs that are identified in the ladder can be associated with any type of actor, whether it be a human worker or machine automation.

This point is important, not just because it is frequently overlooked, but also because it allows us to decouple two types of decisions, namely *what* needs to be done and *who* should do it. The decision ladder provides a generic template for identifying the demands associated with particular control tasks. As a result, it allows us to identify the information requirements that must be satisfied. Regardless of the role allocation architecture that is adopted (e.g., fully automated, fully manual, some hybrid; see Sheridan, 1987), *these requirements must be satisfied if task goals are to be reliably and consistently attained.* As we pointed out in chapter 5, after a control task analysis has been conducted, it is easier for an analyst to make informed decisions about who should be doing what (i.e., how these requirements should be allocated across workers and automation). This topic is addressed in detail in chapter 10.

There is another feature of the decision ladder that is not visible in the graphical depiction in Fig. 8.4, namely that it is used to develop a product description rather than a process description. That is, the goal at this stage of CWA is to find out what needs to be done, not how it can be done. In terms of the concepts introduced in chapter 3, the decision ladder is used to conduct a "black box" input–output task analysis. As we mentioned in chapter 5, after a control task analysis has been conducted, it is easier for an analyst to make informed decisions about how those tasks might be done (i.e., what strategies are useful for achieving the required input–output function). This topic is addressed in detail in chapter 9.

To summarize, the decision ladder can be used to help determine what information-processing activities need to be done, independently of who is to do them and how they can be done.

Formative. In keeping with the overall philosophy of CWA (see chap. 5), the orientation of a control task analysis is formative in nature. The overall objective is to determine how computer-based support systems could be designed to allow workers to effectively meet the challenges they face. Accordingly, it is important that the analyst identify all feasible shortcuts. For example, if it is possible to link the identification of a particular system state with a particular procedure consistently and reliably, then this shunt should be noted in the analysis. The identification of this shortcut then serves as a requirement that must be satisfied during systems design (e.g., by training workers to know that such a shunt is possible, or by building the shunt into the design of the automation or interface). This formative approach should result in the design of a sociotechnical system that will foster and support expert performance.

It is important to emphasize this point, because we could very well conduct the control task analysis from the descriptive viewpoint of a novice, for instance. However, such an analysis would probably lead to the conclusion that every information-processing activity in the decision ladder needs to be undertaken, even for known recurring classes of situations. After all, a novice would not be expected to know that certain shunts or leaps are possible. Although it is possible to use the decision ladder in this descriptive manner (see the two examples presented later in this section), such an analysis would defeat the purpose of CWA. The result of applying the ladder for novice behavior would probably be something that looks very much like the inefficient and usually unnecessary linear sequence in Fig. 8.2. Such an analysis would not lead to a design that is particularly efficient because the requirements that induce and support expert action would not have been identified. Thus, it is important to adopt a formative perspective when using the decision ladder in CWA.

One of the implications of this orientation is that a control task analysis need not utilize all of the elements in the ladder. Because of the shortcuts, several—sometimes, many—information-processing activities can be bypassed. Furthermore, some information-processing activities depicted in the ladder (e.g., evaluate performance criteria) are just irrelevant for some control tasks. In our experience, it is very rare for any one control task analysis to rely on all of the elements listed in the ladder.

Grain of Analysis. Another important factor that needs to be considered when using the decision ladder is the grain of the analysis. For any control task analysis, there are always several different grains of analysis that can be chosen, some finer and others more coarse. For example, in the case of conducting a control task analysis for a hospital, we could adopt a fine level of analysis to identify the requirements associated with low-level activities, such as the administration of drugs to patients. Alternatively, we could adopt a coarse level of analysis to identify the requirements associated with high-level activities, such as human resource management.

Different grains of analysis will lead to different results. For example, in the first case just described, the activation information-processing activity might require detecting that a patient is in pain and needs analgesic. A social-organizational analysis would probably identify such a low-level activity as being the responsibility of a single individual. In contrast, in the second case just described, the activation information-processing activity might require detecting a need to hire more qualified personnel. A social-organizational analysis would probably identify such a high-level activity as being the responsibility of a group of individuals.

As in almost any modeling effort, there are no hard and fast rules for identifying which grain of analysis is most appropriate. There is no privileged grain of analysis. Different levels of resolution are appropriate for different circumstances. To determine which level is most appropriate, we should be very clear about why we are conducting a CWA. Once the purpose of the analysis is explicitly identified, we should be able to determine that certain levels of analysis are clearly more useful than others. In addition, it is also important to consider the boundary that was used to define the work domain. This boundary delimits the scope of the entire CWA, so the scope of a control task analysis cannot require us to go outside the bounds that we defined for the work domain.

Operating Modes. A final factor that needs to be taken into account is that there are usually several qualitatively different operating modes in most complex sociotechnical systems. For example, in process control plants, activities can generally be divided into the following modes: plant start-up, normal operation, disturbance management, and plant shutdown. Similarly, in hospitals, activities may be divided into several modes, including: preoperative, surgery, and recovery. Usually, the control task activities in other complex sociotechnical systems can be divided up into an analogous set of modes.[3]

Because each of these modes will usually impose a qualitatively different set of demands, it is useful to conduct a separate control task analysis for each mode. The decision ladder can be used as a template in each case, but each mode will use a different template to identify the information requirements associated with that mode. Note that a particular information-processing activity may appear in more than one mode. In general, however, the analyses of the different modes will lead to a different set of relevant information-processing activities and shortcuts.

Relationship Between the Decision Ladder and the Abstraction-Decomposition Space

Now that we have described the decision ladder in some detail, we can show you its relationship to the abstraction-decomposition space described in chapter 7. Figure 8.1 shows the relationship between a work domain and a control task. The work domain is the system being acted on. Control tasks, on the other hand, are what needs to be done to the work domain in particular classes of situations. Recall that the abstraction-decomposition space is a modeling tool that can be used to conduct a work domain analysis. As such, it describes the functional structure of the system being controlled. Although the work domain imposes constraints on action, it is not the only source of constraint in a complex sociotechnical system (see chap. 5). For example, there are some constraints on action that arise because of the structure of control tasks, not just the structure of the work domain. It is for this reason that a control task analysis needs to be conducted in addition to a work domain analysis.

---

[3]For those familiar with Rasmussen et al. (1994), it might help to point out that our notion of operating modes corresponds roughly to their notion of prototypical situations or functions. Therefore, our identification of operating modes plays a role that is similar to, albeit less detailed than, Rasmussen et al.'s activity analysis in work domain terms. Our decision ladder analysis corresponds directly to Rasmussen et al.'s activity analysis in decision-making terms.

Decision Ladder

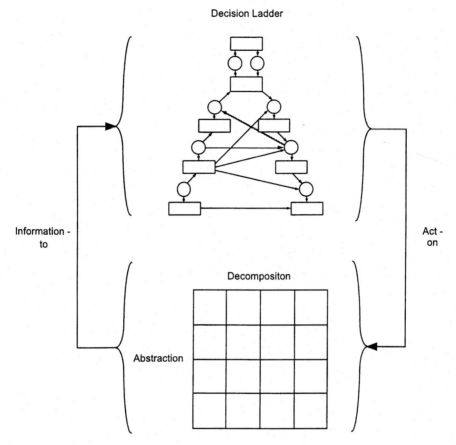

Information -
to

Act -
on

Decompositon

Abstraction

FIG. 8.6. A simplified diagram of the relation between the abstraction-decomposition space and the decision ladder. Compare with Figure 8.1.

As we have pointed out in this section, the decision ladder is a modeling tool that can be used to carry out a control task analysis.

Figure 8.6 combines these insights by graphically showing the relationship between the abstraction-decomposition space and the decision ladder. This figure is an elaborated version of Fig. 8.1 in which the two modeling tools that are used in CWA to conduct a work domain analysis and a control task analysis have been substituted into the figure. Think of the decision ladder as representing the information-processing activities that need to be performed by a hypothetical actor to achieve task goals.[4] These activities are accomplished by: (a) obtaining information about the current state of the work domain, (b) knowing the functional structure of the work domain, (c) knowing the constraints associated with the control task in question, and (d) acting on the work domain to effect a change in its state. Of course, these steps do not occur in such an orderly fashion. As the flexible structure

---

[4]We say "hypothetical" because, following the logic of CWA (see chap. 5), the questions of how many and which actors will be responsible for the various requirements identified in a control task analysis are not dealt with until Phase 4, social-organizational analysis (see chap. 10).

of the decision ladder makes clear, the interaction between actor and work domain is much more opportunistic and iterative than this description would indicate. With this caveat in mind, we hope that you can see from Fig. 8.6 how the abstraction-decomposition space and the decision ladder complement each other in a coherent and systematic fashion. The former describes the constraints imposed by the system being controlled, whereas the latter describes the constraints imposed by the information-processing activities associated with control (see Footnote 1).

Figure 8.6 shows a very simple case of one hypothetical actor operating on one work domain. In almost all complex sociotechnical systems, the control architecture for any one operating mode will result in multiple actors interacting with each other. We could conceptualize such an architecture by using one decision ladder to show the information-processing activities associated with each actor. Typically, each actor would be working on the same work domain.[5] Alternatively, it is also possible to have a hierarchical arrangement where some actors act on other actors. This could be illustrated by having one decision ladder acting on another (see chap. 10). A simple example would be a supervisory control architecture where one worker oversees automation acting on the work domain (Sheridan, 1987). The issue of who (i.e., which actors) are responsible for different control task requirements is dealt with in detail in chapter 10.

Multiple Uses of the Decision Ladder: Some Examples

The decision ladder can be productively used for a number of different purposes. Our focus so far has been, and continues to be, on describing how it can be used to conduct a control task analysis as part of CWA. For instance, in this section, we have described in some detail how the ladder can be used as a template in identifying the requirements associated with particular information-processing activities. In chapter 10, we show how the ladder can also be used as a template in making decisions about how these information requirements can be distributed across multiple actors in an organization. In this subsection, we describe how the ladder can be used for a third purpose that is not directly related to CWA, namely as a template with which to analyze verbal protocols. This use is descriptive rather than formative. However, by showing you a few examples of how the decision ladder can be used in this way, we hope to offer you greater insight into its detailed structure.

Figure 8.7 illustrates the first of two examples taken from Rasmussen (1980), showing how verbal protocols from workers can be mapped onto the ladder in a descriptive fashion. This particular episode begins on the middle right of the ladder with the worker receiving a scheduled order ("We now have to . . ."). The worker does not have to engage in any information-processing activity. He knows what task to perform, which is why the statement is coded using the task state of knowledge circle. Based on experience, the worker knows that the first step that needs to be performed is to see if the work domain is ready to process the order. This shortcut is coded as a leap between two states of knowledge, because there is a direct

---

[5]Strictly speaking, this claim is not true and is made here merely for the sake of simplicity. In practice, there are frequently multiple work domain representations of the same, or at least overlapping, physical worlds (Rasmussen et al., 1994). Each work domain representation is defined by a different set of purposes (see chap. 7, especially Footnote 1). Different actors may thereby be operating in different work domains. Chapter 10 provides an example from engineering design to illustrate this nuance.

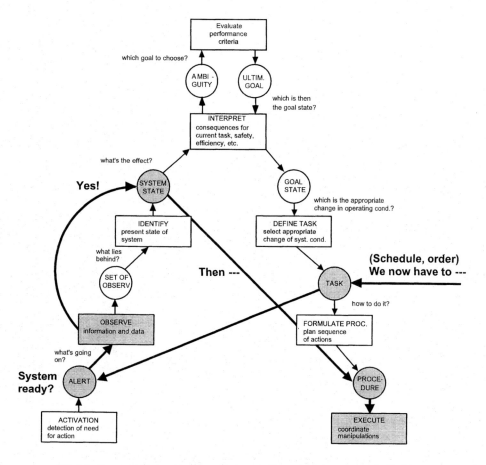

FIG. 8.7. An example showing how verbal protocols can be mapped onto the decision ladder (Rasmussen, 1980). Reprinted from H. T. Smith & T. R. G. Green (Eds.), *Human interaction with computers*, J. Rasmussen, "The human as a systems component," pp. 67–96. Copyright © 1980. Reprinted by permission of the publisher, Academic Press.

association between knowledge of the order and knowledge to check work domain readiness. The worker then observes the work domain, collecting state information. This step in the protocol is coded as an information-processing activity, because the worker is engaged in observing the work domain. Again, based on experience, the worker determines that the work domain is in fact ready. This inference is shown as a shunting path because the worker appears to be able to determine the state of the work domain merely by collecting data and without having to engage in the additional information-processing activity of identification. Once the worker determines that the work domain is ready to process the order, he knows what procedure to use ("Then . . ."), again based on experience. This shortcut is also coded as an associative leap between two knowledge states. Finally, the worker carries out the appropriate actions in the procedure. This last step is coded as an information-processing activity, execute.

The episode of problem-solving activity in Fig. 8.7 reinforces several points we made earlier concerning the use of the decision ladder. First, it is clear that the ladder is being used merely as a template. All of the content comes from the verbal protocols. Nevertheless, the ladder is still useful because it serves as a systematic tool that can be used to parse the verbal protocol into meaningful parts, corresponding to information-processing activities, states of knowledge, shunts, and leaps. Second, the example also shows that activity need not start with the *activation* box, and that activity can sometimes even start on the right side of the ladder. Third, the example clearly shows the opportunistic and flexible nature of skilled cognitive activity. Only a minority of the elements in the ladder were relevant for this particular episode. Furthermore, movement in the ladder did not follow an idealized linear sequence. Instead, the path was more chaotic because of the shortcuts enabled by the one shunt and two leaps.

Figure 8.8 illustrates another example, also taken from Rasmussen (1980). In this case, the episode begins at the bottom left of the ladder with an alarm signal that is processed by the worker. This first step is coded as an activation because the worker detects the need for action. The output of this processing is a state of knowledge, alert ("What's that?"). To answer this question, the worker observes the work domain. This step is coded as observation because the worker collects information about the state of the work domain. As with the previous example, there is a shunt again from observation to system state. Because of his expertise, the worker is able to directly determine the cause of the alarm, without having to integrate the observed data by engaging in identification. Next, the worker knows what task needs to be performed to clear the alarm, again because of experience ("We then have to . . ."). This shortcut is coded as an associative leap between knowledge of system state and knowledge of what task to pursue. Because the task is a familiar one or because a procedure is already available, the worker does not have to formulate a procedure. This step is coded as another associative leap, this time between knowledge of what task to pursue and knowledge of which procedure to use. Finally, the worker carries out the steps in the procedure. This last step is coded as an information-processing activity, execute.

By comparing Figs. 8.7 and 8.8, we can see that these two episodes of cognitive activity are different from each other. Although they are both opportunistic and nonlinear, they represent a different set of information-processing activities and shortcuts. Despite these differences, both episodes can be cleanly mapped onto the elements of the decision ladder. These and other examples (e.g., Pawlak & Vicente, 1996) show that the ladder provides a systematic way of parsing verbal protocol data.

CONTROL TASK ANALYSIS FOR PROCESS CONTROL MICROWORLD

In the previous section, we described the basic structure of the decision ladder, claiming that it can be used as a modeling tool to conduct a control task analysis. In this section, we illustrate that claim by using the ladder to conduct a control task analysis of the DURESS II microworld introduced in chapter 6.

We should warn you up front that this section is very detailed, and perhaps even tedious, in nature. We refer extensively to the variables and equations that describe the state and structure of the DURESS II microworld. This level of detail is necessary to show that the ladder can be used in a systematic fashion and that it can lead to substantial insights. To make it easier for you to follow this section, you should probably first revisit chapter 6, particularly Tables 6.2 and 6.3.

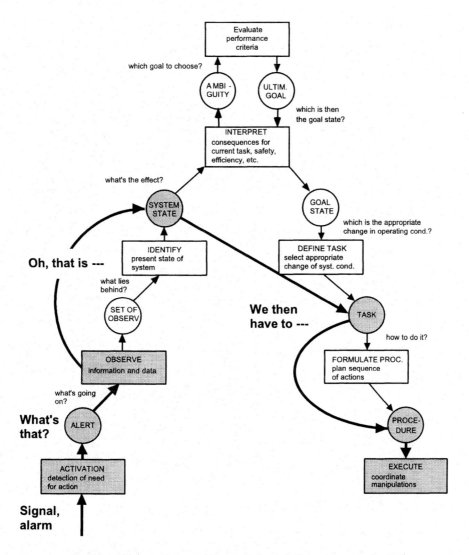

FIG. 8.8. A second example showing how verbal protocols can be mapped onto the decision ladder (Rasmussen, 1980). Reprinted from H. T. Smith & T. R. G. Green (Eds.), *Human interaction with computers*, J. Rasmussen, "The human as a systems component," pp. 67–96, Copyright © 1980. Reprinted by permission of the publisher, Academic Press.

Operating Modes

The operation of DURESS II has been broken down into four modes: start-up, normal operation, system shutdown, and fault management. Each mode is defined as follows:

- *Start-up* deals with meeting the four work domain purposes from a shutdown state.
- *Normal operation* deals with moving the work domain state from one set of demand set point values to another.

• *Shutdown* deals with bringing the work domain to a "zero state" where all components are off and there is no flow or inventory of either heat or liquid.
• *Fault management* deals with coping with equipment failures and other disturbances.

Ordinarily, we would conduct a separate control task analysis for each operating mode, using the decision ladder as a template in each case. However, for the sake of clarity and brevity, only the start-up operating mode analysis is described here.

Start-Up

Graphical and tabular summaries of the control task analysis findings for the start-up operating mode are given in Fig. 8.9 and Table 8.1, respectively. For the sake of

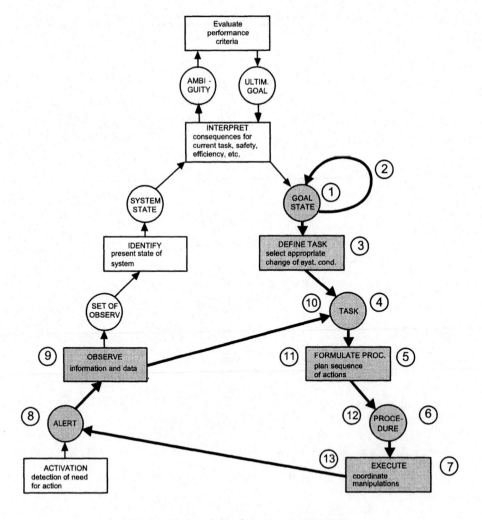

FIG. 8.9. Decision ladder for start-up in DURESS II.

TABLE 8.1 A Summary of the Control Task Analysis for the Start-Up Mode for DURESS II

Step #	Description	Link	Ladder Code	Abstraction Level	Decomposition Level
1.	Achieve Temp Goal → Achieve Demand Goal		Goal State	Functional Purpose	System
2a.	Establish Energy Inv → Establish Mass Inv	Means–Ends/ Part–Whole	Goal State	Abstract Function	Subsystem
2b.	Establish Energy Inflow → Establish Mass Inflow	Topological	Goal State	Abstract Function	Subsystem
3.	Choose Heat Transfer → Choose Input Flow	Means–Ends	Define Task	Generalized Function	Subsystem
4.	Desired Heat Transfer → Desired Flow Rate		Task	Generalized Function	Subsystem
5.	Configure Heater → Configure FWS	Means–Ends/ Part–Whole	Formulate Procedure	Physical Function	Component
6.	Desired Component Settings		Procedure	Physical Function	Component
7.	Change Settings →	Means–Ends	Execute	Physical Form	Component

(Continued)

TABLE 8.1 (Continued)

Step #	Description	Link	Ladder Code	Abstraction Level	Decomposition Level
8.	→ Prepare to Stabilize		Alert	Functional Purpose	System
9.	→ When to Act on Temp		Observe	Functional Purpose	System
10.	→ Time to Act on Temp	Means–Ends/ Part–Whole/ Topological	Task	Functional Purpose	System
11.	→ Stabilize Temp		Formulate Procedure	Physical Function	Component
12.	→ Heater Setting	Means–Ends	Procedure	Physical Function	Component
13.	→ Change Settings		Execute	Physical Form	Component

Note. See also Fig. 8.9.

simplicity, we have exploited the fact that DURESS II has a symmetrical structure by describing only half of its demands (i.e., one reservoir and one FWS) whenever possible. The following account describes the content of the information-processing activities, states of knowledge, and shortcuts identified in Fig. 8.9.

*1. **Goal State.*** The overarching goals in the start-up mode are to stabilize each of the two reservoirs at the required temperature and output demand values. Formally, these requirements can be stated as follows:

$$T1 = \text{set point} \pm \text{tolerance} \qquad MO1 = \text{set point} \pm \text{tolerance}$$

As shown in Table 8.1, this information is represented at the functional purpose level of abstraction and the system level of decomposition in the work domain analysis reported in the previous chapter (see Fig. 7.12). Because the set point values must be given up front, the actor will begin the start-up mode with knowledge of the goal state to be achieved. Accordingly, the path through the decision ladder begins with the *goal state* circle (see Fig. 8.9).

*2. **Goal State Iteration.*** The overall goals just described cannot be achieved directly, but must instead be decomposed into subgoals and sub-subgoals (see Table 8.1). At the first level of recursion, the temperature goal can be achieved only by first establishing a mass inventory and an energy inventory. Similarly, the demand goals can be achieved only by first establishing a mass inventory. At the second level of recursion, to establish an energy inventory, it is first necessary to establish a flow of input energy. Similarly, to establish a mass inventory, it is first necessary to establish a flow of input mass.

These shortcuts steps are shown as associative leaps recursively iterated on the *goal state* node in the decision ladder (see Fig. 8.9). The rationale for this classification is that there are no degrees of freedom associated with this aspect of start-up, and thus no need for information processing. Thus, a skilled actor could automatically associate each goal with the one below it in the goal hierarchy.

Interestingly, this iteration of goal states can be mapped onto the abstraction-decomposition space developed in the previous chapter. As shown in Table 8.1, each step is equivalent to traversing one or more links in the work domain representation. This mapping shows, in more concrete form, the relationship between work analysis and control task analysis that was described abstractly earlier in this chapter (see Fig. 8.6).

*3. **Define Task.*** Given the goals just described, the next step is for actors to define the task that needs to be accomplished to achieve each goal. As shown in Table 8.1, for the establish energy input goal, this step involves choosing a desired heat transfer rate. The goal in doing so is to create an inflow of energy, thereby establishing an inventory of energy, and thereby moving the temperature toward the desired set point. For the mass input goal, this step involves choosing a desired input flow rate of water. The goal in doing so is to create an inflow of mass, thereby establishing an inventory of mass, and thereby moving the output demand toward the desired setpoint. The desired information (i.e., the desired heat transfer and flow rate values) can be found at the generalized function level of abstraction and the subsystem level of decomposition in the work domain representation (see Table 8.1).

These two information-processing steps are shown as *define task* boxes in the decision ladder in Fig. 8.9. The rationale for this classification is that an actor must select an appropriate change in the state of the work domain (i.e., which particular heat transfer and water input flow rate values to adopt). These values must be chosen, at least implicitly, before proceeding further. In addition to knowledge of the set point values described previously, actors must also take into account the following constraints when performing this information-processing activity:

- Conservation of mass in reservoir:

$$\frac{dV1(t)}{dt} = \frac{MI1(t) - MO1(t)}{\rho}.$$

- Conservation of energy in reservoir:

$$\frac{dT1(t)}{dt} = \frac{FH1(t) - [MI1(t)] \, [T1(t) - T_1] \, c_p}{\rho}.$$

The relationships in these equations must be taken into account because they show how the choice of heat transfer (FH1) and water input flow (MI1) values affect the state of the higher order goal variables (T1 and MO1).

The linkages from goal state to define task can also be mapped onto the abstraction-decomposition space developed in the previous chapter. As shown in Table 8.1, each step is equivalent to traversing a link in the work domain representation.

4. Task. The outputs of the two define task activities are two corresponding states of knowledge, represented as *task* circles in the decision ladder in Fig. 8.9. More specifically, the actor should have identified a desired heat transfer rate (FH1) as well as a desired water input flow rate (MI1). This information can be found at the generalized function level of abstraction and the subsystem level of decomposition (see Table 8.1).

5. Formulate Procedure. Given these desired values, an actor must then derive a plan of action that will achieve those values. As shown in Table 8.1, this step involves configuring the heater and configuring the FWS. Choosing an appropriate heater setting will achieve the desired heat transfer rate. Analogously, choosing an appropriate set of valve and pump settings will achieve the desired water input flow rate.

These information-processing activities are shown as *formulate procedure* boxes in the decision ladder in Fig. 8.9. The rationale for this classification is that an actor must plan a sequence of actions (i.e., which component settings to adopt). These settings must be chosen, at least implicitly, before proceeding any further.

When determining the heater configuration, actors must take into account the following constraint.

- Conservation of energy from heater:

$$FH1(t) = HTR1(t).$$

The relationship in this equation must be taken into account because it shows how the choice of heater setting (HTR1) value affects the desired heat transfer (FH1) value.

The FWS configuration is more complex. For this information-processing activity, actors must take into account quite a few constraints:

- Conservation of mass from two feedwater streams:

$$MI1(t) = FA1(t) + FB1(t).$$

- Conservation of mass in feedwater stream:

$$FVA(t) = FA1(t) + FA2(t).$$

- Flow split relation:

$$FA1(t) = \frac{FVA(t)VA1(t)}{VA1(t)+VA2(t)}.$$

- Flow through pump:

> If pump is OFF then FPA(t) = 0,
> otherwise:
> IF [VA1(t) + VA2(t)] > VA(t) THEN
>> FPA(t) = FVA(t)
> ELSE
>> FPA(t) = FVA1(t) + FVA2(t)

- Conservation of mass in pipe:

$$FVA(t) = FPA(t)$$

The relationships in these equations must somehow be taken into account because they show how the choice of component settings (PA, VA, VA1, VA2) values affect the desired water input flow (MI1) value.

There are some sequential constraints that must also be taken into account in the formulation of an appropriate procedure. First, pumps should not be turned on before valves, otherwise there will be no downstream path for the water to follow and the pumps may be damaged. Second, the configure heater step should not be initiated before the configure FWS step, otherwise an empty reservoir will be heated and the reservoir vessel may be damaged. These constraints rule out many action sequences, but they do not uniquely specify a single correct sequence (see chap. 3).

The links between task and formulate procedure can also be mapped onto the abstraction-decomposition space developed in the previous chapter. As shown in Table 8.1, each step is equivalent to traversing several links in the work domain representation.

6. Procedure. The output of the two formulate procedure activities is a state of knowledge, represented as a *procedure* circle in the decision ladder in Fig. 8.9. At this point, the actor will have identified a desired set of component settings organized into an order that does not violate the sequential constraints described earlier (e.g., open the valves, turn on the pumps, ensure that water is going into the reservoirs, and then turn on the heaters).

Although there are two tasks being performed, only one procedure is identified because the plans associated with each task are interrelated by the sequential constraints described previously. Thus, it is useful to consider the two plans together, rather than independently. In this way, the requirements associated with expert action can be identified.

7. Execute. Given an appropriate procedure, actors will have to carry out the changes in component settings (see Table 8.1). This information-processing activity is shown as an *execute* box in the decision ladder in Fig. 8.9. The rationale for this classification is that an actor must coordinate a sequence of manipulations (i.e., actions on the components). These actions must be implemented to begin the start-up task. During this step, actors must know the location of the components. The link between procedure and execution can be mapped onto the abstraction-decomposition space. As shown in Table 8.1, this step is equivalent to traversing means–ends links in the work domain representation. And as shown in Fig. 8.9, this information can be found in the abstraction-decomposition space at the physical form level of abstraction and the component level of decomposition.

8. Alert. After, or even during, the execution of the changes in component settings, it is important for the actor to be in a state of vigilance, being ready to monitor the state of the work domain to determine when it is time to stabilize its state. Otherwise, the influx of water may result in an overflow of the reservoir, and the influx of energy may result in an overshoot of the temperature goal. This step is represented as a shunt between the execute information-processing activity and the alert state of knowledge in the decision ladder in Fig. 8.9. The rationale for this categorization is that the actor must be ready to determine if it is time to stabilize the work domain state.

9. Observe. The next relevant step in the start-up mode is to observe the approach of the goal variables to their respective desired values to determine the appropriate time to act to stabilize the work domain state (see Table 8.1). There are actually two related steps, one associated with determining when to act to stabilize the temperature goal and the other associated with determining when to act to stabilize the output demand goal. These two steps are considered together because, as we see shortly, they are not independent of each other.

This information-processing activity is coded as an *observe* box in the decision ladder in Fig. 8.9. The rationale for this classification is that an actor must gather information about the current state of the work domain to determine when actions are necessary to stabilize the work domain within the desired set point ranges. The temperature and output demand goals are addressed in turn.

For the temperature goal, it should be clear that the current reservoir temperature (T1) and its set point value must be monitored. After all, the difference between these two variables specifies how far away the work domain temperature is from where it should be. But although relevant, this information is actually only of secondary importance. The primary information that is needed to determine when it is time to act to stabilize the work domain state is the *time to contact* (TTC) the goal region (Lee, 1976; Yu et al., 1997). This higher order variable corresponds to the rate of change of the difference between the current and set point temperatures. This variable tells actors how quickly the gap to the goal state is closing. More formally, the TTC can be calculated as follows:

$$TTC = \frac{d(T1goal - T1current)}{dt}.$$

Although this formula describes the TTC, it does not tell an actor exactly when to act. Should action be initiated when TTC is 10 s, 5 s, 1 s? In part, the answer to this question should be determined based on knowledge of the time constants of the components in DURESS II (see chap. 6). Because of the lags in the work domain, it is important that action be taken before TTC = 0 s. Otherwise, by the time the actor's actions have their effect, the temperature will already have overshot the goal region. This constraint rules out certain values of TTC, but it does not identify a single ideal value. A conservative actor may decide to act early, whereas a risky actor may decide to act at the last possible second. In any case, however, TTC is the key variable to monitor and knowledge of the time constants is desirable for expert action.

As for the output demand goal, it is unusual for the following reasons. There is only one component that can be used to directly achieve the output demand goal. Specifically, the output valve (VO1) must be set so that the flow out of the reservoir (MO1) is within the demand set point region. Thus, there are no structural degrees of freedom, as far as this action is concerned. The complication is that putting the output valve at the appropriate setting is necessary but not sufficient to achieve the demand goal. There must also be enough water stored in the reservoir (V1) or enough flow rate being supplied into the reservoir (MI1) to satisfy the demand goal. For example, if the output valve is open to the appropriate setting but there is no volume and no inflow, then the output flow rate will be zero and the goal will not be achieved. Thus, there must be an inventory of volume (V1) and an influx of water (MO1) to stabilize the reservoir at the output demand goal. The interesting point, however, is that the choice of how much water to store in the reservoir is left completely to the actor. The volume can be kept at a low, intermediate, or high level, with different implications for a number of factors including: the time it takes to reach steady state, the likelihood of overflowing the reservoir, and the likelihood of emptying the reservoir (see also Pawlak & Vicente, 1996).

What does this tell us about the *observe* box in Fig. 8.9? The main implication is that, when monitoring the work domain to determine when it is time to act to stabilize the output demand goal, the actor must choose a preferred volume level (V1). This level must be greater than zero but less than the capacity of the reservoir. Given this choice, the key variable to monitor is then the time to contact this preferred level. This information will tell the actor when it is time to act on the work domain to stabilize the volume in the preferred region, thereby creating the conditions that are necessary to satisfy the output demand goal. As with the case of temperature, there are several constraints to take into account (e.g., the time constant on the valves, reservoir capacity) but no single ideal value.

Note that volume (V1) is pertinent to both goals because adding cool water to achieve the output demand goal not only changes volume, but it may also affect temperature as well. In contrast, adding energy by manipulating the heater will change the temperature, but has no effect on volume. Thus, there is an asymmetrical interaction between the means that can be used to satisfy each of these higher order goals. This structural feature of DURESS II was pointed out in the work domain analysis presented in chapter 7. We can use our knowledge of this property to derive implications for control tasks. Specifically, the volume should be stabilized first. This will allow an actor to satisfy the output demand. Once the volume is fixed, then it becomes much easier to stabilize temperature. After all, if volume continues to change, then the actor will continually have to compensate for the resulting changes in temperature. As shown in Table 8.1, the variables pertinent to this *observe* box can be found in the *functional purpose/system* cell of the abstraction-decomposition space.

10. Task. The insights just described allow a skilled actor to go directly from observing the state of the work domain to knowing what task to perform. For example, when the volume is getting close to the preferred level, then it is time to act on the work domain to stabilize the volume and satisfy demand. Similarly, when the temperature is getting close to the set point value, it will be time to act on the work domain to stabilize the temperature in the set point region. These two shunts allow a skilled actor to go directly from observing the state of the work domain to knowing what tasks need to be performed. These shortcuts are represented as a shunt between the *observe* box and the *task* circle in Fig. 8.9. The rationale for this categorization is that intimate knowledge of the properties of the work domain and the control tasks to be accomplished allow a skilled worker to perceive the state of the process in terms of the task to be accomplished next. As shown in Table 8.1, the variables pertinent to this task can be found in the abstraction-decomposition space at the levels of functional purpose/system.

11. Formulate Procedure. Given knowledge of the task to be performed, an actor must then determine what actions need to be performed. As shown in Table 8.1, this corresponds to identifying a procedure that will stabilize the output demand in the goal region and a procedure that will stabilize the temperature in the goal region. These activities are represented in the *formulate procedure* box in Fig. 8.9. The rationale for this categorization is that each of these information-processing activities involves planning a sequence of actions to complete a task. Each of the activities is addressed in turn.

First, to satisfy the output demand goal, the actor must make sure that the output valve (VO1) is set so that the output flow rate (MO1) is in the goal region. Not only is it important to achieve the output demand goal, it is also important to stabilize the work domain state in that goal region. To make sure that the output demand goal remains satisfied, the volume should be stabilized. To determine how this can be accomplished, we can examine the following equation:

• Conservation of mass in reservoir:

$$\frac{dV1(t)}{dt} = \frac{MI1(t) - MO1(t).}{\rho}$$

Stabilizing volume means having $dV1(t)/dt = 0$. To achieve this state, the actor must make sure that the FWS valves are set so that the input flow rate (MI1) equals the output flow rate (MO1). This, in turn, requires knowledge of the same constraints that were described in Step 5 earlier (i.e., the configure FWS task).

Second, to satisfy the temperature goal, the actor must make sure that the heater (HTR1) is set so that the temperature (T1) rests in the goal region. To determine how this can be accomplished, we can examine the following equation:

• Relationship between energy, volume, and temperature:

$$T1(t) = \frac{E1(t)}{V1(t)\, c_p\, \rho}.$$

We can see that the requirement to satisfy the temperature goal is equivalent to a requirement to keep T1 constant. Because c_p and ρ are both constants, this equation tells us that keeping T1 constant is equivalent to maintaining a constant ratio of energy inventory (E1) to volume (V1). But because the volume should be kept stable to satisfy the output demand goal (see previous discussion), the volume (V1) will be a constant. That means that the amount of energy in the reservoir (E1) is uniquely specified, once the temperature set point (T1) is given and once the preferred reservoir volume (V1) is chosen. In other words, to achieve the temperature setpoint, an actor must choose an energy level such that the following constraint is satisfied:

$$E1(t) = T1_{goal}(t)\, V1_{preferred}(t)\, c_p\, \rho.$$

Not only is it important to achieve the temperature goal, it is also important to stabilize the work domain state in that goal region. To determine how this can be accomplished, we can examine the following equation:

• Conservation of energy in reservoir:

$$\frac{dT1(t)}{dt} = \frac{FH1(t) - [MI1(t)]\, [T1(t) - T_I]\, c_p}{V1(t)\, c_p\rho}.$$

Stabilizing temperature means having $dT1(t)/dt = 0$. To achieve this state, the following must be true:

$$FH1(t) = [MI1(t)]\, [T1(t) - T_I]\, c_p.$$

But we already know that the reservoir temperature (T1) is fixed because it has to match the temperature goal value. Similarly, the mass input flow rate (MI1) is fixed because it has to match the output demand goal (MO1). Furthermore, T_I (the inlet water temperature) and c_p are both constants (assuming no disturbances). Consequently, the heat transfer rate from the heater (FH1) is uniquely specified. Not only that, but the heater setting itself (HTR1) is also uniquely specified by the following constraint:

• Conservation of energy from heater:

$$FH1(t) = HTR1(t).$$

Therefore, to stabilize the temperature in the demand region, the actor has to choose the heater setting that will satisfy the constraints just described.

Although it may not be obvious, the intricate reasoning path we have gone through in this subsection can actually be mapped onto the links and representations identified in the work domain representation for DURESS II (see Table 8.1).

12. Procedure. The output of the information-processing activities just described is a sequence of specific actions comprising the valve and heater settings that will achieve the aforementioned task. For reasons that should be self-evident, this step is coded as a procedure state of knowledge in the decision ladder in Fig. 8.9. The information itself (i.e., the component settings) can be found at the physical function level of abstraction and the component level of decomposition (see Table 8.1).

13. Execute. The final step in the start-up sequence involves performing the aforementioned actions to stabilize the work domain in the desired goal states. As mentioned earlier, the actions associated with the demand goal should be performed first to stabilize volume. Then, the heater can be manipulated to stabilize temperature. This information-processing step is coded as an execute activity in the decision ladder in Fig. 8.9. The reason for this categorization is that this activity requires the coordinated manipulation of component settings.

Note that timing is very important in this step. The actor must have and use knowledge of the time constants governing the behavior of the components in deciding when to act. For example, to stabilize temperature, it is important that the heater be manipulated before the upper boundary of the temperature set point is reached. Otherwise, the temperature will overshoot the goal region because of the considerable lag between the time the heater is manipulated and the time it has its effect on the temperature.

Once this step is completed, the actor would be finished with the start-up operating mode, and would enter into the normal operations operating mode (see previous discussion). This new operating mode would entail a new set of constraints that could be identified by a control task analysis analogous to the one that we have presented here.

Discussion: Highlighting the Key Points

This control task analysis for the start-up operating mode for DURESS II allows us to highlight, in a more tangible way, some of the abstract points we made earlier in the book.

Constraint Based. This analysis is constraint based because it merely identifies the constraints on control rather than specifying an idealized flow sequence or a specific timeline trajectory (see chap. 3). You may have noticed this by the absence of a specific set, sequence, or timeline of actions. For example, nowhere did we state that component X should be put on setting Y at time Z. Different actors may choose to adopt different ways of achieving the required tasks. For example, some actors may want to set the heater at its maximum value to start up the process in a fast, aggressive manner. Others may want to set the heater at a lower value to start up the process in a slower, conservative manner. The control task analysis described previously accommodates either of these approaches, and many others as well. At the same time, it provides a great deal of guidance by specifying what constraints must be dealt with to achieve the control tasks reliably and effectively.

The constraint-based nature of control task analysis is illustrated more clearly by Fig. 8.10. Each of the information-processing activities identified in the control task for the start-up mode for DURESS II is listed. By depicting these activities in this way, we can see that each information-processing activity represents a function that transforms a given set of inputs into a given set of outputs according to a specific

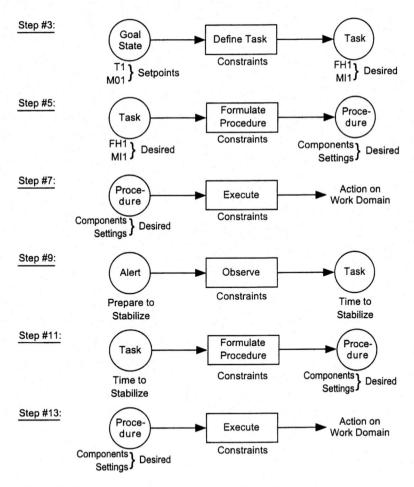

FIG. 8.10. The input–output transformations identified in this control task analysis for the start-up mode for DURESS II. Compare with Table 8.1 and Fig. 8.9. See text for description of constraints.

set of constraints (not listed in the figure for the sake of clarity, but described earlier in the text). These transformations are the requirements that a skilled actor must satisfy, and these constraints are the knowledge that an expert actor must exhibit. This form satisfies the definition of input–output task analysis provided in chapter 3.

An important property of these functions is that they are usually underspecified. That is, they do not usually lead to a unique solution. For example, in Step 3, knowledge of the set points does not result in unique desired values for the heat transfer rate and the mass input flow rate. The constraints rule out some values, but many are still feasible. Following the logic of the constraint-based approach (see chap. 5), the analysis merely identified the degrees of freedom that are available to the actor. By not dictating how work should be done, the control task analysis provides an envelope within which actors can operate. This envelope supports the goal-directed, adaptive flexibility that is the hallmark of context-conditioned variability. And as we mentioned in chapter 3, this type of flexibility is essential to deal effectively with the demands imposed by open systems.

Opportunistic. This analysis also clearly shows the opportunistic aspect of the decision ladder. Several of the information-processing activities and states of knowledge were not used because they were not relevant to this operating mode. Also, the analysis started on the right side of the ladder rather than on the left side because the goal state was given up front. Furthermore, shortcuts were identified wherever feasible.

The opportunistic character of this analysis goes even further, however. Note that the analysis is not intended to imply that an actor will closely follow the 13-step sequence in Fig. 8.9 when conducting a start-up. Actual performance is much messier than this sequence would indicate. It involves backtracking to fix mistakes, iteration to fine-tune the process, and other nonlinear activities. The important point, however, is that an actor will have to deal with all of the requirements identified by the analysis somewhere along the way. Thus, the control task analysis identifies the requirements that must be met by a competent actor, while not only acknowledging, but accommodating, the opportunistic and idiosyncratic nature of actual practice.

Just a Template, but Still Very Useful. This case study should also have made it clear that the decision ladder is just a template. All of the content had to be obtained from sources other than the ladder. Nevertheless, the structure of the ladder provides a very useful basis for conducting a control task analysis. Its constructs provide a systematic basis for parsing information-processing activities and states of knowledge. These constructs also provide an orderly basis for doing a type of conceptual "bookkeeping" to make sure that the analysis is being performed in an orderly manner.[6] Thus, although far from providing all the answers, the decision ladder plays a valuable role as a modeling tool in control task analysis.

---

[6]As with other phases of CWA, the process of conducting a control task analysis could be much easier if appropriate support systems were available to help the analyst. For example, it would not be at all surprising to find that the analysis presented here is not as explicit or as consistent as it should be. Formalizing the elements of control task analysis and using such a formalism to design a computer-based support system could ensure the consistency and rigor of the analysis, as well as making it more efficient to conduct.

"What," Not "How" or "Who". You may have noticed that throughout the analysis, we kept on referring to a hypothetical actor rather than a worker, automation, or some hybrid architecture involving both. The reason for this can be found in the rationale for CWA discussed in chapter 5. The control task analysis merely tries to identify the requirements that must be satisfied by actor(s). The question of which of these requirements should be allocated to automation or workers is dealt with only in a later phase (see chap. 10). The control task analysis does not commit to any one particular scheme.

Similarly, throughout our analysis, we did not make any reference to particular strategies for accomplishing the control tasks. This approach is in keeping with the black box nature of control task analysis modeling. As we see in the next chapter, some control tasks can be accomplished using more than one strategy. However, the question of how tasks can be done should be dealt with only after the question of what tasks need to be done has been thoroughly addressed, as it has been in this chapter.

Formative. Although the decision ladder can also be used in a descriptive fashion (see Figs. 8.7 and 8.8), our partial control task analysis for DURESS II clearly illustrates the formative orientation of CWA. Wherever possible, we tried to exploit knowledge of the work domain and control task constraints to identify efficient, yet reliable, paths through the decision ladder. By squeezing as much as possible out of these constraints, we can identify a set of information requirements that should lead to the design of a sociotechnical system that explicitly supports such expert performance (see the examples in the next section), while retaining flexibility (see previous discussion).

Action Means–Ends, Not Structural Means–Ends. In chapter 7, we introduced a distinction between action means–ends and structural means–ends relations. The former describe the activity of an actor, whereas the latter describe the work domain being acted on. Accordingly, we noted that structural means–ends relations are appropriate for conducting a work domain analysis, whereas action means–ends relations are inappropriate because they do not describe the work domain.

Now that we are conducting a control task analysis, we are interested in identifying the requirements associated with effective control. Consequently, action means–ends relations become relevant. This change in concepts can be seen in Table 8.1. Rather than describing work domain objects and functions, we are now describing what needs to be done with those objects and functions. For example, whereas in the work domain analysis we determined that DURESS II has an FWS as a structural means, in Table 8.1 we noted that the act of configuring the FWS is an action means for achieving the desired input flow rate. Similarly, whereas in the work domain analysis we determined that DURESS II has a heater as a structural means, in Table 8.1 we noted that the act of changing the heater setting is an action means for achieving the desired heat transfer rate.

Not only do these examples illustrate the change from structural means–ends descriptions to action means–ends descriptions, they also illustrate the relationship between these two kinds of descriptions. Structural means–ends descriptions identify the structure of the work domain being acted on, whereas action means–ends descriptions identify how that work domain could be acted on in a particular class of situations. For example, the heater is a structural means. In some situations, it may be necessary to increase the heater setting (e.g., to start up DURESS II), in others it may be necessary to decrease the heater setting (e.g., to avoid overshooting the temperature goal), and in others still it may be necessary to put the heater setting on a particular value (e.g., to stabilize the temperature in the goal region). We can see from this example that action means descriptions are more detailed than structural means descriptions because they describe what must be done with a particular component or function for a particular context, not just that that component or function exists independent of any particular context.

Conversely, we can also see the unique and important value added of work domain descriptions. There are a bounded but infinite number of actions that can be taken on any one component. In open systems, it is impossible to specify precisely what action should be taken for every possible situation. The work domain description gets around this problem by merely specifying the structural means that are available. For example, instead of trying to specify what setting the heater should be set at for all possible situations, the work domain representation tells actors that they have a heater available for each reservoir and that that heater has the function of serving as a heat source. This field description accommodates all possible actions on the heater, although leaving it up to the actor to determine what particular setting to adopt on any one occasion, including events that are not foreseen by the designer. Because the work domain description describes what is possible, not what should be done, it provides a basis of coping with the unanticipated. We hope this discussion has clearly illustrated the complementary nature of structural and action means–ends relations and descriptions.

Relationship to Work Domain Analysis. The complementary nature of structural and action means–ends descriptions highlights perhaps the most important general point in this chapter, namely the complementary relationship that exists between work domain analysis and control task analysis. As Table 8.1 shows, a control task analysis gains a great deal by leveraging the knowledge gathered during a work domain analysis. The abstraction-decomposition representation provides us with knowledge of what functions are required to achieve particular purposes. It also provides us with knowledge of what components can be used to achieve particular functions. As a result, we can systematically determine which variables and relations are relevant for a particular task goal and which can be ignored. Furthermore, the relationships identified in a work domain analysis provide an invaluable source of insight that can be used to identify possible shortcuts or interactions that an actor must consider when performing particular control tasks.

At the same time, the control task analysis adds a great deal of information that cannot be found explicitly anywhere in the work domain analysis. A simple example is information about sequence constraints (e.g., pumps should not be turned on before valves are open). A more complex example is information about which variables and relations in the work domain representation are relevant for a particular task goal. For example, to determine when it is time to act to stabilize temperature, it is sufficient to monitor the time to contact the temperature goal.

The detailed case study we have presented in this chapter should have shown you, in a clear and convincing way, that a control task analysis inherits the layer of constraints identified in a work domain analysis, while simultaneously adding a new nested layer of constraint (see chap. 5, especially Fig. 5.3).

SUMMARY AND IMPLICATIONS

A well-designed sociotechnical system must be based, not just on knowledge of the possibilities offered by the work domain, but also on the requirements that are associated with proficient control. The second phase of CWA, control task analysis, is intended to identify the latter set of requirements. The decision ladder is a modeling tool that can be effectively used for this phase of analysis. Because of its shortcuts, it provides a template that can be used to determine only those activities that are required to exhibit expert action, thereby fostering and supporting efficient and flexible performance. Furthermore, because of its constructs, the decision ladder provides a useful template for breaking down an operating mode into its constituent states of knowledge and information-processing activities.

What are the implications of control task analysis for design? The answer to this question can be obtained by examining Fig. 8.11. Whereas the previous chapter dealt with the first row of this figure, this chapter has dealt with the second row. The concept of a control task can be analyzed using the decision ladder as a modeling tool. We have used this tool to develop a partial control task analysis for our case study. This representation can be used to design procedures, automation, and context-sensitive interfaces. For example, the fact that the pumps should not be turned on before the FWS valves are opened could be built into a procedure that would constrain, but not dictate, workers' actions. Building this knowledge into systems design would make it easier for workers to exhibit expert performance because they would not have to derive this sequential constraint on their own in an ad hoc manner. To take another example, the relationships between FWS valve settings and the resulting flow into the reservoir could be used to design a computer-based aid that would suggest a viable set of component settings, once a worker selects a desired input flow rate. The relationship between the information requirements identified by the control task analysis and the design of automation is discussed in more detail in chapter 10. Finally, there are also several examples of how the results of the control task analysis could be used to design context-sensitive interfaces. To take but one example, the results of the analysis of the monitoring task in Table 8.1 suggest that it might be useful to design a task-specific display to present the time to contact the temperature directly. By collecting together only the information that is required to perform the monitoring task listed in Table 8.1, and by preproc-

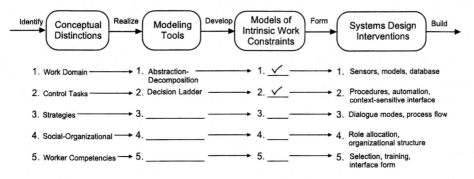

FIG. 8.11. An outline of the CWA framework with the first two phases (i.e., rows) filled in.

essing lower order data to present the required higher order information directly, the monitoring task could be facilitated. Workers could achieve expert performance without having to know and remember what data need to be collected together for this task, what data have to be filtered or ignored, and how the relevant data should be integrated to obtain the higher order information of interest (cf. Woods, 1991). Collectively, these examples show how a control task analysis can identify the requirements associated with expert action. More generally, the examples show the tight connection between CWA and design that is required for analysis to be a worthwhile endeavor.

To conclude, this second phase of CWA is important because it provides support for known classes of tasks in a proficient manner. Nevertheless, as we have mentioned several times already, there are other sources of constraint that need to be considered in CWA. Specifically, a control task analysis does not identify how a particular task can be performed. Furthermore, it also does not address the issue of who could be responsible for different information-processing activities that were identified. The questions of "how" and "who" are addressed in the next two chapters.

Phase 3: Strategies Analysis 9

Heuristic strategies can be highly accurate in some environments, but no single heuristic does well across all contexts. This suggests that if a decision maker wanted to achieve both a reasonably high level of accuracy and low effort, he or she should have to use a repertoire of strategies, with selection contingent upon situational demands.

—J. W. Payne et al. (1993, p. 131)

To support system design we do not need detailed process models of the mental activities which are used by the operators. System design must be based upon higher level models of the structures of effective mental processes which the operators can use and their characteristics with respect to human limitations and preferences so that operators can adapt individually and develop effective strategies. Rather than descriptions of the course and content of actual mental processes, we need descriptions of the structure of possible and effective mental processes.

—Rasmussen (1981, p. 242)

PURPOSE

Our objective in this chapter is to describe the third phase of CWA, strategies analysis, in full detail. As we discussed in chapter 5, the value added by strategies analysis lies in understanding the different ways of accomplishing the activities identified in a control task analysis. Thus, whereas the previous chapter dealt with the question of *what* needs to be done, this chapter deals with the question of *how* it can be done. We describe *information flow maps* (Rasmussen, 1980, 1981) as a modeling tool that can be used to conduct such an analysis. A detailed example of a strategies analysis is then presented for the DURESS II process control microworld. By the end of this chapter, you should understand how strategy constraints inherit, and build on, the work domain and control task constraints discussed in the previous two chapters.

STRATEGIES ANALYSIS: WHAT IS IT AND WHY IS IT IMPORTANT?

We begin by showing the relationship between strategies analysis and control task analysis. Then, we use an example from air traffic control to show why workers often switch between several strategies to cope with task demands. This intuition-building example is followed by a generic definition of the term *strategy*. This definition helps to illustrate the importance of strategies analysis in CWA.

a) Control Task Analysis:

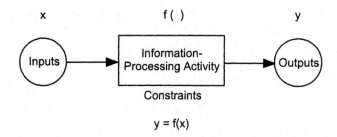

Constraints

$$y = f(x)$$

b) Strategies Analysis:

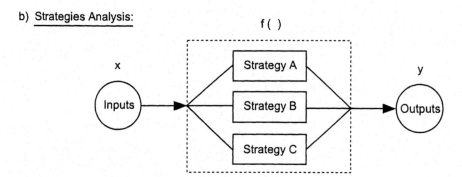

FIG. 9.1. Comparison of control task and strategies analyses. Compare with Fig. 8.10.

Relationship Between Control Task Analysis and Strategies Analysis

Control task analysis and strategy analysis both deal with aspects of action on the work domain (rather than with the work domain itself). However, whereas a control task analysis results in a *product* description of *what* needs to be done, a strategy analysis results in a *process* description of *how* it can be done. The difference between these two kinds of models can be explained with reference to Fig. 9.1 (cf. Luce, 1995). A generic version of one of the boxes in the decision ladder (see chap. 8) is presented in Fig. 9.1a. Connected to the box are two circles representing two states of knowledge, the one on the left being the set of inputs (x) into the activity and the one on the right being the set of outputs (y) of the activity. Viewed in this light, each of the activities identified in the control task analysis (see Fig. 8.10 for a summary) represents a transformation that can be thought of as a function.[1] Accordingly, the mapping of inputs to outputs can be represented as follows:

$$y = f(x).$$

---

[1]Strictly speaking, the use of the term *function* is not mathematically accurate. There can be more than one set of outputs for any one set of inputs, because the constraints associated with an information-processing activity usually underspecify the output.

One of the primary objectives of a control task analysis is to identify the constraints associated with this function. As such, a control task representation is a product, or "black box," model because it focuses on identifying the input–output transformation associated with a particular information-processing activity. No attention is given to the particulars inside the black box (i.e., the process by which the input–output transformation is achieved).

As we tried to show in the previous chapter, many valuable insights can be gained from such a model. However, a control task analysis is limited because it tells us nothing about how the transformation can be achieved. This is a very important point when we consider that, in complex sociotechnical systems, actors are usually faced with multiple degrees of freedom. In other words, there is more than one way to accomplish a given task. These redundant degrees of freedom lead to the possibility of several qualitatively different strategies for accomplishing a single task (Bedny & Meister, 1997; Bruner, Goodnow, & Austin, 1956; Kieras & Meyer, 1998; J. W. Payne et al., 1993; Pejtersen, 1973, 1979, 1988; Rasmussen & Jensen, 1973, 1974; Sperandio, 1978; Venda & Venda, 1995). In other words, the same transformation $y = f(x)$ can be implemented or accomplished in different ways, as shown in Fig. 9.1b. If we can develop a process model for each of the viable and effective strategies, then we will be in a position to design a computer-based system that can provide support tailored to each strategy. A strategies analysis tries to fulfill this objective by identifying the requirements associated with different ways of accomplishing task goals. In other words, a strategies analysis involves going inside the black box to find out what strategies are possible for a given information-processing activity (see Fig. 9.1b).

The Need for Multiple Strategies: An Example From Air Traffic Control

To see why workers usually adopt different strategies to perform any one task, we present a concrete example from the domain of air traffic control. There are certain critical work domain and control task constraints that should always be respected if air traffic controllers (ATCs) are to complete task goals consistently. For example, some of the work domain constraints include the geographical characteristics of the area in and around the airport (e.g., mountains, skyscrapers, bodies of water, length and location of the runways) and the characteristics of the different types of aircraft that need to be controlled (e.g., limits on speed, braking, and maneuverability). One of the more obvious control task constraints is that the physical separation between aircraft in the same airspace must be taken into account. Only by consistently respecting these and other work domain and control task constraints will ATCs be able to perform their job reliably.

Nevertheless, there are still several different strategies that ATCs can, and do, adopt to perform their job while still respecting the aforementioned constraints. Evidence supporting this claim can be found in the insightful work of Sperandio (1978), who reported a number of interesting findings obtained from 10 years of studying ATCs in the field. He found that, for a given situation and a given controller, certain operating procedures are more economic (i.e., require less of a cognitive burden) than others. Moreover, as work load increased (e.g., the number of aircraft to be controlled went up), ATCs would spontaneously relax the performance criteria they were using to perform the task. By doing so, they could adopt a qualitatively different strategy that accomplished the task goals in a less effortful fashion.

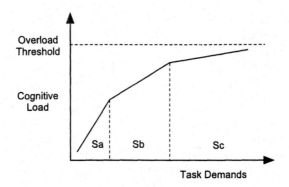

FIG. 9.2. Adaptive regulation of cognitive load via the use of successively less demanding strategies (S_a, S_b, S_c) by air traffic controllers (adapted from Sperandio, 1978).

Figure 9.2 illustrates these findings graphically. As Sperandio (1978) pointed out: "One notices a whole sequence of progressive changes in operating methods, corresponding to different strategies adopted successively by the controllers to avoid crossing the overload threshold and delay disfunctioning" (p. 196). At one particular airport, Sperandio observed the following set of strategies:

1. S_a—When there were about one to three airplanes to control, the ATCs would "nurse" each plane. Because the task demands were relatively low, they would calculate the optimal flight path for each individual plane, based on a number of variables (e.g., speed, course, altitude, and type of aircraft).
2. S_b—When there were about four to six airplanes to control, the ATCs would adopt a less sophisticated, and thus, cognitively more economical strategy. Instead of calculating an optimal path for each airplane, they would adopt uniform speeds and stereotypical flight paths. By relaxing their performance criteria and adopting less effortful strategies, ATCs were able to control a larger number of planes in a manageable fashion.
3. S_c—When there were more than six airplanes to control, the ATCs would create waiting "buffers" that consisted of streams of aircraft. When they were ready, the ATCs would then bring an aircraft off of the buffer and toward the runway at a generally uniform speed and descent path. Because the load is so great with this many aircraft, the performance criteria are once again relaxed so that the only primary concern is safety rather than efficiency.

There are obvious process differences between these three strategies. However, in each case, the control task being performed is the same—land the planes safely.

We might think that these findings suggest that a work analysis should identify the requirements associated with each of these strategies so that contextualized support can be offered for each. For example, the ATCs' displays could be automatically reconfigured as a function of the number of aircraft. In this way, the information needed for each strategy could be provided in a tailored manner, thereby reducing the demands on ATCs. Although this may seem like a good idea, there are a number of strong reasons to believe that such a design would be ineffective. As Sperandio (1978) pointed out, "changes of strategy are not clear-cut and sudden" (p. 199). Instead, they are statistical tendencies that can only be observed over many occasions. Thus, there is no magical number of aircraft that will always lead to a switch in strategy.

There seem to be several reasons for this. First, the point at which the statistical tendency to switch strategies occurs varies according to contextual factors that are hard to anticipate. Even if the number of aircraft is held constant, some situations impose a greater burden on ATCs and are thus more likely to lead to a strategy shift than other situations. Second, the point at which the statistical tendency to switch strategies occurs also varies according to the characteristics (e.g., age, health, motivation, experience) of individual ATCs. Moreover, the same controller may exhibit different tendencies on different days. For example, if a controller happens to be particularly tired or distracted one day, then she may wish to switch to a more economic strategy earlier (i.e., with a lower number of aircraft) than on another day. For all of these reasons, it is very difficult, if not impossible, to reliably predict exactly under what conditions a strategy shift is warranted. Therefore, although it may be possible for a work analysis to identify the range of strategies that ATCs might wish to engage in, it will generally not be possible to determine when a particular controller will want to shift from one particular strategy to another. As Suchman (1987) observed, at a detailed level "every instance of meaningful action must be accounted for separately, with respect to specific, local, contingent determinants of significance" (p. 67).

We can take away several general lessons from this example. First, workers can switch between several different strategies for achieving a particular task goal. Second, each of these strategies is based on a different set of performance criteria, and requires a different kind of information support. Third, workers tend to switch strategies when their cognitive load is getting too great. Fourth, it is very difficult to predict when a strategy shift will occur in practice. As we see later, there is ample empirical evidence for each of these conclusions from a number of other empirical studies.

What Is a Strategy?

The air traffic control example has provided us with some intuitions about the use of strategies. Given this background, we can now turn to defining the term *strategy* in a more systematic, domain-independent manner.

It is important to point out straight away that the term strategy is defined in many different ways, frequently implicitly, by researchers from a wide variety of disciplines. The definition adopted in CWA is best illustrated by contrast. Consider first the definition given by J. W. Payne et al. (1993) in their monograph on adaptive decision making:

> A decision strategy [is] a sequence of mental and effector (actions on the environment) operations used to transform an initial state of knowledge into a final state of knowledge where the decision maker views the particular decision problem as solved. (p. 9)

This definition appears to be consistent with the formulation we provided in Fig. 9.1 because it refers to an initial state of knowledge and a final state of knowledge. In this view, a strategy is a sequence of operations (i.e., a procedure) that takes us from the former to the latter.

Now consider the following alternative definition: A strategy is a category of cognitive task procedures that transforms an initial state of knowledge into a final state of knowledge (Rasmussen, 1981). This definition is the one we use in CWA, and several examples of its use are provided later. For now, we should merely note the primary difference between this definition and that of J. W. Payne et al. (1993). Rasmussen's definition views a strategy as a *category* of procedures, whereas J. W. Payne et al.'s definition views a strategy as an *instance* of a procedure. Each definition may be useful for different purposes, of course. However, it is important to show the nature of this difference precisely because it will shed light on the role of strategies analysis in CWA.

What is the rationale for defining a strategy as a category rather than an exemplar? The answer provided by Rasmussen (1980) is that:

> Data processes in real-life tasks are extremely situation and person dependent. . . . it is [therefore] very difficult to identify general features of the data processes due to frequent spontaneous shifts between different basic process strategies, shifts which are initiated by minute details in the situation. One cannot see the wood for the trees. (p. 78)

In other words, the detailed cognitive procedures used by workers during any one particular situation are idiosyncratic. As a result, there is a great deal of variability across situations. In part, this variability is an expression of the need for context-conditioned variability in open systems (see chap. 3). In part, it is also due to individual differences in behavior, both across workers within a situation and within a worker across situations. Different people may prefer to perform the same task in different ways. Moreover, even the same person can prefer to perform the same task in different ways on different occasions. Sperandio (1978) observed the same kind of idiosyncrasies in his field study of ATCs. The important implication is that, using J. W. Payne et al.'s (1993) definition, strategies are unstable because the precise sequence of operations used to transform an initial state of knowledge into a final state of knowledge varies across situations, at least in complex sociotechnical systems.

Interestingly, the same phenomenon was observed in J. W. Payne et al.'s (1993) laboratory experiments, even though they were conducted under comparatively less complex conditions. J. W. Payne et al. readily acknowledged that cognitive processing is constructive, opportunistic, and, to some extent, ad hoc in nature. The reasons given for these idiosyncrasies are the same as those given by Rasmussen (1980) and by Sperandio (1978): "Individuals make spur-of-the-moment shifts in processing direction rather than merely executing some strategy determined beforehand" (J. W. Payne et al., 1993, p. 172).

These idiosyncratic details present a significant challenge to work analysis. If we describe a strategy as a detailed procedure that is never actually observed in pure form in practice, and if we then use that description to create a computer-based information system, then we are back to the limitations of the normative approaches to work analysis discussed in chapter 3. The design will support "the one right way," but it will not be flexible enough to accommodate the context-conditioned variability that is required to cope effectively with the disturbances found in open systems.

Defining a strategy as a category of cognitive procedures, rather than as an instance, offers a way out of this problem. To see why, have a look at Fig. 9.3. The view of strategies as categories is illustrated in Fig. 9.3a. Each bin represents an

a) Strategies as categories:

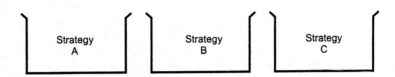

b) Action sequence as switching between instantiated strategies:

c) Timeline of action sequence:

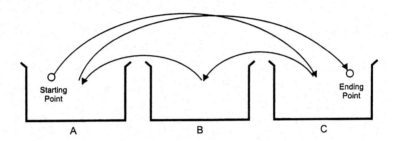

FIG. 9.3. An illustration of the difference between defining a strategy as a category (Rasmussen, 1981) or as an exemplar of an action sequence (Payne et al., 1993).

idealized, abstract description of a strategy, as defined by Rasmussen (1981).[2] Because these descriptions are at the level of a category, they are stable. That is, they subsume all instances (i.e., action sequences) belonging to the respective categories. Each action sequence is an instantiation of a category. Figure 9.3c illustrates a hypothetical action sequence (including cognitive steps) in the form of a timeline. Although only one trajectory is shown, based on the empirical research described previously, we would expect a great deal of variability across trajectories. What specific actions will be taken, in what order, and at what time will be, in part, determined by unforeseeable, idiosyncratic features (e.g., the "spur-of-the-moment shifts in processing direction" observed by J. W. Payne et al., 1993, p. 172). Thus, this detailed level of description is not very useful for design because it is highly variable—it is a moving target (cf. chap. 4).

---

[2]If you are familiar with dynamical systems theory, you can metaphorically think of these strategy bins as attractor basins (Port & van Gelder, 1995). It would be interesting to pursue this connection empirically, because the result may turn out to be far more than a metaphorical link.

However, if we have a description of strategies as categories (i.e., the bins in Fig. 9.3a), then we should be able to accommodate a bounded but infinite number of action sequences, even though we cannot predict any of them. The rationale behind this capability is illustrated in Figs. 9.3b and 9.3c. If we have knowledge of the strategies as categories, then we can use those bins to parse the action sequence timeline. Doing so will reveal that the idiosyncratic trajectory is actually composed of a number of pieces, each representing an instantiation of one of the strategy categories. The action sequence as a whole is composed by shifting between these instantiated strategies, as shown graphically by Fig. 9.3b. It is important to note that this characterization is not a theoretical fiction. Rasmussen and Jensen's (1973) field study data have been coded exactly according to the type of scheme we have described here. An example from their work is provided later in this chapter.

Why Is Strategies Analysis Important?

Although we are getting a little bit ahead of ourselves, it is worthwhile pointing out now that the definition of strategies as categories has important advantages for design. Because the categories are stable constructs, computer-based support can be created to meet the unique needs of each strategy. For example, a different display could be constructed to support the information requirements associated with each viable strategy, and a mechanism could be provided to allow workers to switch easily between displays, and thus strategies (see the design example in chap. 12). Such a design should allow workers to meet task demands in a flexible manner by using whatever strategy they prefer, and by seamlessly switching between strategies as necessary. Therefore, even though we cannot predict what action sequence a worker *will use* on any one occasion, a strategies analysis identifies the strategies that a worker *can use,* thereby providing a stable basis for the design of flexible, yet contextually sensitive, computer-based information systems.

INFORMATION FLOW MAPS: THE CASE OF DIAGNOSING FAILURES IN ELECTRONIC EQUIPMENT

To conduct a strategies analysis, we need a modeling tool that can be used by analysts to achieve the objectives described in the previous section. Rasmussen (1980, 1981) adopted *information flow maps* to describe the categories of cognitive task procedures that constitute strategies. This modeling tool has not reached the same level of maturity as the decision ladder or the abstraction-decomposition space because it has not been comprehensively described as a generic tool, independent of any particular application. Therefore, in this section we introduce information flow maps in the context of a particular example, a strategies analysis of electronic troubleshooting.

A Summary Account of the Strategies

Rasmussen and Jensen (1973, 1974) conducted a field study of professional technicians engaged in troubleshooting electronic equipment. Based on their findings, they identified a number of strategies that can be used to perform this diagnostic task.

We begin by presenting a summary account of the findings by contrasting the engineers' approach to troubleshooting with that of the technicians. In the following subsection, we provide a detailed account of each of the strategies that were identified.

The Engineers' Approach. The experimenters, being engineers, began their study with the expectation that their subjects would adopt the "intelligent" approach to troubleshooting that was advocated by both researchers and textbooks at the time that the study was conducted (the late 1960s). Briefly, that approach exploits knowledge of the functioning of the equipment being repaired so that an equipment-specific plan of attack can be developed. Typically, this knowledge would be acquired, in part, from a documentation manual for the equipment in question. A few carefully chosen measurements could be taken, and then integrated, using a carefully chosen reasoning sequence supported by an understanding of equipment functioning. This approach is elegant in that it requires only a few observations, it is based on a detailed understanding of the equipment, and it produces a logical chain of reasoning that extracts as much information as possible from each individual observation. Furthermore, it also has the benefit of uncovering an explanation for why the fault is producing the observed abnormal symptoms.

The Technicians' Approach. Rasmussen and Jensen (1973, 1974) found that the professional technicians participating in their study tended to adopt a very different approach to electronic troubleshooting. Rather than conduct a small number of observations, the technicians tended to conduct many observations in rapid sequence. Rather than extracting as much information as possible from each observation, the technicians tended to only use each observation in order to determine where to make the next observation. Rather than engage in elaborate, logical reasoning, the technicians tended to make very simple decisions, primarily whether each observation was good (i.e., normal) or bad (i.e., abnormal). Rather than exploit knowledge of the equipment to develop an equipment-specific plan of attack, the technicians tended to use generic methods that were generally independent of both the equipment and the fault. Finally, the technicians also sometimes exhibited a few shortcuts that could be interpreted as signs of laziness. For instance, sometimes they used a circuit diagram or a similar circuit as referents to determine the normal value for a particular observation rather than mentally deducing or remembering that value.

Which Is Better? The Role of Subjective Task Formulation and Performance Criteria. Given the stark contrast in the two approaches, it may seem natural to question the rationality of the technicians' strategies. Compared to the systematic, logical, and calculated engineers' approach, the technicians' approach appears to be very inefficient, ad hoc, and lazy. If nothing else, it is certainly very different from what the textbooks said the technicians should be doing.

Nevertheless, Rasmussen and Jensen (1973, 1974) found that the question of which approach was more rational—or better, more adaptive—could not be answered in isolation. Rather, the question could be answered only by considering two additional factors, namely the technicians' subjective task formulation and performance criteria (cf. Bedny & Meister, 1997; Bruner et al., 1956). The *subjective task formulation* is the task that workers set for themselves. As these field study results show, the very same task can be approached in quite different ways. The engineers set out to develop an explanation for why the equipment exhibited the faulty responses. This is the reason why their approach put a premium on logical deduction. In contrast, the technicians formulated the troubleshooting task in quite a different way. They were merely interested in locating a fault in a piece of equipment that had previously been functioning properly (i.e., to find the root location of the discrepancy between normal and actual state). This is the reason why their approach was much more pragmatic in nature.

The related construct of *performance criteria* was also very important in assessing the comparative adaptiveness of the two approaches to troubleshooting. As shown in Table 9.1, the engineers' and technicians' approaches emphasized a complementary set of criteria. Thus, which approach is more adaptive depends on the importance given to different performance criteria. For example, if the cost of observations is very high, then the engineers' approach may indeed be more adaptive. However, the professional technicians participating in this study worked in an environment where the cost of an observation was usually minimal. In addition, there was no intrinsic reward for using an approach that is more effortful. On the contrary, doing the job efficiently and effectively would suggest that minimizing computational effort would be more desirable. Furthermore, because the technicians' livelihood depended on repairing faulty equipment, it was important that they do so quickly, otherwise a backlog of repairs would begin to accumulate. Doing prompt work also meant minimizing the time spent reading manuals to gather equipment-specific knowledge. The technicians also faced many different faults on many different kinds of equipment on the job, so it would be adaptive to develop a diagnostic approach that was independent of those details. From these conditions, we can see that there is no advantage to using the elegant engineers' approach in this environment. The pragmatic approach exhibited by the technicians was consistent with a set of performance criteria that were adaptive, given the conditions under which the job was being performed.

Diagnostic Strategies

In this subsection, we describe the diagnostic search strategies that were identified in this field study (Rasmussen, 1981). These strategies can be divided into two types, topographic search and symptomatic search.

TABLE 9.1 A Comparison of the Engineers' and Technicians' Troubleshooting Approaches in Terms of Different Performance Criteria

Criterion	Engineer	Technician
Number of observations	few	many
Computational effort	high	low
Dependence on fault	yes	no
Dependence on equipment	yes	no
Use of information	economic	uneconomic
Speed	slow	fast
Knowledge required	lots	little

Topographic Search. An information flow map for the *topographic* search strategy is illustrated in Fig. 9.4. Following the definition of strategy as a category rather than as an instance, this and subsequent maps should not be interpreted as a description of the sequence of cognitive steps that a troubleshooter adopts. Instead, they should be interpreted as an idealized process representation from which particular instances of troubleshooting sequences can be generated.

Referring to the center right side of Fig. 9.4, topographic search requires a model of the normal state of the equipment (i.e., how it should be behaving). This reference information can be obtained in several ways. For example, it is possible to recall the information from memory or read it from a schematic diagram. These sources are generally preferred because they require minimal time and effort. If the reference data are not immediately available from either of these sources, then they will have to be deduced from a model of normal equipment functioning. This method is much more time consuming and effortful because it requires comparatively complex reasoning. The model of normal function is also used to identify the relevant paths or fields in the equipment topology.

Given a model of the normal state of the equipment, a location for the next observation is selected based on a set of tactical search rules (described in more detail in Rasmussen & Jensen, 1974). Based on this selection, the actor will then have to find the desired location. This search step requires knowledge of the visual appearance of different components, information that is provided by a physical

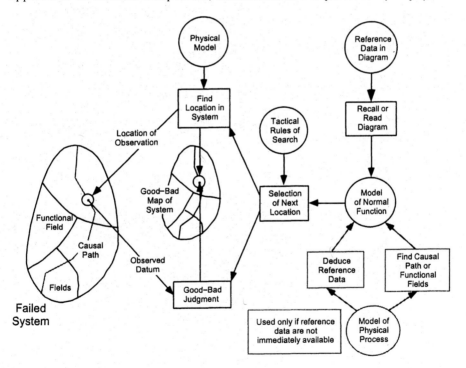

FIG. 9.4. Information flow map for topographic search strategy. Adapted from J. Rasmussen, "Models of mental strategies in process plant diagnosis." In J. Rasmussen & W. B. Rouse (Eds.), *Human detection and diagnosis of system failure*, Plenum Publishing, 1981. Reprinted with permission.

model of the equipment. The actor then takes an observation in the failed equipment, and compares it to the aforementioned reference data. This comparison leads to a judgment as to whether the observation is good (i.e., normal) or bad (i.e., abnormal). This judgment is then used to rule out certain equipment paths or fields as being likely locations of the fault. As more and more observations are taken, the good–bad map is iteratively updated so that the plausible bad area is reduced. This stepwise refinement results in a gradual narrowing of the actor's focus of attention.

Symptomatic Search. There are actually three diagnostic search strategies under the symptomatic search category. The property that they share, and which distinguishes them from topographic search, is that they all use the information content of an observation (rather than its topographic location) to diagnose the equipment state. The three symptomatic search strategies are: pattern recognition, decision table, and hypothesis and test.

An information flow map for the *pattern recognition* strategy is illustrated in Fig. 9.5. We can see that this data-driven strategy involves minimal information processing. The actor recognizes a pattern of data from the failed equipment as being familiar, and attaches a label to that pattern. Note that the label need not be associated with the root cause of the fault. For instance, it is also possible for the actor to recognize that a particular task or action should be performed.

An information flow map for the *decision table* strategy is illustrated in Fig. 9.6. This strategy relies on a library of state models that associate a particular data pattern with a particular state (e.g., IF symptoms S1, S2, and S3 are observed THEN check the power supply; IF symptoms S4 and S5 are observed THEN there is probably a short circuit). Frequently, the decision table is organized in a hierarchical manner based on frequency of occurrence or size of granularity. The library of models is scanned and a particular state model is selected based on tactical search rules. Usually, the selected state will also have a label associated with it (e.g., faulty power supply). The set of symptoms in the state model represents a model of a failed state, and this model is used as a reference pattern for search. The actor then observes the failed equipment to see if there is a match to the pattern. Although this strategy may seem to be similar to pattern recognition, the primary difference is that decision table search is a knowledge-driven (i.e., top-down), rather than a data-driven (i.e., bottom-up), process.

An information flow map for the *hypothesis-and-test* strategy is illustrated in Fig. 9.7. As the name indicates, this strategy begins with a hypothesis about the state of

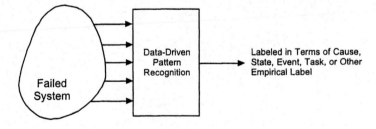

FIG. 9.5. Information flow map for pattern recognition strategy. Adapted from J. Rasmussen, "Models of mental strategies in process plant diagnosis." In J. Rasmussen & W. B. Rouse (Eds.), *Human detection and diagnosis of system failure*, Plenum Publishing, 1981. Reprinted with permission.

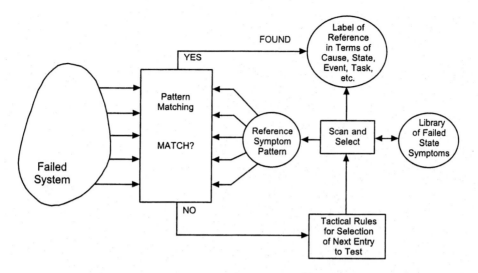

FIG. 9.6. Information flow map for decision table search strategy. Adapted from J. Rasmussen, "Models of mental strategies in process plant diagnosis." In J. Rasmussen & W. B. Rouse (Eds.), *Human detection and diagnosis of system failure,* Plenum Publishing, 1981. Reprinted with permission.

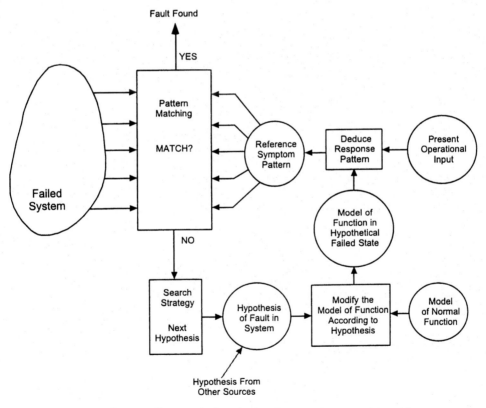

FIG. 9.7. Information flow map for hypothesis and test search strategy. Adapted from J. Rasmussen, "Models of mental strategies in process plant diagnosis." In J. Rasmussen & W. B. Rouse (Eds.), *Human detection and diagnosis of system failure,* Plenum Publishing, 1981. Reprinted with permission.

the equipment. This hypothesis can be derived from an earlier application of the hypothesis-and-test strategy or from one of the other search strategies. A functional model of the normal state of the equipment is also required for this strategy. This normative model is then modified by introducing the aforementioned hypothesis. The result is a functional model of the hypothetical failed condition. The actor then needs to know the input data for the current operating regime. These data are then fed into the model of the failed condition, and a pattern of expected data is deduced. This step involves substantial knowledge and computational effort because the actor must derive the expected pattern from knowledge of how the equipment functions. The result of this deductive reasoning process is a model of the hypothetical failed state. This state model is then compared to observations obtained from the failed equipment. If the patterns match, then the hypothesis is accepted. Otherwise, it is rejected and the actor continues with the troubleshooting task.

Comparing the Strategies. Table 9.2 compares these strategies along various performance criteria. We can see that the various strategies have different resource requirements. Some require more time, whereas some require more observations. Some require more knowledge, whereas some require more memory load. There is no universally best strategy (cf. J. W. Payne et al., 1993).

Rasmussen and Jensen (1973, 1974) found that the technicians exploit these redundant possibilities by switching between strategies to avoid exceeding their resource constraints. Thus, when one strategy is becoming too effortful, technicians would spontaneously switch to another strategy to meet the task demands in a more economic fashion. These findings have been replicated in a number of studies, including Sperandio's (1978) field studies of air traffic control and J. W. Payne et al.'s (1993) laboratory experiments on adaptive decision making. Thus, strategy shifts seem to be a generalizable tactic for dealing with resource-demand conflicts.

As we mentioned earlier, these strategy shifts are largely responsible for the idiosyncratic nature of the detailed procedures exhibited by workers on any one

TABLE 9.2 A Comparison of the Resource Requirements of the Various Diagnostic Search Strategies Identified by Rasmussen (1981)

Performance Factor	Topographic Search	Recognition	Decision Table	Hypothesis and Test
Time spent	——	Low	——	——
Number of observations	High	Low	——	Low
Dependency on pattern perception	——	High	——	——
Load on short-term memory	Low	Low	High	High
Complexity of cognitive processes	Low	Low	——	High
Complexity of functional model	Low	——	——	High
General applicability of tactical rules	High	——	——	Low
Dependency on malfunction experience	Low	High	——	Low
Dependency on malfunction preanalysis	——	——	High	——

Note. Features that are not filled in either are not typical, being dependent on circumstances, or are irrelevant.

occasion. Figure 9.8 provides a concrete example. The figure shows that the sequence followed by a particular technician during a specific troubleshooting episode can be parsed in terms of the strategies just described. This example reinforces the point we made earlier with Fig. 9.3. Action sequence instances are variable and idiosyncratic. However, treating strategies as idealized categories that can be instantiated in combination during particular situations provides a parsimonious way of subsuming this complexity.

Design Implications. This analysis of diagnostic strategies has deep implications for design. As J. W. Payne et al. (1993) pointed out, "we can improve decisions by creating a better match between task demands and the information-processing capabilities and preferences of the decision maker" (p. 218). The first, and perhaps most important, step to take in achieving this goal is to realize that the strategies identified earlier are, in principle at least, actor-independent. Although they were developed based on the findings of a descriptive field study of technicians' actions, there is no reason why the strategies cannot be distributed across workers and automation. This issue is dealt with in greater detail in the next chapter, but it is important that we introduce it here so that we do not lose sight of the connection between work analysis and systems design.

The electronics technicians did not have any special job aids beyond the usual instrumentation that is required to take observation from electronic equipment (e.g., ohmmeter, voltmeter, oscilloscope, etc.). Therefore, aside from the act of measurement itself, the technicians had to perform the tasks essentially on their own. But if the strategies, subjective task formulation, and performance criteria that technicians spontaneously adopt to perform the task have been identified, there is no reason why new forms of information support cannot be introduced to perform the job. In fact, the strategies presented graphically in Figs. 9.4 to 9.7 provide a great deal of information to do exactly that.

If we examine these figures closely, we see that the strategies do not make any reference to a human technician. As we mentioned, they are actor-independent. This is a very important property because it means that different parts of each strategy can be allocated either to a worker or to automation. Figure 9.9 provides a hypothetical example for the hypothesis-and-test strategy. The most resource-demanding aspects of this strategy have been allocated to automation, while still leaving the worker in charge. As the figure shows, it is up to the worker to generate a hypothesis about the state of the equipment. Once this hypothesis is generated, the worker can enter it into the machine, which has a functional model of the normal condition of the equipment stored in it. Using that model and the hypothesis selected by the worker, the automation can modify the normative model to obtain a functional model of the hypothetical failed condition. The automation could also perhaps have input data for different operating regimes stored in it, in which case it could also deduce the data pattern that would be expected for a given hypothetical state. It would then be up to the worker to make the required observations and determine if there is a good match warranting acceptance of the hypothesis. In the case of no match, it would also be up to the worker to generate an alternative hypothesis or to switch to a different strategy. Analogous hypothetical job aids could be proposed based on any of the other diagnostic strategies described earlier.

Note that, with this hypothetical design, the strategy has not changed at all. It remains exactly as it was defined in Fig. 9.7. What has changed are the responsibilities, and thus

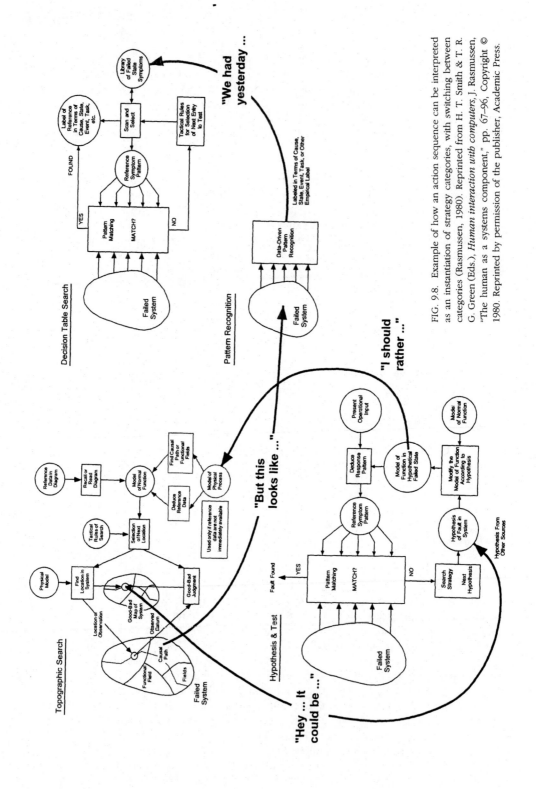

FIG. 9.8. Example of how an action sequence can be interpreted as an instantiation of strategy categories, with switching between categories (Rasmussen, 1980). Reprinted from H. T. Smith & T. R. G. Green (Eds.), *Human interaction with computers*, J. Rasmussen, "The human as a systems component," pp. 67–96, Copyright © 1980. Reprinted by permission of the publisher, Academic Press.

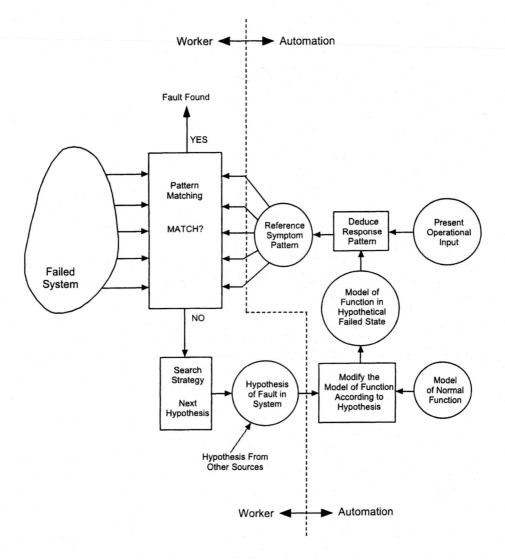

FIG. 9.9. Example showing how the resource-demanding aspects of the hypothesis and test search strategy can be off-loaded to automation, thereby resulting in a distributed cognitive system to perform the very same strategy (adapted from Rasmussen, 1981).

the psychological demands, experienced by workers. This is one of the advantages associated with decoupling how things are done from who will be doing them.

There are two implications that follow from these insights. First, the resource requirements experienced by workers with each strategy can be manipulated through design. For example, in Table 9.2, we noted that the hypothesis and test strategy requires a great deal of equipment knowledge, puts a significant burden on memory, and involves complex reasoning. However, *these resource requirements are properties of the strategy, not the technicians that adopted the strategy.* That is, these characteristics must be dealt with by *any* actor(s) performing the strategy. This insight can be obtained only by separating the question of *how* from that of *who*. This separation makes it clear that whether a strategy is cognitively demanding or not for a worker depends on the information support available. For example, from the aforementioned description, one might think that the hypothesis and test strategy intrinsically requires workers to engage in extensive analytical processing. But as Fig. 9.9 suggests, this need not be the case at all if the most resource-demanding aspects of the strategy are off-loaded to automation. Therefore, it is important to distinguish between the strategy being followed by a worker and the psychological demands experienced by the worker. The mapping between these two constructs is not fixed, but can instead be varied by the information support provided by designers.

Second, and relatedly, it is important to focus on the strategies that *can be* used by workers, not just those that *are being* used (Rasmussen, 1981). If a particular strategy is not being used very frequently by workers, or does not frequently lead to positive outcomes, it does not necessarily follow that the strategy is not a good one. This statement follows from the formative orientation of CWA (see chap. 5). Again, the troubleshooting field study provides an example. Technicians did not adopt the hypothesis-and-test strategy very frequently. Furthermore, when they did use it, the technicians frequently had trouble reaching an accurate diagnosis. It might be tempting to conclude from these descriptive findings that the strategy is not worthwhile supporting with a computer-based information system. Such a conclusion would be premature, however. The strategy may not be used very frequently or very effectively because it is too effortful or too time consuming to do so with the tools at hand (cf. chap. 4). Perhaps, if a new information system were introduced, the strategy could be used much more frequently and to good effect. Again, the hypothesis-and-test strategy provides a nice concrete example to illustrate the point. It is clear that technicians tended to avoid using this strategy because it put too much of a load on their limited cognitive resources. However, if the difficult parts of the strategy were off-loaded (as suggested in Fig. 9.9), then the technicians might be much more likely to use the strategy.

Moreover, there are independent reasons why technicians could use this strategy to greater advantage than was observed. Specifically, the hypothesis-and-test strategy has the advantage of being capable of dealing with unfamiliar and complex faults. When the technicians encountered such faults, they frequently became stuck because their preferred strategies were no longer very effective. The experimenters instructed the technicians to use the more broadly applicable hypothesis-and-test strategy under these situations, but the technicians were not able to do so in the normal repair shop environment. Thus, it seems that providing information support for this strategy would be very useful. Not only would it make the strategy less cognitively demanding for technicians, but it might also allow them to use that strategy to cope effectively with unfamiliar and complex faults. The bottom line is that CWA should try to identify the strategies that can be effectively used, not just those that are currently being used.

In summary, we do not know whether it would be feasible to build a job aid like the one suggested in Fig. 9.9, but that is not important for our purposes. Our point is that, because strategies are actor-independent, a strategies analysis provides a basis for designing information support that: (a) allows workers to use strategies that they might spontaneously adopt anyway, (b) takes on the burden of dealing with the most resource-intensive aspects of the strategies, (c) makes it easier for workers to adopt effective strategies that might otherwise not be used, (d) keeps workers in charge of the task, and (e) gives workers the flexibility to switch between strategies. In short, such a design would provide the flexible support that would enable the context-conditioned variability that is required to cope effectively with open systems.

Abstracting From the Particulars

We have introduced information flow maps by example, showing how they can be used as a tool to model fault diagnosis strategies (see Figs. 9.4 to 9.7). We conclude this section by discussing this modeling tool in its generic form and by briefly discussing how we might go about identifying strategies in practice.

Information Flow Maps. It should be clear from the examples and from our discussion that information flow maps are not intended to be sequence descriptions of cognitive activity. Rather, they are idealized categories of task procedures (see Figs. 9.3 and 9.8). It should be equally clear that the elements used to construct information flow maps will be context-specific. For example, the strategies analysis described in the next section uses a different set of elements because the task being represented is completely different in nature from the one described in this section.

Nevertheless, it would be useful to develop some sort of taxonomy or formalism that could be used to construct information flow maps for various contexts. What we need is something like the elementary information processes (EIPs) developed by J. W. Payne et al. (1993). These EIPs are like a periodic table that can be used to describe the processing activities required by different strategies. Any one strategy can be described by a unique sequence of EIPs. Examples of the EIPs developed by Payne et al. include: READ, COMPARE, DIFFERENCE, ADD, PRODUCT, ELIMINATE, MOVE, and CHOOSE. These particular EIPs happened to be the ones that were required to describe the strategies investigated by Payne et al. The list would have to be expanded to cover the strategies encountered in a wide range of application domains.

More important, however, a shift in the definition of the EIPs is also required. Recall that J. W. Payne et al. (1993) defined a strategy as a sequence of cognitive operators. In contrast, in CWA a strategy is a category of cognitive procedures. A refinement of the information flow map modeling tool would involve accommodating this shift from exemplar to category. This remains an important topic for future research (see Rasmussen, 1986a, for a first pass).

Where Do Strategies Come From? How do we come up with descriptions of strategies like those in Figs. 9.4 to 9.7? As we mentioned in the Preface, this is not a "how to" book so a detailed answer to this question is beyond our scope. Nevertheless, we can point to three ways of collecting the information that is needed to perform a strategies analysis. The most obvious tactic is to conduct a *descriptive field study* of current practice, like that conducted by Rasmussen and Jensen (1973, 1974). As we discussed in chapter 5, the data obtained from such an empirical study could provide a basis for the formative modeling of CWA. Accordingly, the objective of such a study would be to determine, not just what strategies are being used, but also what strategies could be used, and the performance criteria and subjective task formulation exhibited by workers.

A second way of getting information that can be used to identify strategies is to conduct an *a priori analysis*. The first step along this path would be to identify the possibilities for action provided by the work domain. Then, the degrees of freedom associated with particular classes of situations could be identified by conducting a control task analysis. These remaining degrees of freedom can then suggest several qualitatively different strategies for performing the task. The case study presented later in this chapter provides a detailed example of this approach. Note, however, that this tactic can probably only be used effectively in highly structured application domains that are well understood (i.e., for which there are useful models).

Finally, it is also possible to use *operations research models* to identify strategies. For example, Dessouky et al. (1995) used scheduling theory to identify strategies that could be used by actors to perform planning tasks in various application domains. Although this approach benefits from an extensive and sophisticated set of formal methods, it is important to ensure that psychological criteria and constraints are not ignored in the modeling process. Otherwise, operations research models may provide nothing more than yet another unrealistic "one right way" to get the job done (cf. chap. 3).

In summary, there are different ways of collecting information that can be used to conduct a strategies analysis. Which approach is best depends on the application domain.

STRATEGIES ANALYSIS FOR PROCESS CONTROL MICROWORLD

In this section, we use information flow maps as a modeling tool to conduct a strategies analysis for DURESS II. Ordinarily, we would identify the strategies associated with each of the information-processing activities identified during a control task analysis. For example, six activities were identified in the previous chapter for the start-up mode of DURESS II (see Fig. 8.10 for a summary). A comprehensive strategies analysis would ordinarily be conducted for each of these activities. However, for the sake of clarity and brevity, we focus on only one of those activities, namely the *formulate procedure* box that requires an actor to configure the FWS to achieve a particular water input flow rate into the reservoir (see Step 5 in Table 8.1).

Product Constraints: Control Task Analysis Results Revisited

Figure 9.10 summarizes the results of the control task analysis for the FWS configuration task. Following the schematic diagram in Fig. 9.1a, we have identified the inputs, outputs, and constraints associated with this information-processing activity. The input state of knowledge is a desired water input flow rate into each reservoir (MI1, MI2). The output state of knowledge is a set of valve and pump settings that will achieve these desired flow rates. The steady state constraints associated with this activity are as follows:

- Conservation of mass from two feedwater streams:

$$MI1(t) = FA1(t) + FB1(t).$$

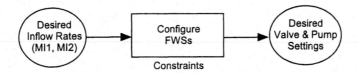

FIG. 9.10. A schematic diagram of the configure FWS activity. Compare with the generic diagram in Fig. 9.1.

• Conservation of mass in feedwater stream:

$$FVA(t) = FA1(t) + FA2(t).$$

• Flow split relation:

$$FA1(t) = \frac{FVA(t)\ VA1(t)}{VA1(t) + VA2(t)}.$$

• Flow through pump:

If pump is OFF then FPA(t) = 0,
otherwise:

IF [VA1(t) + VA2(t)] > VA(t) THEN

 FPA(t) = FVA(t)

ELSE

 FPA(t) = FVA1(t) + FVA2(t).

• Conservation of mass in pipe:

$$FVA(t) = FPA(t).$$

In addition, it is important that the pumps not be turned on before the valves. Otherwise, there will be no downstream path for the water to flow through.

Although there are many constraints to take into account, there is no unique solution to this problem. For any given pair of desired input flow rates (MI1, MI2), there is a bounded but infinite set of component settings that will achieve the desired values. In other words, the actor has degrees of freedom available in choosing how to accomplish the task.

Process Constraints: Feasible Strategies

Because there are degrees of freedom available after the product constraints have been considered, it follows that different strategies can be used to configure the FWSs. Each strategy has associated with it a set of process constraints that can be mapped out as an information flow. Following the logic depicted in Fig. 9.1b, the objective of a strategies analysis is to identify the process constraints for each viable and effective strategy. For the most part, the use of the pumps is nondiscretionary (they must be on for there to be any flow at all), so we focus our analysis on the use of the valves.

We have identified three structurally distinct strategies for the FWS configuration task in DURESS II:

- Single—Uses only one of the two FWSs and thus three input valves.
- Decoupled—Uses both FWSs but only four of the input valves.
- Full—Uses both FWSs and five or six of the input valves.

Figure 9.11 provides graphical examples of each of these strategies.[3]

Single. Figure 9.12a shows an information flow map for the single FWS strategy. The first step is to select one of the FWSs. For example, an actor may choose to fill up both reservoirs from FWS A. This choice requires knowledge of the topological properties of the work domain. Given this choice, the next step is to determine what the flows should be through the secondary valves. This decision will be made based on knowledge of the flows to be achieved (MI1, MI2) and a functional model of the work domain. For example, if MI1 = 4 and MI2 = 6, the actor may choose to establish a flow of four units/second through valve VA1 to feed Reservoir 1 and a flow of six units/second through Valve VA2 to feed Reservoir 2 (see Fig. 9.11a). Finally, the actor must then select three valve settings (VA, VA1, VA2) that will achieve these goals. This decision must be made based on knowledge of the relationship between a functional model of flow rates and a physical model of component settings. Continuing with the same example, VA could be set to 10, VA1 to 4, and VA2 to 6. The output of the strategy is a desired set of three valve settings.

Decoupled. Figure 9.12b shows an information flow map for the decoupled FWS strategy. Right away we see an obvious difference from the map for the single FWS strategy in Fig. 9.12a. For the decoupled strategy, there are two parallel information flows, one dealing with formulating the procedure to achieve the desired inflow into each reservoir. As we see later, the information flow maps are essentially independent because the two FWSs are decoupled in this strategy.

Beginning with the map for Reservoir 1, the desired input flow rate (MI1) is an input. Then, the actor must choose one of the two available flow paths into that reservoir. For example, one way to feed Reservoir 1 is to use VA1 (see Fig. 9.11b). This choice requires knowledge of the topological properties of the work domain. Referring back to Fig. 9.12b, the choice of flow path and the desired inflow can be used as inputs to determine a desired flow through the chosen flow path. For example, if MI1 = 10, then the actor might choose to establish a flow of 10 units/second through VA1. Next, the actor must select a pair of valve settings that will achieve this flow. This decision must be made based on knowledge of the relationship between a functional model of flow rates and a physical model of component settings. Continuing with the same example, VA could be set to 10, and VA1 to 10. The output is a desired pair of valve settings to satisfy the flow to Reservoir 1.

---

[3]Note that the diagrams in the figure are only examples. Because each strategy is a category, there are multiple instances of each. For example, the single FWS strategy could use FWS B instead of FWS A.

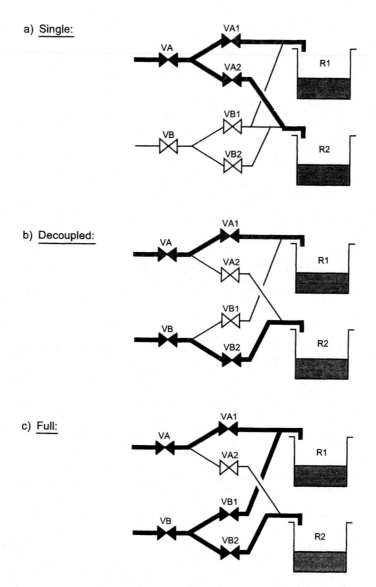

a) Single:

b) Decoupled:

c) Full:

FIG. 9.11. Graphical examples of the strategies that can be used to configure the FWSs in DURESS II. Filled-in valves represent valves that are used (i.e., open) for that strategy. Thick flow lines indicate paths through which water should flow for that strategy. See Fig. 9.12 for an information flow map for each strategy.

The second parallel information flow map in Fig. 9.12b deals with the flow to Reservoir 2 and is almost identical. The only difference is that there is no need to choose a flow path because there are no remaining degrees of freedom. For example, if VA1 is used to feed Reservoir 1, then VB2 must be used to feed Reservoir 2 if a decoupling strategy is to be implemented.

As shown in Fig. 9.11b, the defining feature of this strategy is that the two FWSs are decoupled from each other. Each FWS feeds only one reservoir. In the case of

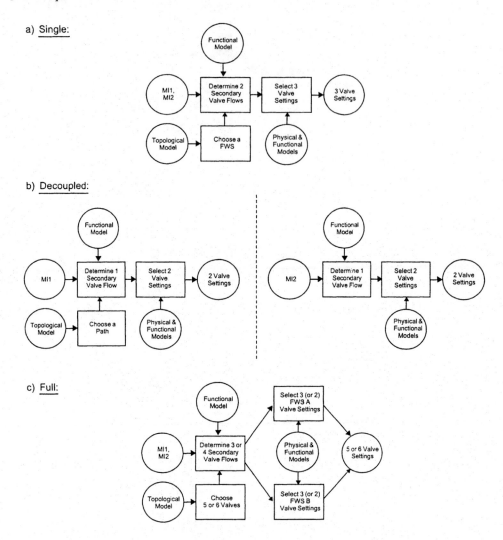

FIG. 9.12. Information flow map for each of the FWS configuration strategies summarized graphically in Fig. 9.11.

the example in Fig. 9.11b, the two crossover valves (VA2 and VB1) are not used, thereby achieving the decoupling. As a result, the FWS are structurally less complex because there is a simple one-to-one mapping between FWSs and reservoirs.

Full. Figure 9.12c shows an information flow map for the full FWS strategy. The first choice to be made is whether to use all six FWS valves or only five. This choice is somewhat arbitrary in that five- and six-valve configurations are equally capable of satisfying the same range of input flow rate pairs (MI1, MI2) under normal (i.e., nonfault) conditions. Nevertheless, the choice must be made because it constrains subsequent steps in the strategy. To make the decision, a topological model of the work domain is required. As an example, if MI1 = 12 and MI2 = 3, then one possibility is to use five valves: VA, VA1, VB, VB1, and VB2 (see Fig. 9.11c). The next step, which requires a functional model, is to determine what the flows through the secondary valves will be. If we stick with the same example, the actor may choose to get a flow of 10 units/second through VA1, 2 units/second through VB1, and 3 units/second through VB2. These flows would combine to satisfy the desired reservoir input flow rates (MI1 = 12, MI2 = 3). As shown in Fig. 9.12c, the next two steps involve selecting the valve settings that will achieve these flows through the secondary valves. Each FWS must be dealt with in turn, because the flows within a FWS are affected by all of the valve settings in that FWS. These decisions require physical and functional models of the work domain. The output is a desired set of five or six valve settings.

Note that this strategy results in a many-to-many mapping between FWSs and reservoirs (see Fig. 9.11c). This relationship is more complex than that achieved in the decoupling strategy.

Performance Criteria: Comparing the Strategies

Just as the various diagnostic strategies in the troubleshooting field study differed in terms of various performance criteria (see Table 9.2), the three strategies for configuring the FWS differ in terms of the performance criteria listed in Table 9.3.

The first criterion that distinguishes between the strategies is their respective capacity limits. Recall from chapter 6 that each pump has a maximum flow rate capacity of 10 units/second of flow. As a result, each FWS cannot produce more than 10 units/second of flow. Knowing this, if we examine Fig. 9.11, we can see that the strategies cannot all be used under any circumstances. The single FWS strategy uses only one FWS, so it cannot be used if the sum of the required input flow rates (i.e., MI1 + MI2) is greater than 10. The decoupled FWS strategy uses one FWS to feed each strategy, so it cannot be used if either input flow rate is required to be greater than 10. The full FWS strategy reaches its limit of applicability when the sum of the required input flow rates (i.e., MI1 + MI2) is greater than 20 because the two FWSs cannot collectively produce more than 20 units/second of flow.

Table 9.4 summarizes these constraints in a different way, by listing the feasible strategies for different combinations of desired input flow rates (MI1, MI2). We can see that the full strategy is the most broadly applicable of all, the single strategy the most narrow, and the decoupled strategy in between.

TABLE 9.3 A Comparison of the FWS Configuration Strategies in Terms of Different Performance Criteria			
Criterion	**Single**	**Decoupled**	**Full**
Cannot be used if . . .	(MI1 + MI2) > 10	MI1 or MI2 > 10	(MI1 + MI2) > 20
Likelihood that change in demand will result in need to reconfigure	likely	less likely	zero
Facility in dealing with a major fault	most cumbersome	less cumbersome	least cumbersome
Memory load	lowest	intermediate	highest
Computational effort	intermediate	lowest	highest

TABLE 9.4 A Comparison of the Three FWS Strategies in Terms of Their Capability to Satisfy Different Pairs of Desired Input Flow Rates (MI1, MI2)		
(MI1 + MI2) ≤ 10	**(MI1 ≤ 10) & (MI2 ≤ 10)**	**(MI1 + MI2) ≤ 20**
Single		
Decoupled	Decoupled	
Full	Full	Full

Returning to Table 9.3, the second row indicates that the strategies also differ in terms of the likelihood that a change in demand will require a reconfiguration of the FWSs. For example, if MI1 and MI2 change, can the same strategy be used or will a different strategy have to be adopted? This criterion follows directly from the capacity limitations constraints. The more narrow a strategy, the more likely the need for reconfiguration. For example, if a single FWS strategy is being used and then there is a need to supply more than 10 units/s of flow to both reservoirs or even to just one reservoir, then the single strategy must be relinquished. In contrast, as long as the sum of the required flows is less than or equal to 20, there will never be a need to switch from the full strategy.

In the third row of Table 9.3, we see that the strategies also differ in terms of the ease with which they allow an actor to cope with a major fault. The single FWS is the most cumbersome of all. If a fault occurs in the FWS under these conditions, then it will be necessary to start up from scratch the FWS that is not being used. This task will require some time, a fair number of actions, and some mental processing. In contrast, if a fault occurs when the full FWS strategy is being used, then there is no need to reconfigure the FWSs structurally. The worst that could happen is that the settings of the valves that are already being used would have to be changed. Thus, this strategy is the least cumbersome in terms of resilience to faults. For reasons that should be obvious, the decoupled strategy lies somewhere in between.

The fourth row of Table 9.3 shows how the strategies differ in terms of memory load. Recall that the single, decoupled, and full FWS strategies use three, four, and five (or six) valves, respectively. The number of items that an actor would have to keep track of would likely be proportional to the number of valves used. Thus, the single FWS strategy should impose the least memory load, the full FWS strategy the highest, and the decoupled FWS strategy somewhere in between.

The fifth row of Table 9.3 lists how the three strategies differ in terms of computational effort. A relevant factor here is the mapping between FWSs and reservoirs. As shown in Fig. 9.11, the single FWS strategy results in a one-to-many mapping, the decoupled FWS Strategy in a one-to-one mapping, and the full FWS strategy in a many-to-many mapping. The implications of these different mappings can be deduced by examining the information flow maps in Fig. 9.12. With the single FWS strategy, an actor must determine two secondary flows and then compute the relationship between three valve settings and their respective flow rates. Given the nature of the relevant constraints (see the equations at the beginning of this section), these tasks can represent a nontrivial computational burden. The full FWS strategy is even more complex computationally. It requires that an actor determine three or four secondary valve flows and then compute the relationship between three valve settings and their respective flow rates twice, once for each FWS. The decoupled FWS strategy is actually the least effortful computationally. Because there is a one-to-one mapping between FWSs and reservoirs, the number of interactions that have to be taken into account is minimized. This property is illustrated in Fig. 9.12b by the fact that the information flow maps for the two input flow rates are essentially independent of each other. Moreover, this simple mapping also means that the actor needs to consider fewer relationships between valves and flows. As shown in Fig. 9.12b, only one secondary valve flow needs to be considered at a time and the relationship between only two valve settings and their respective flow rates needs to be computed at a time. Decoupling allows actors to accomplish the overall task by breaking it up into a set of simpler parts that can be dealt with one at a time, thereby reducing computational effort.

In summary, Table 9.3 clearly shows that no strategy is globally superior. Which strategy is best depends on the actor's criteria and the current demands. For example, if the actor formulates the task as configuring the FWS to achieve the desired inflows and any future eventuality (e.g., fault or change in demand), then the full FWS strategy may be preferred. On the other hand, if the actor formulates the task as configuring the FWS to achieve the desired inflows with the least amount of effort, then the decoupling strategy may be preferred. The relationship between strategies, performance criteria, and subjective task formulation is, at a general level, very similar to that observed in the electronic troubleshooting study described earlier in this chapter.

Stepping Back From the Details: A Few General Points

The strategies analysis for the FWS configuration task in DURESS II is graphically summarized in Fig. 9.13. A control task analysis had identified the inputs and outputs associated with this task, as well as the product constraints that had to be taken into account. The objective of the strategies analysis was to try to identify various ways in which the transformation between inputs and outputs might be accomplished. As shown in the figure, three such strategies were identified, each with a unique set of resource requirements (see Table 9.3). Now that we have described these strategies in great detail, it is useful to step back to discuss a few general points that transcend the details of the analysis.

The first point is that the strategies for the FWS configuration (see Fig. 9.12) and those for the fault diagnosis task (see Figs. 9.4 to 9.7) seem quite different from each other. In part, these differences can be explained by the fact that the strategies are dealing with entirely different control tasks (procedure formulation vs. state identification) in different work domains. Thus, it should not be surprising to find that

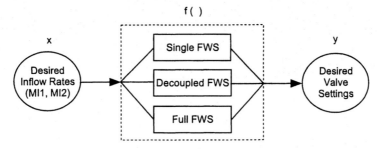

FIG. 9.13. Three strategies for performing the FWS configuration task (compare with Fig. 9.1b). See Figs. 9.11 and 9.12 for detailed accounts of each strategy.

the details of the two analyses differ substantially. Having said that, we believe that part of the difference may also be due to the fact that information flow maps have not been standardized as a modeling tool. If this tool was more systematically defined in an application-independent manner, then perhaps we would be better able to see the relationship between strategies analysis for different applications. As we mentioned earlier, the development of information flow maps remains as an important topic for future research.

A second important point is that, despite their apparent level of detail, the strategies for the FWS configuration task are categories. As such, they must be instantiated by an actor for a particular situation. These instantiations would represent specific action sequences that are exemplars of each category (see Fig. 9.3). One way to see that the strategies are categories is to note that no specific valve settings are listed in the information flow maps in Fig. 9.12. Similarly, the specific flow paths to be used are also not specified. To take a concrete example, the single FWS strategy can be instantiated for FWS A or FWS B. All of these missing details would have to be added when a particular strategy category is realized as an exemplar. These details would be chosen as a function of the context of the moment. As we mentioned earlier in the chapter, these local decisions are frequently opportunistic and idiosyncratic in nature. Modeling strategies as categories allows us to abstract away these messy details and thereby capture the features that remain invariant across instances of action.

A third related point is that, because they are categories rather than exemplars, the FWS configuration strategies are idealized. An actor need not strictly follow one strategy. On the contrary, given the different resource requirements of the various strategies (see Table 9.3), it would not be surprising to find that actors would switch between strategies to keep their work load at a manageable level, just as in the other studies we reviewed earlier. In fact, an analysis of subjects' behavior from an experiment using DURESS II shows that proficient subjects do exactly that (Yu et al., 1997).

A final general point that we should address is whether the same strategies can be used in different contexts. Recall from the previous chapter that a separate control task analysis should be conducted for each operating mode (e.g., start-up, normal operations, shutdown, fault management). For the sake of brevity, we presented an analysis for only one operating mode. As a result, we were not able to show that, in some cases, the same information-processing activity can appear in more than one operating mode. For example, in the case of DURESS II, the FWS configuration task is relevant, not just to start-up, but to normal operations as well. In cases such as this one, there is no need to duplicate the strategies analysis. The input and output states of knowledge are the same for the FWS configuration task, regardless of whether it takes place during start-up or normal operations. Consequently, the set of strategies that can be used is the same in either case. It is worthwhile noting such opportunities for consolidation because they can lead to substantial savings in analysis effort.

SUMMARY AND IMPLICATIONS

The first phase of CWA, work domain analysis, was concerned with describing the field on which action takes place. The second phase, control task analysis, was concerned with describing what tasks need to be accomplished on that field. In the third phase of CWA, strategies analysis, we have been concerned with how these control tasks can be accomplished. In complex sociotechnical systems, there is usually more than one way to achieve any given task goal. Field studies and laboratory experiments both show that people exploit this redundancy by using different strategies to perform the same task, and by switching between strategies. These opportunistic actions allow people to resolve resource-demand conflicts, thereby achieving task goals without exceeding their information-processing limitations. The findings of Sperandio (1978) in air traffic control provide a nice prototypical example. ATCs changed their strategies to keep their work load at a manageable level. A computer-based information system should support, not inhibit, these opportunistic actions. It should be comprehensive enough to provide tailored support for each strategy, and at the same time, it should be flexible enough to allow workers to switch between strategies seamlessly. The objective of a strategies analysis is to identify the requirements that can lead to the design of a support system with these characteristics.

What are the implications of strategies analysis for design? The answer to this question can be obtained by examining Fig. 9.14. This chapter has dealt with the third row in this figure. The concept of a strategy can be analyzed using an information flow map as a modeling tool, although this tool is still relatively immature and is in need of further development. The strategies representations can then be used to design dialogue modes. For example, each of the FWS configuration strategies for DURESS II could be supported by a different display that tailored information presentation to the unique demands of each strategy. One possibility would be to provide information support for the portion of the strategies that require actors to determine the quantitative valve settings that will achieve the desired flow rates. As shown in Fig. 9.12, this step is demanding because it requires both physical and functional models. Note also that the details of this step differ for each strategy. Thus, it would be useful to develop contextualized displays that could take on this resource-demanding step for each of the three strategies. Similarly, each of the diagnostic strategies described by Rasmussen (1981) could serve as the basis for a different interaction mode that was tailored to the needs of each strategy. Figure 9.9 provides one hypothetical example for the particular case of the hypothesis and test strategy. In chapter 12, we provide a detailed design example showing how a strategies analysis for library information retrieval led to the development of a computer-based device that presents different dialogue modes for different strategies.

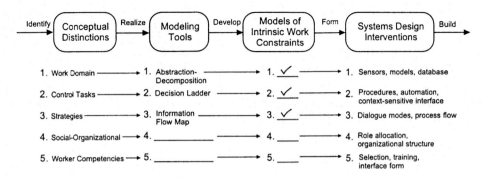

FIG. 9.14. An outline of the CWA framework with the first three phases (i.e., rows) filled in.

In addition, a strategies analysis can also be used to design process flows within a dialogue mode. For example, each of the FWS configuration strategies for DURESS II defines a different process flow sequence that could be exploited by a computer-based aid. Similarly, the diagnostic strategies described by Rasmussen (1981) also have different flow sequence constraints that could be used productively during the design of an information system. Note, however, that the objective is not to enforce a particular way of doing things. Because strategies are defined as categories, they accommodate many different action sequences. Thus, an information system designed according to such strategy definitions should be flexible enough to support many different action sequences. And as mentioned earlier, it is also important to allow seamless strategy switching if we are to design an aid that is flexible enough to support, rather than restrict, the opportunistic character of action in complex sociotechnical systems. The library information retrieval example that we present in chapter 12 shows how these ambitious objectives can be accomplished in practice.

To conclude, this third phase of CWA is important because it provides support for different ways of accomplishing task goals. But as we mentioned earlier in this chapter, there are several ways to implement any one strategy. For example, in Fig. 9.9, we provided a hypothetical example to show how a strategy can be distributed across human workers and machine automation. A strategies analysis does not explicitly address the question of *who* could be responsible for different aspects of a strategy. This issue falls in the realm of social organization and cooperation, which is the topic of the next chapter.

Phase 4: 10
Social Organization and
Cooperation Analysis

For years, Human Factors research has been focusing on the interaction between the individual user and a computer-based system. On the basis of this paradigm, considerable progress has been made in the design of human-computer interfaces. The individualistic presupposition of this paradigm is obviously invalid, however. Work is [an] inherently social phenomenon.
—Schmidt (1991b, p. 75)

It is possible for the team to organize its behavior in an appropriate sequence without there being a global script or plan anywhere in the system. Each crew member only needs to know what to do when certain conditions are produced in the environment.
—Hutchins (1995a, p. 199)

PURPOSE

The purpose of this chapter is to discuss the fourth phase of CWA, social organization and cooperation analysis, in detail. Now that we have described the field of activity (i.e., the work domain), what needs to be done (i.e., control tasks), and how it can be done (i.e., strategies), we can address the important issues of how these requirements can be distributed across human workers and machine automation, and how such actors could communicate and cooperate. The objective is to determine how the social and technical factors in a sociotechnical system can work together in a way that enhances the performance of the system as a whole. This phase of CWA is particularly important because organizational factors are being increasingly recognized as having the most pervasive influence in complex, sociotechnical systems (e.g., Leveson, 1995; Moray & Huey, 1988; Perrow, 1984; Pool, 1997; Reason, 1990). We show how the modeling tools introduced in the last three chapters (i.e., abstraction-decomposition space, decision ladder, and information flow maps) can all be productively used in this phase of analysis as well. These concepts cannot directly give us the answers we need, but they can be used as tools that help us obtain those answers. Several miniexamples are presented to illustrate the application of these tools.[1] By the end of this chapter, you should understand how social-organizational constraints inherit, and build on, the layers of constraint discussed in the previous three chapters.

---

[1]In contrast with the previous three chapters, the concepts in this chapter are not illustrated using the DURESS II process control microworld as a case study. As we pointed out in chapter 6, DURESS II is only a microworld so it does not have a social dimension to it. Consequently, it is not rich enough to illustrate the social organization and cooperation analysis phase of CWA. Accordingly, in this chapter, we use a number of illustrative examples from a diverse set of other domains.

To avoid any false expectations, it is important to point out at the outset that the concepts that we present in this chapter are not meant to replace the knowledge of disciplines that are concerned with social-organizational factors (e.g., management science, organizational behavior, sociology, etc.). Instead, the goal is to provide a set of concepts that can be used to leverage the findings from these disciplines. In particular, we are concerned with integrating social-organizational factors and technical factors into a common framework so that implications for systems design can be made more clear. A corollary of this approach is that a proper analysis cannot be conducted using only the concepts we describe. A great deal of knowledge about social-organizational factors is also required.

THE NATURE OF SOCIAL ORGANIZATION AND COOPERATION

In chapter 7, we built intuitions about work domain analysis by discussing Simon's (1981) parable about an ant navigating on a beach. In this chapter, we build intuitions about social organization and cooperation analysis by first discussing another very simple example, namely the case of ants collectively foraging for food (Resnick, 1994). You may be asking yourself why we are discussing ant foraging in a book on work analysis for complex sociotechnical systems. Needless to say, we do not mean to suggest that the actions of workers in such systems is the same as the behavior of the ants in the following example. Nevertheless, this example is useful because it helps illustrate the generic properties of self-organization, a phenomenon that is particularly important in complex sociotechnical systems. We then present a more complex example of social coordination, namely the organizational structure found on navy aircraft carriers during flight operations at sea (Rochlin, La Porte, & Roberts, 1987). This example shows that self-organization phenomena can be observed, not just in groups of ants, but in human organizations as well. We go on to argue that this type of flexible, adaptive structure is essential for meeting the unanticipated disturbances found in complex, sociotechnical systems.[2]

Example #1: The Case of Ant Foraging

The behavior of social insects provides a fascinating example of social organization and cooperation (Kugler & Turvey, 1987). We borrow from Resnick's (1994) lucid description of the specific case of ants foraging for food. In particular, we focus on the computer simulation that Resnick constructed using existing empirical research on ant colonies. First, we describe the behavior of the simulated ants, and then we describe the mechanism that is responsible for generating that behavior.

---

[2]As we pointed out in chapter 5, CWA does not rule out more centralized forms of organizational structure. There may be situations in complex sociotechnical systems that require centralized control (see the examples in chap. 5). CWA can be used to identify such cases because the analysis will show that the domain demands are comparatively closed in certain situations (i.e., there are few unanticipated disturbances). Having said that, such cases will be the exception rather than the norm in complex sociotechnical systems, as the examples in this chapter illustrate.

Figure 10.1 shows a simulated, hypothetical progression of ant foraging with the ants' nest in the center and three sources of food in the periphery. The first frame shows the initial state of the simulation with all of the ants in the nest. The second frame shows a number of ants (each represented by a small dot) wandering randomly in space. The third frame shows a qualitative shift in behavior. The ants are now forming a trail to the closest source of food, to the right of the nest. The fourth frame shows another qualitative shift. In this case, the food from the first source on the right is depleted, while a new trail of ants forms to the next closest source of food, to the lower left of the nest. In the fifth frame, this second trail grows even stronger as more ants travel to the second source. In the sixth frame, this second food source also becomes depleted. The seventh frame shows yet another qualitative shift in behavior. This time, a new trail of ants forms to the most distant source of food, to the upper left of the ants' nest. In the eighth frame, this trail also disappears as the food in that source is depleted. Finally, in the ninth frame, the behavior of the collective returns to its original mode with the ants searching randomly, this time in vain because there is no more food to be found in this simulated environment. If we look across all of the nuances depicted in Fig. 10.1, we should be struck by the fact that this behavior seems surprisingly sophisticated for a group of mere insects.

What mechanism is responsible for these organized patterns? On the surface, the behavior of the ants seems very purposeful, suggesting that it is logically planned a priori through some sort of centralized decision-making mechanism. After all, the ants were smart enough to "decide" to pursue the nearest food source first. In addition, once they made this decision, they acted in a coordinated fashion, progressing en masse to the food site. This organized collective behavior suggests that the ants have some centralized means of communicating with each other, so that they can "tell" each other where the nearest food source is. Moreover, the ants only "chose" to seek out the next nearest food source after the first source was being depleted, and so on. This logical serial order suggests some kind of rational criterion for making decisions (e.g., minimal distance to food). Perhaps the ants have some mechanism for determining how far away they are from different food sources, a clever means of communicating with each other, and some effective planning process for marshaling their collective behavior in a rational manner. Maybe there is even a leader ant—analogous to a military commander—who is the central focus of all of this activity.

As Resnick (1994) pointed out, however, the mechanisms that are actually responsible for the ants' behavior are not as "rational" or as complex as the preceding paragraph would suggest. In fact, the ants' behavior is not centralized or planned at all—there is no leader. Furthermore, the mechanism responsible for the behavior of each ant is actually quite simple. Each simulated ant's behavior can be described by a rather simple set of rules (see Resnick, 1994, for more details):

1. If a pheromone trail is not sensed (see later discussion), travel away from the nest in a random path.
2. When a food source is encountered, take some food, return back to the nest, and drop a chemical pheromone on the way.
3. When a pheromone trail is sensed, follow it to the food source.

Thus, we see that there is no centralized plan or decision-making entity that is governing the behavior of the ants as a group. Instead, the behavior of each individual

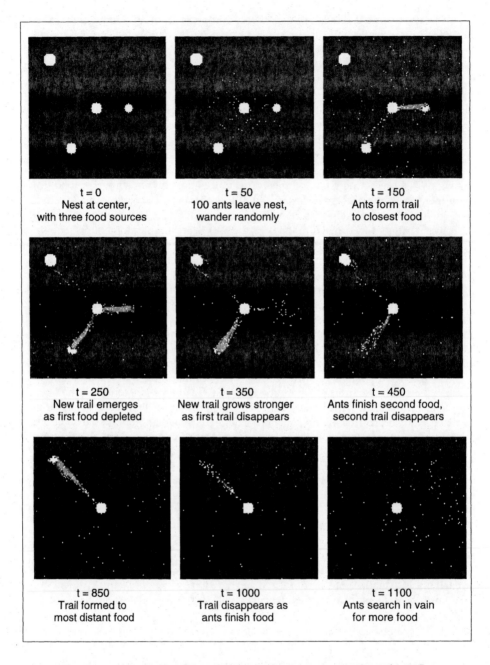

FIG. 10.1. Ant foraging with three food sources and nest at center (Resnick, 1994). Reprinted from M. Resnick (1994), *Turtles, termites, and traffic jams: Explorations in massively parallel microworlds*, published by MIT Press. Copyright © 1994 Massachusetts Institute of Technology. Reprinted by permission.

ant is described by relatively simple, local rules.[3] Nevertheless, rather complex and purposeful behavior emerges on a global (i.e., group) scale (cf. the quote from Hutchins, 1995a, at the start of this chapter).

Resnick (1994) went on to explain that one of the key contributors to this emergent complexity is the pheromone trail in the environment, shown as a light gray path in Fig. 10.1:

> When other ants detect the pheromone trail, they follow it to the food source. Then they, too, return to the nest, reinforcing the pheromone trail. Before long, there is a strong trail between the food and the nest, with hundreds of ants walking back and forth. (p. 61)

The reason why the closest food source is accessed first is simply due to the interaction between its location and random chance; because it is closer to the ants' nest, the nearest source is more likely to be encountered by the randomly exploring ants. This explanation accounts for the fact that the sources are accessed serially according to their proximity (see Resnick, 1994, for a more detailed explanation). Two other important features of the pheromone are that ants drop it only when they are carrying food, and that it gradually decays with time. Therefore, when a particular food source is depleted, the pheromone trail is not reinforced because there are no longer any ants bringing back food to the nest. The pheromone trail then becomes weaker over time due to diffusion and evaporation. As a result, the existing pheromone trail is eventually extinguished as it decays over time. Thus, not only does the pheromone concentration in the environment serve as a natural accelerator for the ants, drawing it to a food source, but it also serves as a natural decelerator as well, making it less likely that ants will travel to extinguished food sources.

Remarkable as it may seem, the patterns of organized group behavior depicted in Fig. 10.1 arise from the simple mechanisms just described, without any planning, complex decision making, or explicit communication.[4] In the words of Resnick (1994):

> The pheromone trail is an example of a large-scale, orderly structure, created entirely through local interactions. It lasts only as long as the food source. Once the food source is fully exploited, the [pheromone] trail gradually fades away . . . and the ants wander around aimlessly. (p. 64)

In short, this example of ants foraging for food is a paradigmatic case of a self-organizing system.

Self-Organization Phenomena: Emergent, Distributed, and Nonspecific

The case of insect foraging is valuable because it illustrates several general properties that are common to self-organizing systems (Kugler, Kelso, & Turvey, 1982). First, the

---

[3]The fact that real (i.e., nonsimulated) ants' behavior can be *described by* these rules does not mean that the ants *are using* rules to generate their behavior (i.e., that the rules play any causal role).

[4]According to Braitenberg (1984), it is usually very difficult for people to guess inductively the internal structure of a system just by observing its behavior. There seems to be a natural tendency to overestimate the complexity that is actually involved. Thus, rational, centralized mechanisms tend to come to mind much more easily than less "intelligent," distributed mechanisms. In the case of ant foraging, this general pattern seems to explain why it is easy for us to attribute more "intelligence" to the ants than is actually warranted. Braitenberg's artificial "vehicles" provide some additional simple, yet powerful, examples of this phenomenon (see also Resnick, 1994).

behavior of the system is temporally *emergent* rather than being completely planned a priori. The insects are not using a plan or procedure (see Footnote 3). Instead, their behavior is generated online, in real time as a function of the current context. In a very rudimentary but very concrete way, their actions are situated (Suchman, 1987).

Second, the behavior of the system is also *distributed* in several senses. The most obvious sense is that the behavior of the colony as a whole is distributed across multiple actors (i.e., all of the participating ants). It does not reside in, nor is it dictated by, any one actor. Perhaps less obviously, the behavior of the system is also distributed over the actors and their environment, particularly the information in the environment. The informative, spatial structure of the pheromone field plays as strong a role as the behavioral capabilities of the ants themselves. Just as there could be no self-organization without ants, there could be no self-organization without information in the environment. In the absence of a pheromone trail, the ants would merely wander around aimlessly.

Finally, changes in the qualitative structure of the system (e.g., the transition from the initial trial-and-error search mode to the subsequent directed foraging mode) are *nonspecific*. According to Haken (1988), this property is the hallmark of self-organizing behavior:

> A system is self-organizing if it acquires a spatial, temporal or functional structure without specific interference from the outside. By "specific" we mean that the structure or functioning is not impressed on the system, but that the system is acted upon from the outside in a nonspecific fashion. (p. 11)

In other words, for a system to be self-organizing, it must exhibit a qualitative change in structure. Moreover, that change cannot be completely specified by factors outside of the system (e.g., by another actor or by a prespecified plan). Although a self-organizing system will be influenced by its environment, the way in which the system changes its structure is partly determined by factors that are internal to the system itself. In summary, not only does the behavior of a self-organizing system spontaneously emerge over time, but so does its very structure (Kugler et al., 1982).

Of course, ant foraging is not the only example of self-organization. Biological evolution is a much more complex example (see Weiner, 1994, for a highly readable and fascinating account). And as we see shortly, self-organization is also very important in complex sociotechnical systems.

Example #2: Aircraft Carrier Flight Operations

Just as we went beyond the ant and the beach parable in chapter 7, it is important to go beyond the case of ant foraging to show the relevance of this simple example to the more complex situations in which we are most interested. With this goal in mind, we discuss the work of Rochlin et al. (1987) who conducted a field study of how U.S. Navy personnel coordinate their work activities on aircraft carriers at sea.

This application domain is very complex, as evidenced by the following set of characteristics: There is a potential for high hazard; the domain is highly technical; the domain is tightly coupled; the actors face severe time pressure; there is a great deal of personnel turnover; and training is conducted largely on the job. The following quote from a senior navy officer gives a concrete feel for the complexity involved:

So you want to understand an aircraft carrier? Well, just imagine that it's a busy day, and you shrink San Francisco airport to only one short runway and one ramp and gate. Make planes take off and land at the same time, at half the present time interval, rock the runway from side to side, and require that everyone who leaves in the morning returns that same day. Make sure the equipment is so close to the edge of the envelope that it's fragile. Then turn off the radar to avoid detection, impose strict controls on radios, fuel the aircraft in place with their engines running, put an enemy in the air, and scatter live bombs and rockets around. Now wet the whole thing down with salt water and oil, and man it with 20-year old's [sic], half of whom have never seen an airplane close-up. Oh, and by the way, try not to kill anyone. (Rochlin et al., 1987, p. 78)

Clearly, aircraft operation at sea meets many of the criteria of complexity that we identified in chapter 1. Despite this level of complexity, the navy has been able to achieve a very high level of reliability. Accidents with loss of life or equipment are extremely rare. How can such a high level of reliability be achieved in the face of such daunting demands?

Rochlin et al. (1987) argued that the secret lies in the organizational structure that is adopted to coordinate the work activities. As one would expect from a military organization, the formal organization (i.e., the one documented on an organizational chart) is rigid and centralized: "On paper, the ship is formally organized in a steep hierarchy by rank with clear chains of command and means to enforce authority far beyond that of any civilian organization" (p. 83). During much of the time, this hierarchical organizational structure does in fact govern the operations of the ship. However, Rochlin et al. found that a very different type of organizational structure is adopted during the complex flight operations described earlier. This organizational structure is informal in that it is not officially documented. Furthermore, the informal work organization is flat and distributed rather than hierarchical and centralized. Workers who are at the bottom of the command chain will make important decisions without first obtaining approval from their superiors.

Why would the organizational structure of the ship change in such a strong way when it is time to conduct flight operations? According to Rochlin et al. (1987):

This [organization] contributes greatly to the ability to seek the proper, immediate balance between the drive for safety and reliability and that for combat effectiveness. Events on the flight deck, for example, can happen too quickly to allow for appeals through a chain of command. Even the lowest rating on the deck has not only the authority, but the obligation to suspend flight operations immediately, under the proper circumstances and without first clearing it with superiors. (pp. 83–84)

This level of autonomy provides a stark contrast with the strongly hierarchical formal organization. Rochlin et al. argued that the remarkable degree of flexibility that they observed is essential for coping with the challenging demands of the job. Thus, the change in organizational structure during flight operations represents an adaptive response to a complex work environment.

Self-Organization Revisited

In the title of their article, Rochlin et al. (1987) referred to the aircraft carrier as a "self-designing organization," which suggests that there may be a parallel between their field study findings and the generic characteristics of self-organization phenomena described earlier.

Beginning with the first characteristic, the informal organizational structure exhibited on the aircraft carriers is emergent on two different time scales. First, the adoption of a flat and distributed, rather than hierarchical and centralized, structure just emerged with experience on the job. Nobody sat down beforehand to determine what organizational structure would be the most reliable and productive for aircraft operations. Thus, the switch from the formal to the informal organization during flight operations was a result of an exploratory process. Second, the specific organizational structure that is exhibited on any one occasion is also emergent. For example, people frequently change roles to help others who are being overloaded (Rochlin et al., 1987). There is a great deal of redundancy in the organization, and thus no simple or fixed mapping between people and roles. The organizational structure literally adapts locally to changes in circumstances.

The distributed nature of the organizational structure observed by Rochlin et al. (1987) should be evident from our description of their findings. Changes can occur very quickly, requiring rapid responses. As a result, the same authority and responsibility have been allocated to a number of workers in parallel because you never know who will be in the best position to respond. Furthermore, these workers are frequently near the bottom of the official organization chart. But because they have the most immediate access to the information that is required to deal effectively with life-and-death situations, they are given the autonomy to make decisions. Therefore, during flight operations, decision making is not completely planned in a centralized fashion. Instead, it is largely distributed across a number of different actors in parallel.

Finally, the informal organizational structure on aircraft carriers is nonspecific as well. The qualitative change in structure that is observed during flight operations is not imposed on the organization from the outside. Instead, it has developed from within. Through experience, the workers have found that a flat, distributed organization provides an effective structure for balancing the challenging demands of the job with the limited capacity of the resources available. In fact, reading Rochlin et al.'s (1987) article, we get the strong impression that the collective actions of the workers are highly "organic" in nature. The organization is like a living, breathing entity with a life of its own. Its actions are an adaptive response that is shaped, but not dictated, by changes in its environment.

We can reinterpret these findings in terms of the concepts we introduced in chapters 3 and 5. Flight operation on an aircraft carrier at sea is a very open system. There are substantial disturbances that cannot be anticipated up front by designers. Furthermore, these disturbances occur at a very rapid pace, making it impossible to go up the chain of command to obtain orders as to what should be done. Consequently, a centralized organizational structure cannot possibly deal effectively with the nature of the demands. Instead, a distributed organizational structure has evolved over time. Although this flat structure differs markedly from the typical hierarchical military command structure, it provides the flexibility that is required to cope effectively with unanticipated disturbances. The people who have access to the information locally are given the autonomy to make important decisions, even if those people are at the bottom of the chain of command. It is up to these workers to "finish the design." Thus, the Rochlin et al. (1987) field study provides a great example of why a distributed control architecture is required to deal with the unique demands of open systems (see chap. 5).

General Lesson: The Importance of Adaptive, Self-Organization

Rochlin et al's (1987) findings show the importance of adaptive, self-organization at a social, cooperative level. Moreover, their lessons generalize well beyond aircraft carrier operations at sea. For example, Schmidt (1991a) made the same points in the broad context of advanced manufacturing. This application domain is also teeming with unanticipated disturbances caused by a number of factors, including the rapid pace of technological change and increasing global competition. Because of the open nature of this application domain, there is a need for versatility and flexibility to adapt effectively to volatile markets. As a result, work coordination cannot be completely planned in detail up front. Instead, there is a strong organizational need for finishing the design (i.e., self-organization) through local adaptation to the situated context. According to Schmidt, such adaptation is best supported by a distributed organizational structure.

Gerson and Star (1986) took this point to the extreme, claiming that it applies to every application domain:

> Every real-world system is an open system: It is impossible, both in practice and in theory, to anticipate and provide for every contingency which might arise in carrying out a series of tasks. No formal description of a system (or plan for its work) can thus be complete. Moreover, there is no way of guaranteeing that some contingency arising in the world will not be inconsistent with a formal description or plan for the system. . . . *Every real-world system thus requires articulation* to deal with the unanticipated contingencies that arise. Articulation resolves these inconsistencies by packaging a compromise that "gets the job done," that is, closes the system locally and temporarily so that work can go on. (p. 266, emphasis in original)

The scope of these conclusions is debatable (see chap. 3), but there can be little doubt that complex sociotechnical systems are largely open and thereby require distributed, adaptive organizational structures. This conclusion is not a new one (Thompson, 1967), but it is one with which information systems designers have not yet fully come to terms. How do we deliberately design to support adaptive, self-organization?

To improve on the existing situation, a framework for work analysis must somehow take into account the fact that complex sociotechnical systems require a flexible, distributed organizational structure. This is an onerous demand because it means that the organizational structure may evolve slowly in response to long-term pressures, and may adapt very quickly to suit the needs of the moment. CWA tries to accommodate this dynamic flux by capturing the criteria behind organizational adaptation to change (Rasmussen et al., 1994).

DISTINGUISHING CONTENT AND FORM OF COOPERATION

In the previous phases of CWA, we have been primarily concerned with identifying the requirements that must be satisfied for a complex sociotechnical system to function effectively. Somehow, these requirements must be integrated with existing knowledge of social-organizational factors so that a viable organizational structure can emerge. One set of factors that needs to be considered is the set of criteria by which the previously identified requirements can be distributed across actors or organizational entities. In this section, we describe some of these criteria. Note that the discussion is intended to be illustrative, rather than exhaustive. We distinguish between two aspects of social organization and cooperation analysis, namely the *content* and the *form* of the organizational structure (Rasmussen et al., 1994). Each of these aspects is briefly described.

Content: Division and Coordination of Work

The content dimension of social-organizational analysis pertains to the division and coordination of work. Simply put, what are the criteria governing the allocation of roles among actors? As we show in the following section, it is possible to use the abstraction-decomposition space as a tool to make and represent such decisions. Thus, different actors would be responsible for working in different areas of the work domain representation. Similarly, it is also possible to use the decision ladder as a tool to make and represent such decisions. From this perspective, different actors would be responsible for different information-processing activities (i.e., boxes in the ladder). Finally, it is also possible to use information flow maps in an analogous fashion. In this case, different actors would be responsible for different aspects of a strategy. Examples of each of these cases are provided in the following section. In each case, the concepts of CWA are being used as tools, but a great deal of other knowledge is required to make decisions about what allocation schemes are feasible and desirable. Note that, throughout our discussion, multiple levels of resolution are possible. Thus, in some cases, it makes sense to divide work among individual actors, whereas in other cases it makes more sense to divide work among groups or teams of actors (Rasmussen et al., 1994).

One important point to keep in mind for now is that the division and coordination of work determines what information content actors need to perform their duties. For example, if one group of actors is responsible for a particular level of abstraction in the work domain, then the information contained in that level defines the information content that should be presented to that particular group. Similarly, if a team is responsible for a particular box in the decision ladder, then the information (e.g., data and constraints) associated with that box should be presented to that team. Also, if an actor is responsible for a particular aspect of a strategy, then the resources required to implement those steps should be made available to that actor. In each case, making decisions about how work demands should be divided up has important implications for the identification of relevant information content.

A second important point is that the division and coordination of work is a dynamic process in sociotechnical systems. As the aircraft carrier example shows, the organizational structure can change markedly depending on the circumstances. As a result, it is important for us to consider some of the criteria that can govern such shifts. Rasmussen et al. (1994) identified six criteria for dividing work demands, although others are probably also possible.

The first criterion is *actor competency*. Work demands can impose a heterogeneous set of demands on actors in terms of competency, in which case it would make sense to divide up work accordingly. For example, during surgery, demands are distributed across surgeons, anesthesiologists, and nurses, in part because each organizational role requires a different set of competencies.

A second criterion for dividing work demands is *access to information or action means.* In some application domains, actors will not have equal access to information. In such cases, it would make sense to distribute work demands so that the actors who have the most immediate access to a particular information set are responsible for making decisions associated with that information. For example, in Rochlin et al.'s (1987) field study, the low-ranking workers on the deck of the aircraft carrier were given the responsibility for making important decisions because they had the most immediate access to the relevant information. Similar arguments can be made for access to action means. For example, surgeons have the most direct access to the surgical field (i.e., the part of the body being operated on), so it makes sense that they have responsibility for decisions pertaining to anatomical interventions.

A third criterion for dividing work demands is *facilitating the communication needed for coordination.* In some application domains, dealing with work demands effectively requires coordination of multiple decisions or constraints. In such cases, it would make sense to allocate responsibility for these decisions or constraints to one actor or to a group of actors. In this way, the interdependencies that need to be coordinated can be factored in without requiring too much communication across actors or groups, respectively. For example, in engineering design, it is useful to create a single team that includes end users and human factors engineers, as well as hardware engineers, software engineers, and marketing professionals. Such an organization allows multiple criteria to be coordinated within a group rather than across groups. This scheme facilitates the communication required for coordination compared to the case where different interests are each represented by a separate group (e.g., user representative group, human factors group, software group, hardware group, marketing group).

A fourth criterion for dividing work demands is *work load sharing.* In many cases, work is distributed across actors because the demands are too great for a single actor to cope with effectively. An example can be found in some nuclear power plants when an emergency occurs on one unit in a multiunit control room. Under these circumstances, an operator of a different unit is required to come over and help the operator of the unit who is experiencing problems. The reason for this change in division of work is that the demands associated with emergency operations are too great for one actor to shoulder alone.

A fifth criterion for dividing work demands is *safety and reliability.* In many application domains, redundancy is deliberately built into the organizational structure to improve safety or reliability. This redundancy can take many different forms. For example, in commercial aircraft cockpits, the pilot and copilot are supposed to verify each other's actions when following procedures. The rationale is that two people are more reliable than one because one individual can detect the mistakes of another. In some process control plants, there are multiple automatic safety systems, each with the capability of shutting down the plant and bringing it to a safe state. By having multiple systems, a failure in one will not jeopardize safety. Thus, redundancy can involve multiple workers or multiple automated systems.

A final criterion for dividing work demands is *regulation compliance.* In some application domains, there are corporate or industry regulations that constrain how roles can be distributed across actors. In these situations, such regulations could shape the organizational structure. A simple example is the case where there is a legal agreement between union and management that specifies the job responsibilities of particular classes of workers.

It should be clear that these criteria are not independent of each other. In any one case, decisions about how work is to be divided can be governed by multiple criteria. For example, surgeons, anesthesiologists, and nurses have different roles, not just because different competencies are required for each role, but also because the work load associated with the demands of surgery could not be taken on by any one worker alone.

By adopting criteria like those just listed, we can make decisions about how work demands can be distributed across actors. Note, however, that the criteria are defined with respect to particular classes of situations, so the organizational structure can change accordingly. By using such criteria to identify the *boundaries* of responsibility of different actors, we leave room for the flexibility and adaptation that are required in complex sociotechnical systems. At the same time, however, the definition of such boundaries allows us to identify the dynamic communication needs of the various actors in the organization. These insights should become more tangible in the following section when we present various examples of division and coordination of work.

Form: Social Organization

Once work demands have been allocated across actors or groups of actors, we are still left with the question of how the various actors should communicate with each other. This issue takes us into the realm of social organization. As Rasmussen et al. (1994) pointed out, various structures are possible. For example, one possibility is to have an *autocratic* organization where one worker is in charge of all of the others. We might find this type of social organization in a small, private company. Another possibility is to have an *authority hierarchy* where there is a clearly defined chain of command. The official social organization in the military would be one example. A third category of possibilities is to adopt a *heterarchical* social organization. There are various alternatives within this category. For example, a democracy consisting of decision making by committee is common. A more extreme alternative is to have an anarchy where individuals act independently and communicate implicitly by noticing changes in the state and structure of the work domain caused by others' actions.

Of course, there are many other factors that need to be considered. We have just barely scratched the surface. But as we pointed out at the beginning of the chapter, our goal is to leverage, not replace, existing knowledge in disciplines concerned with social-organizational factors so that the implications for systems design can be made more apparent. Accordingly, the most important point to take away from this discussion is that the social organization determines the form of the communication among actors. Depending on what type of organization is adopted, the very same information can be communicated as neutral facts, as advice, as instructions, or as orders. Thus, decisions about social organization have implications for information system design.

APPLICATION EXAMPLES

Each of the modeling tools presented in the last three chapters can be used in social organization and cooperation analysis as well. In this section, we review a number of miniexamples from a diverse set of application domains to illustrate how the CWA concepts can be used. These examples show that criteria like those described in the previous section can be combined with knowledge of social-organizational factors to build an organizational structure "on top of" the requirements identified in previous phases of CWA. In this way, the abstraction-decomposition space, the decision ladder, and the information flow maps can be collectively used to define an inventory of requirements that organizational structures must be capable of satisfying.

Mapping Actors Onto the Abstraction-Decomposition Space

The first set of examples show how the abstraction-decomposition space used in work domain analysis (see chap. 7) can also be used to divide up work. Recall that this two-dimensional space identifies the information set that should be contained in a database for the entire organization. If we can distribute actors or groups of actors across different areas of the space, then we can identify what subset of the overall database would be useful for different actors or groups.

Example #1: Medicine. Hajdukiewicz (1998) identified the levels for a work domain analysis of the human body. An overview of the representation he developed is presented in Fig. 10.2a in the form of an abstraction-decomposition space. Figure 10.2b shows how this work domain representation can be used to divide up responsibilities across workers for a class of situations, in this case surgery. The darker overlay shows the areas of the abstraction-decomposition space that have been allocated to an anesthesiologist, whereas the lighter overlay shows the areas of the space that have been allocated to a surgeon. Whereas the anesthesiologist is primarily responsible for maintaining higher level bodily functions, the surgeon is primarily responsible for managing the patient's anatomy and physiology. It is important to keep in mind that Fig. 10.2b describes the cooperative structure during surgery only. Based on the results of Rochlin et al. (1987), we would expect that the structure will change to adapt to meet the unique demands of other classes of situations, such as preoperation (e.g., Rasmussen et al., 1994; Xiao, 1994; Xiao et al., 1997).

Several points are worth noting. First, the division of work is likely based on multiple criteria, including actor competency and work load sharing. Second, each worker is responsible for different but overlapping areas in the space. This partitioning leaves considerable room for situated adaptation as a function of context. In fact, the actual cooperative structure observed at any time can change from one moment to the next. Third, the mapping of actors onto the abstraction-decomposition space has obvious implications for systems design. The information and relationships identified in the work analysis can be parsed into two overlapping sets, one that is most relevant to the anesthesiologist and another that is most relevant to the surgeon. This partitioning would enable us to create displays that are tailored to the information needs of each worker. It would also allow us to identify the unique knowledge that a training program would need to induce for each worker. Furthermore, the work analysis would also allow us to create an overarching database that would provide a global, shared representation that could support communication and coordination between surgeon and anesthesiologist. Although they are responsible for different areas of the space, each worker would be able to see the relationships between their own area of concern and that of the other worker. Not only would this shared, global representation facilitate communication, but it would facilitate coordination as well. The links between levels

a) Work Domain Representation for the Human Body:

Level of Decomposition

	Whole Body	System	Organ	Tissue	Cell
Purposes	Homeostasis (Maintenance of Internal Environment)	Adequate Circulation, Blood Volume, Oxygenation, Ventilation	Adequate Organ Perfusion, Blood Flow	Adequate Tissue Oxygenation and Perfusion	
Balances	Balances: Mass and Energy Inflow, Storage, and Outflow *	System Balances: Mass and Energy Inflow, Storage, Outflow, and Transfer *	Organ Balances: Mass and Energy Inflow, Storage, Outflow, and Transfer *	Tissue Balances: Mass and Energy Inflow, Storage, Outflow, and Transfer *	
Processes	Total Volume of Body Fluid, Body Temperature, Supply: O_2, Fluids, Nutrients, Sink: CO_2, Fluids, Wastes	Circulation, Oxygenation, Ventilation, Circulating Volume	Perfusion Pressure, Organ Blood Flow, Vascular Resistance	Tissue Oxygenation, Respiration, Metabolism	Cell Metabolism, Chemical Reaction, Binding, Inflow, Outflow
Physiology			Organ Function	Tissue Function	Cellular Function
Anatomy			Organ Anatomy	Tissue Anatomy	Cellular Anatomy

(left axis label: **Level of Abstraction**)

* Balances include: Water, Salt, Electrolytes, pH, O_2, CO_2

b) Mapping Responsibilities onto the Work Domain:

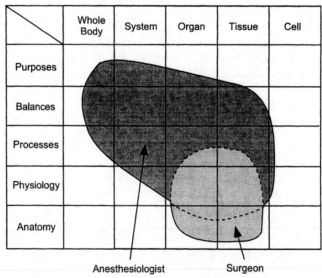

Anesthesiologist   Surgeon

FIG. 10.2. Abstraction-decomposition space (a) for the human body, (b) showing responsibilities of anesthesiologist and surgeon during surgery (Hajdukiewicz, 1998). Reprinted with permission of the author.

in the abstraction-decomposition space should allow the surgeon and anesthesiologist to observe the relevance of their own actions to the concerns of their colleague. Thus, social organization and cooperation analysis can be used to identify many important information needs for different actors in an organization.

Figure 10.2 can also help us reinforce two additional points that were made in earlier chapters. First, note that the levels of abstraction (purposes, balances, processes, physiology, and anatomy) are different from the ones that we used to represent DURESS II (see chap. 7). The reason for the difference is straightforward—the categories of constraints that govern the human cardiovascular system are different from those that govern a process control microworld. Therefore, the definition of the levels of the AH change accordingly. Second, note that only the cells that are roughly along the diagonal of the space in Fig. 9.2 were found to be of use. This finding confirms those obtained in other domains (see chap. 7).

Example #2: Engineering Design. The abstraction-decomposition modeling tool can also be used to illustrate another factor that is extremely important for social organization and cooperation analysis. In chapter 7, we pointed out that a work domain representation is purpose-dependent. Just as different types of maps are required for different purposes (e.g., landscape elevation vs. average annual rainfall), different representations of the same work domain can be developed for different purposes. The implications of this observation come to the fore now that we are discussing the social-organizational phase of CWA. Many complex sociotechnical systems require coordination across different sets of constraints, each united under a different overarching purpose (Rasmussen, 1990b; Rasmussen et al., 1994). As a result, actors must coordinate their activities across multiple "object worlds" (Bucciarelli, 1994), each constituting a different, but usually overlapping, work domain.

These ideas are perhaps best illustrated by an example from engineering design, an application domain that clearly reveals the relevance of multiple work domain representations to social organization and cooperation analysis (Rasmussen, 1990b). Burns and Vicente (1995) conducted a participant-observer field study of engineering design in industry. The observer worked as a full-time human factors consultant for a large U.S. company over a period of 4 months. The project studied was the design of an industrial control room, a large-scale effort involving over 150 people in many departments for several years. The observations documented here draw only on experiences in the design of a control panel for that control room.

Figure 10.3 shows the various representations that were developed to understand the coordination processes in the project.[5] The second through the sixth columns each represent a different AH defined by a unique set of purposes. The domains chosen were as follows:

1. Human factors design—the human factors and ergonomics constraints relevant to this project.

2. Structural design—constraints relevant to physically detailing the panel, choosing materials, manufacturers, and ensuring that the panels would be of adequate strength and meet structural requirements such as seismic qualification.

3. Implementers—constraints relevant to implementation concerns such as the manufacture, the shipping, and the construction of the design artifact.

4. Customer—the constraints imposed by the party that initiates the project and ultimately decides if the product is satisfactory.

5. Management—constraints associated with project management.

---

[5]The decomposition dimension of the space has been omitted for the sake of clarity.

These design domains were chosen because they all had different views and purposes. Yet, to reach an acceptable design solution, the constraints associated with all of these design domains had to be coordinated somehow.

The top two rows in Fig. 10.3 describe the viewpoint represented by each of these object worlds. Each design domain has a different perspective on the design problem and approaches the solution differently. Human factors designers viewed the panel as a work space to display indicators and to allow operators to use controls. For structural designers, a control panel is a physical housing for electronic equipment that must meet strength and stability requirements. The implementers view the panel as something that must be built, shipped, and installed and therefore must be made of feasible materials and meet shipping requirements and installation needs. For the customer, the panel is a highly visible part of their control room. The design of the panel is also something they have contracted out and are paying for. For management, the panel represents a project that must be completed and a contracted obligation that must be met. Management may also view the project as an opportunity to please a customer and earn future contracts. *These are all different views of the same physical design artifact*, in this case a control room panel. All of these views are correct, but only partial, views of the design problem.

The bottom three rows in Fig. 10.3 describe the levels of abstraction that comprise an AH representation in this case. Each level of abstraction is relevant to each design domain, but the contents of each level will obviously differ for each design domain (see later discussion). The three levels of abstraction were defined as follows:

1. *Objectives.* Objectives are the overall purposes of the different design domains. These are the highest level constraints that a design solution must meet to be considered acceptable by that design discipline.

Domain	HF Design	Structural Design	Implementers	Customer (Utility)	Upper Management
View	display surface for indicators and controls	physical housing for indicators and controls	something they have to produce	furnishings in their control room	contract completion
Objectives	visibility, operability	strength, stability	feasibility	image, cost	marketshare, on time, within budget
Processes	viewing angles, reach envelopes	seismic testing	manufacturing processes, shipping, installation, on-site modifications	approval process	scheduling, resource allocation
Physical Components	panel dimensions, panel geometry, room configuration				schedule, personnel, $, resources
	anthropometric data	construction materials		room dimensions, building dimensions, plant staffing	

FIG. 10.3. The various work domain representations identified in a field study of engineering design (Burns & Vicente, 1995). Each representation identifies the constraints relevant to the purposes associated with a particular view or "object world". See text for description. Reprinted from Burns and Vicente (1995), Copyright © 1995 ACM, Inc. Reprinted by permission.

2. *Functional Processes.* Functional processes are a lower level of constraints that identify the means by which the higher level objectives will be met. In a design domain, the functional processes are usually substantiated in the tests and methods used by designers of each domain. In many cases, these processes represent criteria by which the design is judged to ensure that it is meeting the design objectives.

3. *Physical Components.* The physical components constraints include such things as the raw data sources that designers use and the physical materials, sizes, and shapes of the design solution itself. The physical component constraints are those elemental constraints that are used in the functional processes. An interesting aspect of this level of abstraction is that the same physical components are often shared across different design domains.

In each column in Fig. 10.3, the contents of each of these three levels of abstraction are shown for each design domain. For the human factors designers, the objectives were to create a panel that allowed displays to be seen and controls to be operated. The processes used by human factors designers to reach these objectives were evaluations of human reach limits, or reach envelopes, and evaluations of viewing angles and distances. Human factors designers used anthropometric data to support these processes and were concerned with the size and shape of the panel in reference to the configuration of the control room.

The structural designers needed to create a housing for equipment and so their objectives were to have a panel of a certain strength and stability. These designers evaluated the design of the panel under seismic conditions and had to ensure that the panel would support the weight and size of instruments and controls chosen. These designers worked with the dimensions and geometry of the panel as well as the internal construction and the construction materials.

The implementers of the panel wanted a panel that was easily manufactured, shipped, and installed. The functional processes in this case were manufacturing processes, shipping methods, and the installation process. The physical components of the panel that these designers worked with were the construction materials, dimensions, geometry, and sectioning. For installation, these designers had to consider the dimensions of the room, the dimensions of the building (e.g., hallways and doors), and the skill levels of the installers.

The objectives of the customer were to have an attractive control panel within a budgeted cost. Their functional processes are represented in the process by which the customer approves the design work of the contracted company. In addition to the design of the panel, the customer had control over the dimensions of the building and staffing and use of the control room.

The management objectives were to complete the contract for the panel successfully. Specifically, they wanted a satisfied customer that might consider additional contracts. They also wanted the panel completed on time and within budget. Their functional processes included scheduling processes, assignment of designers to the project, and allocation of resources such as computers, consultants, and budgets to the designers. The physical components that the managers worked with were the schedule, money, personnel, and equipment.

It is interesting to note that, at the level of physical components, these domains shared concerns over many of the physical details of the design. At the higher levels though, their objectives were very different. By providing an overarching, unified representation of the space, we can provide workers with a shared, global representation that they can use as a basis for effective cooperation (e.g., to make decisions regarding design trade-offs).

From a social-organizational perspective, the most important feature of Fig. 10.3 is that it represents the various constraints that must be taken into account as designers from various disciplines try to coordinate their respective actions. For the design solution to be considered successful by any one party, it must satisfy their objectives, successfully mesh with their functional processes, and be feasible in physical components. For the design solution to be considered successful in a global sense, it will have to satisfy the constraints of all of the represented design domains. These insights can be illustrated by describing a few of the events that Burns and Vicente (1995) observed during their field study, and by mapping those events onto the representations in Fig. 10.3.

Figure 10.4 shows the trajectory representing the initial activities in the design project. The first solution concept for the panel was proposed by the human factors designers. The initial design specified the dimensions and shape of the panel (1). Anthropometric data for a population of operators was obtained (2). Using the data, the early design was checked against expected reach envelopes and viewing angles for the operators (3). Adjustments to the dimensions and geometry were made (4). This entire design sequence was performed by the human factors designers.

Figure 10.5 shows the trajectory representing a subsequent sequence of design activities. This chain of events began with a discovery of the hallway sizes in the building belonging to the utility where the control room was to be installed (1). The panel implementers compared this information with the current panel dimensions and detected a conflict with the shipping and installation processes (2). The panels, as designed, could not be moved in through the hallways of the building! The panel dimensions were modified again (3). The human factors concerns were checked (4).

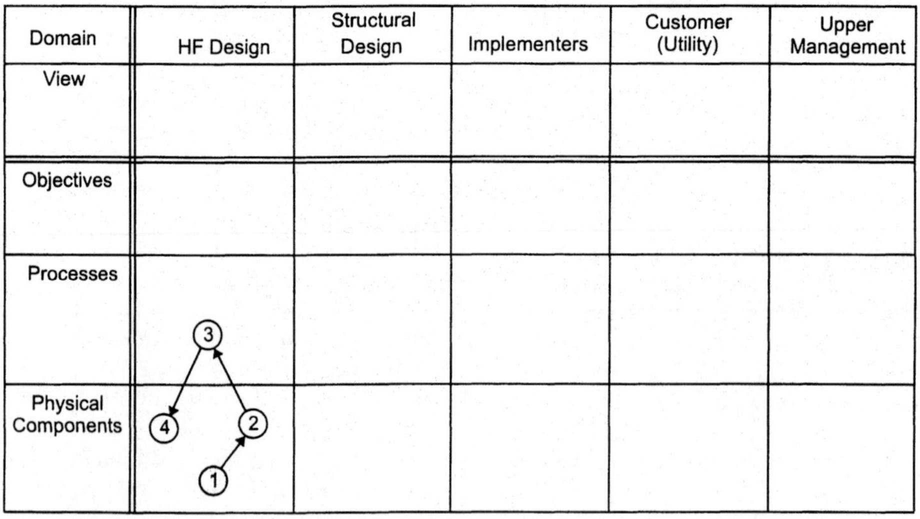

Domain	HF Design	Structural Design	Implementers	Customer (Utility)	Upper Management
View					
Objectives					
Processes					
Physical Components					

FIG. 10.4. A trace of design activity mapped as a trajectory on the representation defined in Fig. 10.3 (Burns & Vicente, 1995). See text for description. Reprinted from Burns & Vicente (1995), Copyright © 1995 ACM, Inc. Reprinted by permission.

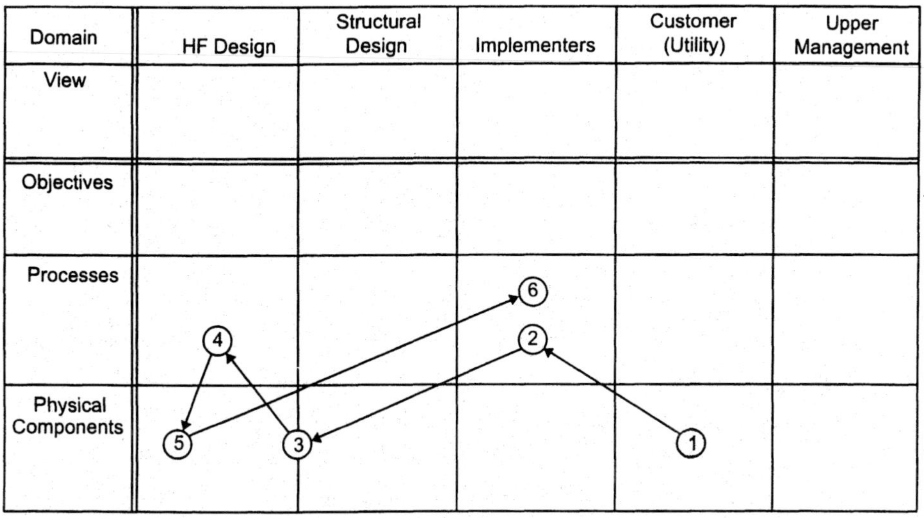

Domain	HF Design	Structural Design	Implementers	Customer (Utility)	Upper Management
View					
Objectives					
Processes					
Physical Components					

FIG. 10.5. A second trace of design activity mapped as a trajectory on the representation defined in Fig. 10.3 (Burns & Vicente, 1995). See text for description. Reprinted from Burns & Vicente (1995), Copyright © 1995 ACM, Inc. Reprinted by permission.

The decision was made to build the panels as designed but in smaller segments (5), affecting their manufacturing processes (6). Unlike the previous trajectory, this trace shows how social coordination and cooperation activities arise because of the interactions between different object worlds.

Figure 10.6 shows a final trajectory representing another trace of design activities. This episode started with a high-level constraint on the part of the customer. When they were shown the proposed control room, the customer was concerned because the panels "looked too small" (1). The engineers explained that the dimensions were based on anthropometric data and were chosen to suit a full range of operator heights. Nevertheless, the customer thought that the panels should be taller because a power plant control room should look impressive! Consequently, this higher level constraint resulted in a need to redesign the panel to be consistent with the image expectations of the customer. To resolve the problem, the customer decided to place a minimum height limit on the operators who would be allowed to work in this plant. In other words, the option the customer took was to manipulate one of their own lower level constraints—plant staffing (2)—which set up a "domino effect" (Fox, 1992). The anthropometric data set changed (3), the human factors analyses are performed again (4), and the panel was modified once more (5).

What can we learn from these examples? First, different purposes lead to different work domain representations, even if the objects being represented are largely the same. The reason for this is that the properties that are relevant, and thereby worthwhile representing, change depending on the purposes. As a result, each set of purposes leads to a different object world (Bucciarelli, 1994).

Second, different actors tend to work in different object worlds (i.e., different areas of Fig. 10.3). To take an obvious example, the human factors engineers work largely within the human factors object world. Note, however, that it is possible for actors to cross object worlds and for the same actors to be responsible for more

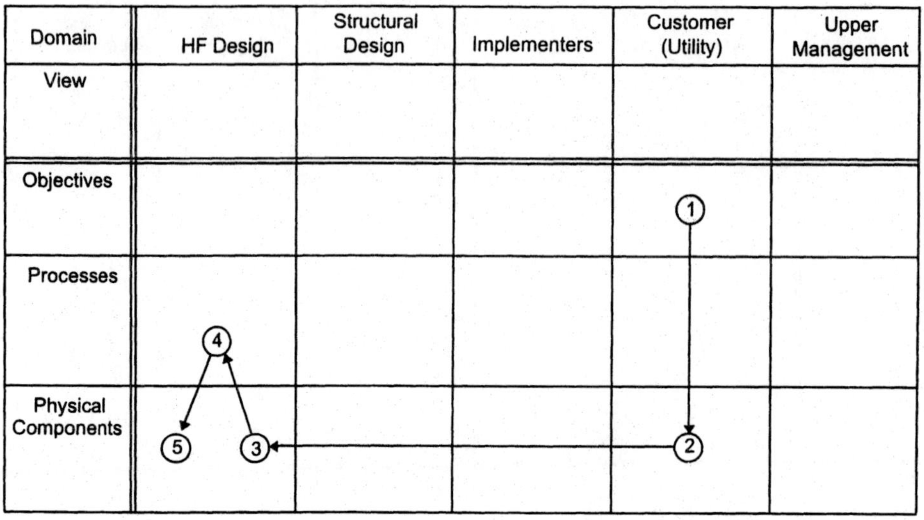

Domain	HF Design	Structural Design	Implementers	Customer (Utility)	Upper Management
View					
Objectives				①	
Processes	④				
Physical Components	⑤ ③			②	

FIG. 10.6. A third trace of design activity mapped as a trajectory on the representation defined in Fig. 10.3 (Burns & Vicente, 1995). See text for description. Reprinted from Burns & Vicente (1995), Copyright © 1995 ACM, Inc. Reprinted by permission.

than one object world. For example, in some engineering design projects, the structural design and implementation object worlds may both fall under the purview of the same group of engineers.

Third, many design decisions require the resolution of constraints that fall under different object worlds. And because different people are usually responsible for different objective worlds, there is a strong need for cooperation among actors.

Fourth, the traces of design activity appear to illustrate the properties of self-organization described earlier in this chapter (i.e., emergent, distributed, and nonspecific). The informal organizational structure changed in a nonspecific manner, depending on what was going on in the design project. In the first case shown in Fig. 10.4, the organizational structure is compartmentalized according to design disciplines. All of the activity took place in one group, namely the human factors design team. Contrast this situation with that in the second case shown in Fig. 10.5. Here, there is a flurry of emergent activity that cuts across several object worlds and that was distributed across several design teams. The designers could no longer work solely within their disciplinary groups. Instead, they spontaneously had to create a very transient organizational structure that supported the intergroup cooperation that was required to deal with the problem of the moment. These findings are reminiscent of those of Rochlin et al. (1987) in the aircraft carrier domain.

Fifth, by developing multiple work domain representations like those shown in Fig. 10.3, we can provide the information support that is required to accommodate this self-organization process. That is, by building a global, shared representation of the constraints in the various object worlds into a computer-based information system, we provide the various organizational groups (in this case, the design teams) with a basis for coordinating their activities in a flexible and adaptive manner (Pejtersen, Sonnenwald, Buur, Govindaraj, & Vicente, 1997). Therefore, the idea of multiple object worlds, each defined by an AH representation, has important implications for systems design.

Summary. These examples show that the abstraction-decomposition space can be used to divide up work demands among various actors in a flexible, adaptive fashion. The information contained in the work domain representation represents a set of requirements that the organization must face. Using some of the criteria that we identified in the previous section (e.g., competency), we can map different workers (e.g., surgeon vs. anesthesiologist) or groups (e.g., various design teams) onto different areas of the space. Doing so provides important implications for systems design because we can identify the information content that different actors will need to fulfill their organizational responsibilities effectively. For instance, the unique training and information display requirements associated with each role can be identified. In addition, a global, shared representation of the entire work domain can be created to foster fluent, dynamic cooperation.

Mapping Actors Onto the Decision Ladder

The decision ladder, first introduced in chapter 8 as a tool for modeling control task requirements, can also be used in social-organizational analysis. The basic idea is that the tasks that fall under the purview of different actors (or groups) can be identified by partitioning the requirements identified during a control task analysis. The idea is very similar to the one described in the previous subsection. The primary difference is that, instead of allocating different actors to different areas of the abstraction-decomposition space, we are allocating different actors to different in-formation-processing activities (i.e., boxes) in the decision ladder.

An informative example of how this can be done can be found in the work of Rasmussen and Goodstein (1987) on role allocation in the nuclear industry. One of the important insights in their work is that decision activities can be allocated among three types of actors: computer automation, operators, and designers. The notion of distributing information-processing activities to a designer is unusual (although, see Hutchins, 1995a), but the following examples show the rationale behind this approach.

Example #1: Automated Shutdown. Figure 10.7 provides a hypothetical example of how the decision ladder can be used to distribute control tasks across actors. The situation being analyzed is the automatic safety shutdown of a nuclear reactor. There is one decision ladder for the designer, another for the computer, and another for the control room operator. The shaded circles and boxes identify the control tasks for which each actor is responsible. Beginning at the top of the designer's ladder, the designer evaluates company policy and legal requirements to determine what level of safety is unacceptable and thereby warrants an emergency shutdown. Based on this evaluation, a loss of control of the energy balance in the reactor core is assumed. Working backwards, the designer then determines what symptoms would be expected under these circumstances. Moving now to the right of the ladder, the designer determines that an emergency shutdown can be achieved by inserting the rods in the reactor and by maintaining cooling. Then, the designer formulates a procedure in the form of a valving and switching sequence. Note that all of these information-processing activities are conducted offline during design, not online during operation (cf. chap. 5).

Moving now to the shaded boxes in the lower left of Fig. 10.7, we can see what the role of the computer is to be continually vigilant, monitoring the relevant

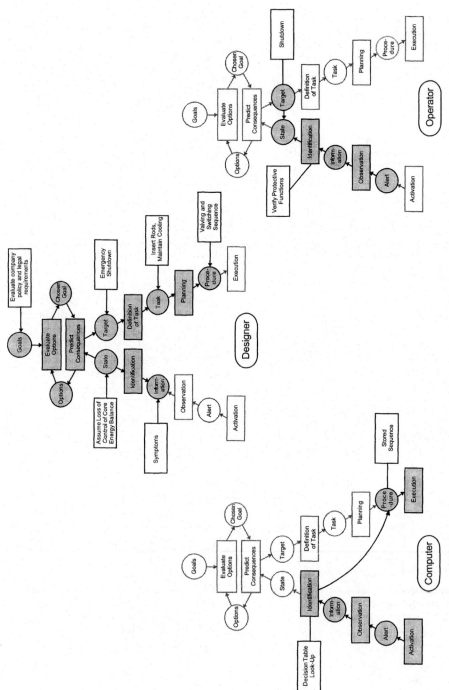

FIG. 10.7. Distribution of control tasks across designer (top), computer (left), and operator (right) for automatic safety shutdown of nuclear reactor (Rasmussen, 1986a). Reprinted with permission of the author.

shutdown variables. These observations are to be compared with the symptoms in a decision table that has been defined by the designer, a priori. The decision table consists of a set of threshold values that can trigger a shutdown. As soon as the current state of any relevant variable exceeds its respective threshold value, an automatic shutdown sequence is initiated. This leap in logic is represented as a shunt in the computer's decision ladder. To carry out the shutdown procedure, the computer then uses the sequence of actions that has been stored in it by the designer. Note that, because the designer has taken on the responsibility of conducting much of the analysis beforehand and has built the results of that analysis into the computer, there is comparatively little left for the computer to do online.

Finally, moving over to the shaded boxes in the bottom right of Fig. 10.7, the activities for which the operator is responsible are shown. Once the threshold values in the decision table are exceeded, the operator is notified that the target state to be achieved is a shut down reactor. The operator is then required to monitor the relevant work domain variables and to verify that the automated safety system is functioning as designed. Thus, in this particular case, the operator's role is that of a supervisor of a completely automated process.[6]

Example #2: Operator Intervention. Figure 10.8 illustrates another example from Rasmussen and Goodstein (1987), showing how actors can be mapped onto the decision ladder. In this case, the situation being analyzed is a loss of coolant accident requiring protection of a nuclear reactor via operator intervention. Again, there are three decision ladders, one for each actor. The information-processing activities taken on by the designer are roughly the same as those in the previous example. A significant difference, however, appears when we look at the decision ladder for the computer in the bottom left of Fig. 10.8. Instead of checking to see if the symptoms in a decision table are satisfied and initiating an automated shutdown sequence, in this example the computer plays a more modest role. It collects information regarding the status of the relevant parameters and displays that information to the operator. Thus, the computer is supporting the operator rather than telling him or her what to do.

If we move to the operator's decision ladder in the bottom right of Fig. 10.8, the shaded boxes illustrate that it is the operator's responsibility to identify the event using the information presented by the computer. Based on the results of the designer's analysis (see top of Fig. 10.8), the operator would be trained to know what symptoms justify a shutdown sequence. If these symptoms are observed, then the operator knows that the task is to shut down the reactor. This shortcut is shown as a shunt in the bottom right of Fig. 10.8. However, rather than carrying out that intention directly, the operator gives an order to the computer to start the shutdown sequence. Returning to the computer's decision ladder in the bottom left of Fig. 10.8, we see that, once this order is received, the computer can automatically carry out

---

[6]It might look like that this centralized organization is inconsistent with the philosophy of implementing a distributed organization so that workers can finish the design. However, in chapter 5, we mentioned that there are situations where a particular task should be fully automated or proceduralized. The examples we gave included situations where there are no degrees of freedom available for action, the time to react is beyond human capabilities, and the consequences of failure are extremely severe and the designers are confident that their preplanned solution will work in every conceivable situation. The example in Fig. 10.7 is consistent with these criteria.

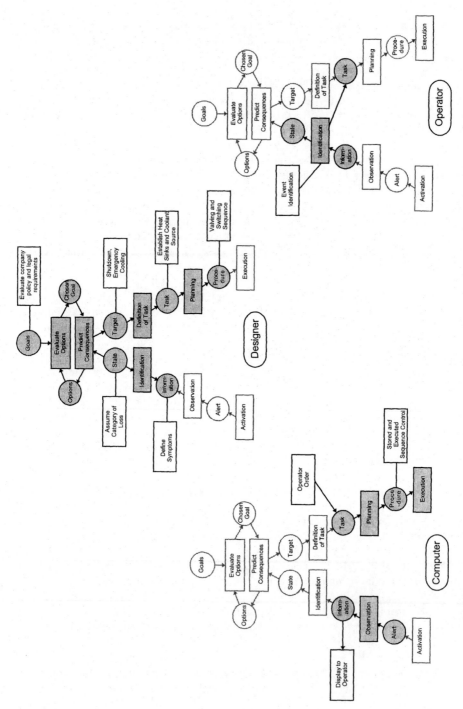

FIG. 10.8. Distribution of control tasks across designer (top), computer (left), and operator (right) for protection of nuclear reactor by operator intervention (Rasmussen, 1986a). Reprinted with permission of the author.

the task because it has a stored sequence of control actions that have been identified a priori by the designer.

This example differs from the previous one because the operator is left in charge of diagnosing the event and issuing the shutdown order. Such an organization would be justifiable in cases where the designer is not completely confident that their preplanned solution will work in every conceivable situation.

Summary. These two examples show how the decision ladder can be used to address social-organizational issues. A control task analysis identifies what needs to be done for different operating modes. Given these requirements, a social organization and cooperation analysis can then distribute these demands across actors or groups. The examples we presented involved a designer, computer, and operator. However, it is possible to use the same modeling technique to represent alternative control architectures. For example, an operator responsible for supervising several automatic controllers that act directly on the work domain could be represented by a hierarchical set of decision ladders, one for the operator and one for each of the automatic controllers. Note that, regardless of the particular architecture, the decision ladder is merely being used as a tool. A great deal of other knowledge is required to make decisions about what kind of architecture is feasible and desirable.

Nevertheless, the ladder serves as a very useful and effective means of (a) summarizing what responsibilities there are to distribute across actors or groups, including designers, (b) documenting the results of such decisions on a well-defined template, (c) ensuring that all responsibilities have been allocated, (d) illustrating the communication and coordination that is required across actors to perform certain control tasks, and (e) identifying the information needs of each actor, once an allocation decision has been made.

Mapping Actors Onto Information Flow Maps

The information flow maps that we used to conduct a strategies analysis (see chap. 9) can also be used to address cooperation issues. We already gave one hypothetical example in the previous chapter, where the steps in the hypothesis-and-test troubleshooting strategy were distributed across a human worker and computer automation (see Fig. 9.9). In this subsection, we provide a second hypothetical example for another of the troubleshooting strategies described by Rasmussen (1981).

Recall from chapter 9 that one of the strategies that the technicians were observed to use was a decision table search. Figure 10.9 shows the information flow map for this strategy. The technicians in Rasmussen and Jensen's (1973, 1974) field study performed this strategy without any computer support. Thus, they had to have a library of state models stored in their memory. Then, they had to scan and select a state model to see if its reference patterns matched the current set of symptoms. However, now that the strategy has been identified, it is possible to distribute its demands over worker and automation. Following the philosophy of having workers finish the design, we can keep the worker in charge of strategy selection while having the computer take on the burden of dealing with the more resource-intensive aspects of the strategy.

Figure 10.9 shows how the decision table search strategy could be partitioned to achieve these objectives. The workers would be kept in charge of the strategy.

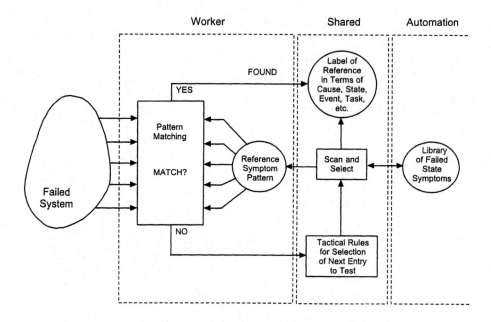

FIG. 10.9. Example showing how the resource-demanding aspects of the decision table search strategy can be off-loaded to automation, thereby resulting in a distributed cognitive system to perform the very same strategy (adapted from Rasmussen, 1981).

However, a computer aid could be constructed to allow workers to browse a library of state models in a way that is consistent with the workers' tactical search rules. Moreover, the computer could take on the responsibility of storing and labeling the state models themselves. These models could be arranged hierarchically to allow workers to narrow down their focus of attention in an efficient manner. In addition, state models could be shown according to their probability of occurrence. A more sophisticated aid might even have the capability to acquire and restructure new state model entries, adding to the library as more and more experience is gained with different types of equipment failures. With such an aid, workers could still adopt the very same strategy, but the amount of effort and knowledge that they would require to carry out that strategy effectively would be considerably reduced. After all, a great deal of information-processing capability and knowledge would already be built into the aid.

Summary. A strategies analysis identifies the various strategies that can be used by actors to perform the required control tasks. However, it leaves open the issue of who is responsible for performing different steps in the strategy. Recent research in cognitive science has shown that human performance can be improved by creating external representations to aid workers (e.g., Hutchins, 1990, 1995a, 1995b; Zhang & Norman, 1994). The information flow maps used in strategies analysis can be reused in the social organization and cooperation analysis phase of CWA to address this issue. As shown by the example in Fig. 10.9, information flow maps provide a basis for creating distributed cognitive systems that functionally integrate workers and automation (see also Fig. 9.9). This provides yet another example of how social-organizational analysis has important implications for systems design.

Mapping Actors Onto a Social Organization

So far in this section, we have been concerned with identifying the content of social organization and cooperation. In this final subsection, we very briefly discuss how to identify the form of social communication. The literature pertinent to this issue is vast, spanning various disciplines, and is thus largely beyond the scope of this book. Consequently, we limit ourselves to a cursory introduction to the topic.

Figure 10.10 illustrates how the style of organizational communication and interaction builds on the content of communication. The bottom layer of the figure shows that different actors have been mapped onto different areas of the work domain. For example, actors A and C work in two independent areas of the work domain representation, whereas actors A and D work in two different but overlapping areas. Several examples of this mapping between actors and work domain were given earlier in this chapter. The middle level of Fig. 10.10 shows how the various actors coordinate their communication patterns. In part, the exchange of information will be shaped by the work demands (i.e., what is required to get the job done). However, as shown at the top of the figure, the form of the communication patterns is also constrained by the social organization that is chosen. For example, if an autocratic structure is adopted, then information will be exchanged in one way; whereas if a heterarchic structure is adopted, then the very same information will be exchanged in a different way. In the example in Fig. 10.10, an autocratic structure is shown,

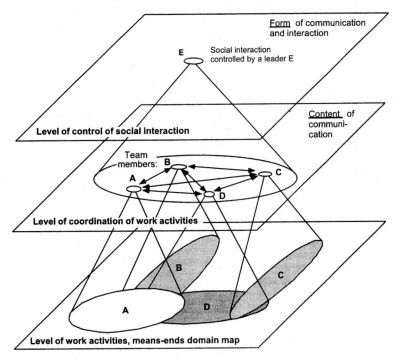

FIG. 10.10. Hypothetical example showing how management style (top layer) affects communication between actors. Adapted from J. Rasmussen, A. M. Pejtersen, & L. P. Goodstein, *Cognitive systems engineering*, Copyright © (1994, John Wiley & Sons, Inc). Reprinted by permission of John Wiley & Sons, Inc.

with actor E controlling the social interaction between team members A, B, C, and D.

Summary. Figure 10.10 shows the relationship between the style of the social organization and the form of communication in an organization. However, it is clear that our very brief description does not do justice to the complexity and importance of this topic. This aspect of CWA needs to be fleshed out in much more detail by drawing on research from organizational behavior, management science, sociology, and industrial/organizational psychology.

SUMMARY AND IMPLICATIONS

Experience in the form of field studies and incident analyses have shown that social-organizational factors play a crucial role in complex sociotechnical systems. Thus, CWA should be concerned with identifying requirements, not just for the design of devices, but for the design of corresponding organizational structures as well. Any mismatch between the two is bound to lead to significant operational problems.

This chapter has addressed these issues by describing the social organization and cooperation phase of CWA. As shown in Fig. 10.11, all of the modeling tools that were used in previous phases of CWA can be used in this phase as well. The abstraction-decomposition space, the decision ladder, and the information flow maps collectively provide complementary frames of reference that can be used to distribute the requirements identified in previous analyses across multiple actors or groups. It is very important to note that the resulting organizational structure can involve multiple machine actors and multiple workers coordinated together. And because the demands in complex sociotechnical systems are open, the organizational structure will shift to satisfy the demands of the moment, as in Rochlin et al.'s (1987) aircraft carrier example.

These modeling tools have important implications for systems design. By using them in a social organization and cooperation analysis, we should be able to identify how responsibility for different areas of the work domain may be allocated among actors, how control tasks may be allocated among actors, and how strategies may

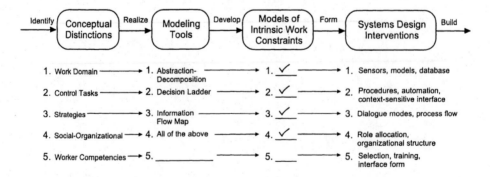

FIG. 10.11. An outline of the CWA framework with the first four phases (i.e., rows) filled in.

be distributed across actors. Thus, a social-organizational analysis describes how actors may be organized into groups or teams, how they may communicate and cooperate with each other, and what authority relationships may govern their cooperation. These general implications can be translated into more specific design products, including (a) a shared, global representation to support coordination and communication, (b) training requirements that are tailored to the responsibilities of each actor, (c) displays that are tailored to the information requirements of each actor, and (d) functional integration across workers, designers, and automation.

The final question that remains to be answered by CWA is what types of competencies do workers need to function effectively. Now that we know what demands must be satisfied, and who will be responsible for satisfying them, we can determine the constraints imposed on the skills, rules, and knowledge of individual workers. This topic is addressed in the next chapter.

Phase 5: Worker Competencies Analysis 11

What we need is not a global quantitative model of human performance but
a set of models which is reliable for defined categories of work conditions
together with a qualitative framework describing and defining their coverage
and relationships.

—Rasmussen (1983, p. 257)

PURPOSE

The purpose of this chapter is to discuss the fifth and final phase of CWA, worker
competencies analysis, in detail. Only at this point do we finally address the traditional
core concerns of the human factors and HCI communities (i.e., the relevance of
human strengths and weaknesses to systems design). Following the logic of the
formative approach described in chapter 5, the overall objective in this phase is to
identify the competencies that an ideal worker should exhibit. That is, rather than
assuming a particular set of worker competencies up front, the analysis lets the
requirements of the application domain tell us what kinds of competencies workers
need to have to function effectively.[1] This overall goal is achieved in two steps. First,
the analysis consolidates the requirements imposed by the previous phases of CWA.
Decisions made in those phases affect the competencies that workers need to function
effectively. For example, if certain responsibilities are off-loaded to computer automation, then workers' responsibilities may be lessened. Therefore, knowledge of
how demands are to be partitioned allows us to identify the competencies that
workers need to cope effectively with the demands that they are responsible for
satisfying. Second, the worker competencies analysis helps us determine how those
requirements can be met in a way that is consistent with human limitations and
capabilities. Note, however, that the goal is to leverage, not replace, existing knowledge of human cognition (e.g., Wickens, 1992).

We introduce the skills, rules, knowledge taxonomy (Rasmussen, 1983), a modeling tool that can be used to conduct such an analysis. This taxonomy provides a
framework within which domain requirements and existing knowledge of human
cognition can be integrated so as to make their collective implications for systems
design more obvious. We again use the DURESS II process control microworld as
a detailed example by which to illustrate this final layer of CWA. By the end of this
chapter, you should be able to understand how the constraints analyzed in the

---

[1]As we mentioned in chapter 5, CWA can also be used in a descriptive rather than a formative fashion.
For example, if we are designing a device for an existing set of workers, then we can identify the
competencies that workers currently exhibit (as opposed to the ones that they need to exhibit). The
results of this analysis can then be compared to the requirements identified in the earlier phases of CWA.
This comparison will tell us whether the existing workers are currently capable of meeting the demands
imposed by the application domain.

previous phases of CWA collectively influence the competencies that workers need to function effectively.

WORKER COMPETENCIES: CONTEXT AND OBJECTIVES

How does worker competencies analysis fit into CWA, and what do we hope to accomplish with such an analysis? To answer these questions, we briefly review the relationship between the five layers in the CWA framework. This review helps us put worker competencies analysis into context. Then, we briefly discuss the challenges that are involved in deriving design implications from knowledge of human strengths and weaknesses.

The Ecological Approach Revisited

In chapter 2, we devoted considerable effort to describing why a framework for work analysis should incorporate both cognitive constraints and environment constraints. Ignoring either one will result in significant problems, albeit of a different type. For instance, failing to take into account cognitive constraints will lead to a design that is difficult for workers to use. The control room design of the TMI nuclear power plant control room provides one well-known example (Rubinstein, 1979). There were many human factors design deficiencies, including poorly labeled displays and controls, unclear procedures, and counterintuitive layout of displays. Failing to take into account environment constraints, on the other hand, will likely lead to a design that does not have the functionality that is required to meet the control requirements of the domain. Interestingly, the TMI control room provides some examples of this type of deficiency as well. Several variables that were relevant to the purposes of the work domain were not sensed and thus not displayed to workers (e.g., the flow rate through the auxiliary feedwater subsystem and the actual state of the pressurizer relief valve).

It is important that the difference between these two types of deficiencies be crystal clear. The first, caused by ignoring cognitive constraints, reflects a lack of appreciation for the capabilities and limitations of human information processing. The second, caused by ignoring environment constraints, is completely different because it has very little directly to do with human capabilities. Instead, it reflects a lack of appreciation for the demands imposed by the application domain being analyzed (i.e., the information and functionality that it takes to get the job done). Environment constraints should be incorporated into work analysis because they represent the demands that have to be taken into account by any actor. Even if the entire system were to be completely automated, the very same environment constraints still have to be identified and built into the design, otherwise the system will not always function properly. Therefore, as we pointed out in chapter 2, it is important that a framework for work analysis incorporate both cognitive and environment constraints.

The other point that we made in chapter 2 was that work analysis should be based on an ecological approach (Flach et al., 1995; Hancock et al., 1995). That is, work analysis should begin by studying the environment constraints and only then

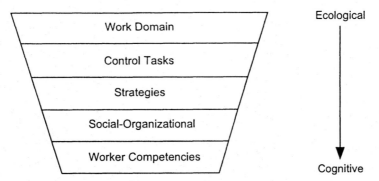

FIG. 11.1. The transition from ecological to cognitive considerations in CWA.

study the cognitive constraints. Why is it important to start and drive work analysis from environment constraints? The rationale we provided in chapter 2 was that cognitive compatibility is of little use unless ecological compatibility has already been established. For example, making an interface compatible with a worker's mental model will not do much good if that model does not correspond with the way in which the environment actually behaves. It is for this reason that environment constraints need to be given priority. Several concrete examples were provided in chapter 2 to support this conclusion.

However, this does not mean that there is no room for cognitive constraints. On the contrary, attention to human characteristics is also important and should be given a significant place in work analysis. Once it has been established that environment constraints have been respected, knowledge of human characteristics provides an important basis for making the remaining design decisions.[2] In this way, work analysis can make sure that physical and social reality are not ignored, while simultaneously ensuring that cognitive characteristics are not overlooked either. The end result should be computer-based systems that present workers with the information they need to develop accurate mental models in a form that is compatible with existing knowledge of human cognition.

How are these insights built into the CWA framework? A preliminary answer to this question was provided in chapter 5, particularly in Figs. 5.2 and 5.3. Now that we have seen in detail what four of the five phases of CWA look like, it may be useful to revisit this issue. Doing so not only consolidates our knowledge of CWA, but it also places the fifth and final phase, worker competencies analysis, in its proper context. Figure 11.1, a reproduction of Fig. 5.2, shows the ecological orientation of CWA. The figure depicts a funneling-in with each successive phase because the degrees of freedom available for action are reduced as we gradually transition from environment constraints to cognitive constraints.

As we pointed out in chapter 7, work domain analysis identifies the field of action. It delimits the inventory of structural means that are available to any actor, and shows how those means are connected to higher level functions and purposes. These

---

[2]Admittedly, in some application domains (e.g., the information retrieval example discussed in the next chapter), the environment does not have as much structure, in which case cognitive constraints will play a predominant role. In other application domains (e.g., DURESS II), the environment has a great deal of structure, in which case cognitive constraints will play a less prominent role. Regardless of the application domain, however, each of the five phases of CWA should be addressed roughly in the stated order, although in practice there will be extensive iteration between phases.

constraints must be obeyed by any actor. For example, ATCs cannot make an airplane go faster than it is capable of, nor can they move the geographical location of the landing strip or any nearby mountains or buildings. The set of work domain constraints needs to be considered by any actor, if there is to be any chance of consistently achieving the purposes for which the work domain was designed. Thus, work domain analysis provides a foundational bedrock layer of constraint in CWA.

But as we pointed out in chapter 8, the work domain is not the entire story. There are additional constraints imposed by the control tasks that need to be performed. These constraints should also be obeyed by any actor. More important for the present purposes, control task constraints go beyond work domain constraints in the sense that they add additional requirements on action. For example, the landing gear on airplanes should be down while landing but up while cruising. A work domain analysis will not explicitly identify these relationships because it merely represents the possibilities for action, independent of any particular situation. Thus, task-dependent information can be found only in a control task analysis. This simple example shows how control task constraints narrow down the meaningful possibilities for action beyond the boundaries identified by the work domain constraints.

Strategies analysis, described in detail in chapter 9, narrows down the possibilities even further. For each control task that needs to be performed, there may be several alternative strategies that can be used as generative mechanisms by which to satisfy the required input–output function. Each strategy represents a different category of processes, and thus rules out particular ways of performing the task. For example, each of the strategies identified by Sperandio (1978) for air traffic control includes some action sequences but excludes others. These constraints cannot be found in a control task analysis, because it only provides a black box representation of what needs to be done. From the perspective of that layer, each strategy looks the same because it satisfies the same input–output function (e.g., land the planes safely). Thus, the constraints added by strategies lie in the processes that can be used, not in the product that is to be achieved. As with the other layers, the requirements identified by a strategies analysis are relevant to any actor because each strategy can, in principle at least, be distributed across a wide number and configuration of human workers or computer automation.

By now, you should be able to see how the social organization and cooperation constraints discussed in chapter 10 fit into this progression. Each strategy can be implemented in different organizational structures. By choosing one structure over another, we add a new set of constraints that cannot be found in the strategy descriptions themselves. For example, the decision table search strategy identified by Rasmussen and Jensen (1973, 1974) could be implemented by the troubleshooting technicians on their own, or it could be distributed across technicians and a computer support tool (see Fig. 10.9). These two structures define two different sets of constraints because the actions in which a worker can engage differ in the two cases. Thus, a social organization and cooperation analysis identifies a unique layer of constraint that adds to the requirements identified by previous layers of CWA.

As shown in Fig. 11.1, these four layers percolate down to have a direct impact on the demands placed on the workers in a complex sociotechnical system. Collectively, they affect the competencies that are required of individual workers. For example, in the case of the medical domain, a surgeon should be capable of dealing with one set of demands, whereas an anesthesiologist should be capable of a different (although overlapping) set. But just as we found in each of the other layers of CWA, a particular phase of analysis does not just inherit the constraints of previous phases. It also adds new requirements that are not explicitly represented in previous phases. For the particular case of worker competencies analysis, the new set of constraints are the capabilities and limitations of human cognition. There are some things that people do well, and other things that they do not do well. Research and practical experience in human factors and HCI has shown that systems design should be based on an understanding of these constraints. Otherwise, the resulting design will suffer from significant human factors deficiencies.

In summary, the purpose of a worker competencies analysis is twofold. First, it should help us consolidate the requirements imposed by the preceding phases of CWA. Second, it should help us determine how those requirements can be dealt with in a fashion that is consistent with what we know about human cognition. Thus, the goal is not to replace existing knowledge of human capabilities and limitations, but rather to leverage that knowledge to derive important practical implications for systems design.

From Cognitive Constraints to Systems Design

Based on the description just given, it should be clear that worker competencies analysis must draw on existing knowledge of human capabilities and limitations. The challenge, however, is to collect and organize this knowledge, integrate it with the requirements identified in previous phases of CWA, and derive implications for systems design. This is a complex problem because the psychological literature is highly fragmented (de Groot, 1990; A. Newell, 1973). There are many theories and models that are potentially relevant to systems design, but each is applicable only to narrow psychological phenomena. Table 11.1 lists a small sample of the many theories and models that are available to choose from. On the positive side, we should feel fortunate that there is so much potentially relevant knowledge to draw on. On the negative side, it is not at all obvious how to identify the knowledge relevant to any one design problem and then package this knowledge into a form that makes the practical implications for systems design more obvious.

The challenge is exemplified by the experience of one of our colleagues in the nuclear industry. Based on the extensive analyses that were conducted shortly after the TMI incident, this colleague thought that he and his coworkers had a good

TABLE 11.1   Examples of Models and Theories That Describe Human Capabilities and Limitations for Particular Types of Psychological Phenomena

Attribution theory	Decision theory
Estimation theory	Fuzzy set models
Manual control models	Optimal control models
Production systems	Psychological problem-solving models
Queuing theory	Sampling theory
Scripts	Signal detection theory
Social judgment theory	

Note. See Rasmussen (1986c) and Rouse (1980) for descriptions of each example.

understanding of the demands that nuclear power plant operators had to face. Thus, he reasoned, it should be a relatively straightforward job to consult the psychology and human factors research literatures to determine how best to design for these demands. Much to his surprise, our anonymous colleague found it very difficult to uncover research findings that were useful for design. As he put it:

> We found that one researcher was working on the eye, another was working on the ear, another was working on memory, and so on. Meanwhile, we had to design a control room for an entire human being. We just couldn't put all the little pieces together in a way that would address the overall demands we had identified.

Thus, the challenge in conducting a worker competencies analysis is to pull together the requirements of the application domain and the relevant subset of existing knowledge about human cognition in an integrated way. This integration should allow us to derive practical implications for systems design. In the next section, we present a modeling tool that can be used to help meet this objective.

THE SKILLS, RULES, KNOWLEDGE TAXONOMY

The skills, rules, knowledge (SRK) taxonomy was developed by Rasmussen (1983) as a way of addressing the needs identified in the previous section. It is not intended to replace existing knowledge of human cognition, but rather to help organize that knowledge into a form that is more useful for systems design. Although the SRK taxonomy is widely cited, reaching the status of a "market standard" in some research communities (Reason, 1990; Sanderson & Harwood, 1988), there are still many misconceptions about its form and role. Thus, it is important that we describe its basic structure, as well as several illustrative examples of its application to research and design problems.

Basic Structure

As Sanderson and Harwood (1988) pointed out, it is very easy to misinterpret the SRK taxonomy because the role that it is intended to play is an unusual one. To avoid any misconceptions, we begin by describing the foundational premises on which the taxonomy is built, and then go on to describe some of its basic properties.

Foundational Premises. The first general point that needs to be made is that the SRK hierarchy is a taxonomy, not a model. More specifically, the SRK taxonomy provides "some basic distinctions which are useful in defining the categories of human performance for which separate development of models is feasible" (Rasmussen, 1983, p. 257). There are several aspects to this quotation that are worth highlighting. First, the taxonomy provides a set of basic distinctions, not a detailed model of psychological processes. Second, one of the primary criteria in the development of the taxonomy is usefulness, not necessarily "truth." Third, each level in the taxonomy corresponds to a category of human performance. The idea is to distinguish things that are of a different nature—to keep from "mixing apples and oranges." Fourth, respecting these distinctions should make it easier to develop models for each category of human performance. Otherwise, the modeling process will be contaminated by the fact that we are trying to model fundamentally different things at the same time. By cleaving apart apples from oranges, we should be better able to develop models for each category.

What criteria should we use to distinguish between different categories of human performance? Many criteria are possible, of course, and different criteria are useful for different purposes. But according to the purpose just stated, we should select criteria that make it easier to develop separate models of human performance. Rasmussen (1983) translated this purpose into the following premises:

> Meaningful interaction with an environment depends upon the existence of a set of invariate constraints in the relationships among events in the environment and between human actions and their effects. . . . The constraints can be defined and represented in various different ways which in turn can serve to characterize the different categories of human behavior. (p. 258)

In other words, the SRK taxonomy is defined by distinguishing categories of human behavior according to fundamentally different ways of representing the constraints in the environment.

We can unpack these premises into a series of logical steps. The starting point in the logical sequence is that goal-directed interaction between a worker and an environment depends on constraints. If workers are to achieve their goals reliably and consistently, then they must somehow take these constraints into account. Take the simple example of catching a ball in midair. One of the constraints that needs to be taken into account in this act is the force exerted by gravity. If a person is to catch the ball reliably and consistently, then they have no choice but to take this constraint into account. If they ignore gravity, then they will not be able to place their hand in the position that is required to intercept the ball. The same logic applies to the different types of constraints that are relevant to the worker's current goals (e.g., work domain constraints, control task constraints).

The second step in this logical sequence is that the constraints that need to be taken into account can be represented by workers in fundamentally different ways. Returning to the same example, one (unlikely) possibility is to take gravity into account by reasoning analytically using a symbolic representation of Newton's second law. A second (also unlikely) alternative is to use a set of instructions that would specify what actions should be taken (e.g., if the ball is X seconds away from you, then begin to stretch out your hand). A third possibility is to use an internal embodiment (or implicit model) of the relevant dynamics to continuously guide actions, much like an analog computer. In each of these three cases, the very same environmental constraint—the force of gravity—is being taken account. However, that constraint is being represented internally in three very different ways.

The third and final step in the logic behind the SRK taxonomy is that each of these ways of taking into account action-relevant constraints is linked with, or defines, a different category of human performance. This relationship is shown in Fig. 11.2 (Rasmussen, 1990a). Each of the three levels in the SRK taxonomy defines a different level of cognitive control (i.e., a different category of human action). Beginning at the top, the knowledge-based behavior (KBB) level of cognitive control is defined by serial, analytical reasoning based on a symbolic representation of the relevant constraints in the environment. Thus, KBB guides action by representing the goal-

Behavior	Representation of Problem Space	Process-Rules
Knowledge-Based	Mental model; explicit representation of relational structures; part–whole, means–end, causal, generic, episodic, etc. relation	Heuristics and rules for model creation and transformation; mapping between abstraction levels; heuristics for thought experiments
Rule-Based	Implicit in terms of cue-action mapping; black-box action-response models	Situation-related rules for operation on the task environment, i.e., on its physical or symbolic objects
Skill-Based	Internal, dynamic world model representing the behavior of the environment and the body in real time	Not relevant - an active simulation model is controlled by laws of nature, not by rules

FIG. 11.2. Relation between levels of cognitive control in SRK taxonomy and the way in which constraints in the environment are represented and processed internally (Rasmussen, 1990a). Reprinted from D. Ackermann & M. J. Tauber (Eds.), *Mental models and human-computer interaction 1*, Copyright © 1990, p. 65. Reprinted with permission from Elsevier Science.

relevant constraints in the environment as a *mental model*.[3] The middle row in Fig. 11.2 shows the rule-based behavior (RBB) level of cognitive control. This level is defined by an if–then mapping between a familiar perceptual cue in the environment and an appropriate action. No reasoning is required. Instead, there is a direct association between a cue and an action. Thus, RBB guides action by representing the goal-relevant constraints in the environment in terms of perceptually grounded rules. Finally, the skill-based behavior (SBB) level of cognitive control involves real-time, direct coupling to the environment via what Rasmussen (1983, 1990a) referred to as a *dynamic world model*. In some ways, this label is misleading because the dynamic world model is only an implicit model of the environment (Mizutani & S. E. Dreyfus, 1998), much like an analog computer (cf. Norman, 1998). For example, if we are skilled in motor control, experience with which actions are successful and which are not allows us to learn to exploit, or compensate for, gravity without having to represent gravity explicitly internally. Similarly, a physician learns through experience to recognize breathing sounds that indicate pneumonia without having to represent those sounds explicitly internally. Just as the representation of information in an analog computer corresponds to its physical structure, the dynamic world model is an embodiment (rather than an explicit model) of the goal-relevant constraints of the environment. As such, it enables SBB by providing a basis for direct coupling and parallel, continuous interaction with the world.

Because each level of cognitive control is based on a different type of internal representation, it represents a different category of human performance. The result

---

[3]The term *mental model* is frequently used as a catch-all phrase that includes various types of knowledge. In this framework, the term takes on a more specific meaning, referring to an internal, symbolic representation of the relational structures in the environment. Consequently, purely procedural knowledge is not encompassed by this definition because such knowledge does not contain an explicit model of the environment (Rasmussen, 1990a).

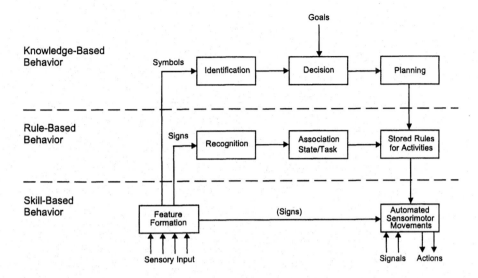

FIG. 11.3. The SRK taxonomy of human performance categories. Adapted from Rasmussen, J. (1983). Skills, rules, and knowledge; signals, signs, and symbols, and other distinctions in human performance models. *IEEE Transactions on Systems, Man, and Cybernetics, SMC-13,* 257–266. Copyright © 1983 IEEE. Reprinted with permission.

is a three-level taxonomy, graphically depicted in a simplified fashion in Fig. 11.3. SBB consists of smooth, automated, and highly integrated patterns of action that are performed without conscious attention. A prototypical example is an automated psychomotor activity (e.g., walking) that is driven by a continuous perception-action loop. Action is usually based on feedforward (i.e., prospective), rather than feedback (i.e., reactive), control. That is, SBB typically consists of anticipatory actions, rather than waiting for the state of the world to change and only then reacting to it. Because it involves direct coupling to the environment, SBB is based on prototypical temporal-spatial patterns, a point that we return to later. Because it does not require conscious attention, SBB is not verbalizable.

RBB, shown in Fig. 11.3, consists of stored rules derived from procedures, experience, instruction, or previous problem-solving activities. Action is goal oriented but goals are not explicitly represented. Workers may know the goals that the rules can achieve, but they are not thinking about those goals when they are following the rules. Instead, goals can only implicitly be found in the structure of the rules. Typically, the rule will reflect the functional properties of the environment that constrain action. Thus, successful RBB can be explained in analytical terms by an observer, but such a description is a rational reconstruction that has no psychological validity (see the example described later). In the RBB mode, people are not reasoning; they are merely using familiar perceptual cues in the environment to trigger actions directly. In contrast with SBB, workers are usually aware of their cognitive activities at the RBB level, and thus, can verbalize their thoughts.

Finally, KBB consists of deliberate, serial search based on an explicit representation of the goal and a mental model of the functional properties of the environment (see Fig. 11.3). Thus, in contrast to RBB, in the KBB mode goals are considered explicitly rather than implicitly. Furthermore, KBB is slow, serial, and effortful because it requires conscious, focal attention. KBB is frequently, although not exclusively, used in unfamiliar situations where previous experience is no longer valid and a solution must be improvised by reasoning.

The Role of Information Interpretation: Signals, Signs, and Symbols. As we already mentioned, the three levels in the SRK taxonomy are distinguished by the kinds of internal representations on which they are based. However, the three levels are also distinguished according to the way in which workers interpret information from the environment. To understand this relationship, we need to introduce a distinction between signals, signs, and symbols defined by Rasmussen (1983). These three concepts represent three ways in which workers can interpret information from the environment. *Signals* have a strong perceptual basis because they are continuous quantitative indicators of the time-space behavior of the environment. For example, when you are driving a car, the changing distance between your car and the lane markers may be a time-space signal. *Signs,* on the other hand, are arbitrary but familiar perceptual cues in the environment. They refer to the state of the world by convention or by prior experience. For example, when you are driving your car, a red octagon with the word *stop* painted on it may be a sign. *Symbols* are meaningful formal structures that represent the functional properties of the environment. For example, when you are driving, you may use your knowledge of the causal relationships between the car's components to diagnose a malfunction. In this case, your knowledge of the car's components are represented as symbols in a mental model.

Several points need to be made regarding the signal, sign, symbol distinctions. First, the concepts refer to the way in which an observer interprets information, not just to the form in which information is presented. The very same display may be interpreted as a signal, as a sign, or as a symbol, depending on the intentions, expectations, and expertise of the observer (see the example discussed later). Thus, we cannot say that a particular item is a symbol, for instance, unless we have access to the thought processes of the observer. Second, signals and signs have a perceptual basis, whereas symbols have a semantic basis. That is, both signals and signs refer to perceptual properties of the environment, such as distance, color, size, shape, configural patterns, and so on. In contrast, symbols, although they may take a particular form, refer to the meaning of the information, not its form.

We can now explain the relationship between signals, signs, and symbols and the three levels in the SRK taxonomy. As shown in Fig. 11.3, the way in which information is interpreted is linked to the level of cognitive control that is activated. During SBB, information is interpreted in terms of signals. The time-space signals obtained from the environment feed into the dynamic world model which, in turn, guides action. During RBB, however, information is interpreted in terms of signs. The familiar perceptual cues obtained from the environment trigger the rules that directly specify an appropriate action to take. Finally, during KBB, information is interpreted in terms of symbols. The meaning of entities or relationships in the environment is interpreted in the context of a mental model that supports analytical reasoning. In summary, there is a nonarbitrary, one-to-one relationship between the way in which information is interpreted and the three levels of cognitive control in the SRK taxonomy.

Figure 11.4 provides a simple example to show this linkage in a more concrete way. Three cases are shown in which a worker is observing a flowmeter. Despite the fact that the object being observed is exactly the same in each case, the information from

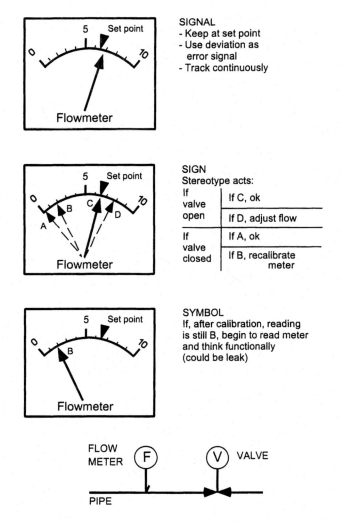

FIG. 11.4. A hypothetical example showing the difference between signals, signs, and symbols. Adapted from Rasmussen, J. (1983). Skills, rules, and knowledge; signals, signs, and symbols, and other distinctions in human performance models. *IEEE Transactions on Systems, Man, and Cybernetics, SMC-13*, 257–266. Copyright © 1983 IEEE. Reprinted with permission.

the environment is being interpreted in three different ways. The first case, shown at the top of Fig. 11.4, shows the relationship between signals and SBB. In this context, the worker's goal is to keep the flowmeter at a particular set point value. Consequently, the deviation between the current flow rate and the set point value is interpreted as a continuous, quantitative error signal. Based on this time-space signal, the worker can rely on SBB to continuously adjust the flow.

The second case, shown in the middle of Fig. 11.4, shows the relationship between signs and RBB. Here, the worker's actions are guided by a set of rules. If the valve is open and the indicator is at position C (the set point), then everything is okay and no action is warranted. If the valve is open and the indicator is at position D (not at the set point), then the flow needs to be adjusted to match the set point. In contrast, if the valve is closed and the indicator is at position A (the zero point), then everything is okay and no action is warranted. If the valve is closed and the indicator is at position B (not at zero), then the meter needs to be recalibrated. Note that it is possible to derive the very same set of actions by reasoning about what should be done in each case. However, such a rational reconstruction has no psychological validity in this case. Because these are familiar, recurring contexts, the worker has developed stereotypical rules with experience. Thus, rather than interpreting the environment in terms of what it means and engaging in analytical reasoning, the worker is instead interpreting the environment in terms of familiar perceptual signs that are directly associated with appropriate actions.

Finally, the third case, shown at the bottom of Fig. 11.4, shows the relationship between symbols and KBB. In this context, the worker has tried to calibrate the flowmeter (see the previously stated second rule), but the indicator is still at position B. As a result, the worker may choose to interpret the information on the flowmeter as a symbol. Then, the worker can use her mental model of the work domain (see the schematic diagram at the bottom of Fig. 11.4) to support a functional problem-solving process. Doing so may result in the conclusion that there is a leak in the valve, which would explain why there is flow despite the fact that the valve is closed and properly calibrated.

In summary, although the form in which information is presented to workers may affect how information is interpreted (see later discussion), in the end, whether something is a signal, sign, or symbol ultimately depends on how workers decide to interpret information in the environment. The very same item or relation can be interpreted in different ways by different workers (e.g., if some workers are more experienced than others) or by the same worker on different occasions (e.g., if the worker's goals or expectations change across occasions).

People Are Complicated: Interactions Between Levels. So far, we have described the SRK taxonomy as if the three levels of cognitive control are independent of each other. This impression is misleading because, in fact, there are significant interactions between levels. Figure 11.5 tries to capture some of these interactions, albeit in a highly simplified manner. The interactions between levels are best appreciated by adopting a temporal perspective. Accordingly, four different types of temporal activities (synchronous, synchronic, diachronic, and achronic) have been mapped onto the three levels in the taxonomy. We illustrate each of these concepts by drawing on examples from automobile driving.

Synchronous activities are those that occur online and in real time and deal with the current situation. In these cases, the worker is directly coupled to spatial-temporal properties in the environment. As a result, such activities fall into the realm of SBB. A simple example would be continuously steering your car to make sure that it is within the lane markers.

Synchronic activities also occur online and in real time, but they deal with the signs that are used to trigger new SBB activities (see the line in Fig. 11.3 connecting "stored rules for activities" at the RBB level to "automated sensorimotor patterns" at the SBB level). Thus, such activities fall into the realm of RBB. A simple example would be that you hear your engine revving up to a familiar reference pitch, so you shift gears. Note, however, that only the choice to change gears occurs at the RBB level (if engine = familiar pitch, then change gears). For an experienced driver, the

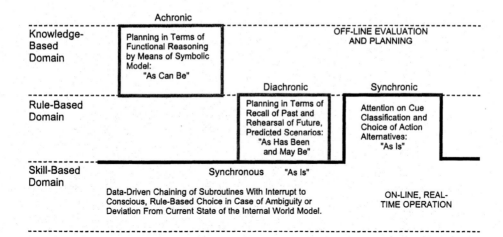

FIG. 11.5. Interactions between the different levels of cognitive control (Rasmussen & Vicente, 1989). Reprinted from *International Journal of Man-Machine Studies, 31*, J. Rasmussen & K. J. Vicente, "Coping with human errors through system design: Implications for ecological interface design," pp. 517–534, Copyright © 1989. Reprinted by permission of the publisher, Academic Press.

act of changing gears would likely take place at the SBB level because it would require direct coupling to the time-space properties of the environment.

Diachronic activities occur offline and require evaluation or planning. They involve using previous experience to interpret the world in terms of signs that could trigger SBB activities that may be required in the near future. Thus, unlike synchronic activities, diachronic activities deal with what has occurred in the past and may occur again in the future, not with what is occurring in the here and now. A simple example would be recognizing a familiar stretch of highway that, in the past, has been followed by an icy patch of road. Using this act of RBB, the driver will be prepared to slow down up ahead if ice is in fact encountered.

Finally, *achronic* activities also occur offline and require evaluation or planning. The difference is that they involve reasoning in symbolic terms, not associating actions based on familiar perceptual signs. As a result, achronic activities belong to the KBB level. A simple example might be that the driver, being a university professor, is thinking about what he can do to stop students from complaining for higher grades after each exam. In this case, the object of activity is not related to the here and now, nor to what has happened in the past and is likely to happen in the near future. Instead, the professor is reasoning about what can be.

Although the terminology might be unfamiliar, these distinctions are important because they show that the three levels of cognitive control interact. For example, it would not be surprising to find a driver who is involved in more than one of these activities (i.e., steering the car, shifting gears, watching for upcoming ice patches, and thinking about problems at work) at the same time. Thus, the SRK taxonomy is richer and more complex than it might first appear to be.

Avoiding Some Misconceptions. As we mentioned, it is not uncommon to find statements in the literature that reflect misconceptions about the SRK taxonomy. We have tried to avoid most of these by describing the basic structure of the taxonomy in detail. However, there are a few other misconceptions that are common, and thus worthwhile addressing explicitly.

First, it may be tempting to interpret the three levels as representing points along a continuum with completely preattentive, automatic processing on the skill-based end and completely attentive, controlled processing on the knowledge-based end (cf. Schneider & Shiffrin, 1977; Shiffrin & Schneider, 1977). On closer examination, however, this interpretation is incoherent. After all, the three ways of interpreting information (i.e., signals, signs, and symbols) are discrete concepts that do not lie on a continuum. For instance, information can be interpreted as a signal or as a sign, but there is nothing in between the two. Thus, the three levels of cognitive control should be interpreted as qualitatively distinct, not as points on a continuum.

Second, actions directly on the world are only possible at the SBB level. This point is frequently overlooked, but it is graphically represented on the right side of Fig. 11.3. We can see that actions on the world are driven by automated sensorimotor patterns at the SBB level. Thus, the three levels of cognitive control are nested on the output side because the output arrows of the KBB and RBB levels both go through the SBB level.

Third, because of the intuitive appeal of the distinctions, it is also tempting to classify tasks as belonging solely to one of the three levels in the taxonomy. For example, one might think that solving a logic problem may be a knowledge-based task and that driving a car may be a skill-based task. Such an interpretation would be overly simplistic. For one, the three levels interact, as we discussed previously. For some people, driving has rule-based components (e.g., if stop sign then put on the brakes), and even the deepest of reasoning tasks has skill-based components. After all, as we just described, actions on the world are only possible at the SBB level. Moreover, there are interactions between levels that are not depicted in this simplified diagram. For example, the SBB level is always active and is responsible for directing attention, activating higher levels, controlling information gathering, as well as transferring intentions into control of movements (see Rasmussen, 1990a, for more details). Thus, any one molar task will typically draw on more than one level of cognitive control.

Fourth, and relatedly, the relationship between a particular task and levels of cognitive control is not fixed because it can be mediated by several variables. One mediating factor is the worker's level of expertise. For example, novices may have to resort to KBB to determine what action is appropriate for the current context because the situation may be unfamiliar to them. In contrast, experts may recognize familiar perceptual cues that enable RBB. A second mediating factor is the form in which information is presented. For example, if data are presented in a graphical form, then it may be easier to extract signs or time-space signals from the display, thereby encouraging lower levels of cognitive control (see later discussion). In contrast, if the same data are presented in an alphanumeric form, then it may be easier to extract symbols from the display, thereby encouraging KBB. A third mediating variable is the degree to which workers reflect on their performance. For example, an unreflective worker may execute a procedure simply by using RBB and SBB. In contrast, a reflective worker may also periodically rely on KBB (e.g., knowledge-driven monitoring) to ensure that the intentions behind the procedure are being satisfied, thereby making it more likely that errors will be detected (Olsen & Rasmussen, 1989; Vicente, Burns et al., 1996). This latter example also highlights another important but underappreciated point, namely that KBB is not used only during unfamiliar situations.

Examples of Its Usage

Now that we have described the basic structure of the SRK taxonomy in detail, it is useful to provide a few examples of how it has been used in the past. As we see, the primary value of the SRK taxonomy is in putting forth conceptual distinctions that can help us avoid muddled thinking and thereby see regularities that might not be observable from other perspectives.

A Taxonomy of Human Performance Models. We mentioned at the outset that one of the driving forces behind the development of the SRK taxonomy is to facilitate the development of models of human performance. Because there are so many different models and theories available in the psychological literature (see Table 11.1), it would be useful to organize those findings in a parsimonious and coherent way.

Figure 11.6 shows how the models and theories listed in Table 11.1 can be mapped onto the three levels of cognitive control. The SRK taxonomy provides an overarching qualitative framework that provides us with some direction in identifying categories of human cognition. Once we are aware of the distinctions provided by the taxonomy, we can develop (or map on) more detailed, quantitative models for each category. These models can then provide more detailed guidance regarding specific systems design questions.

This integrative quality applies not just to models, but to empirical findings as well. As the remaining examples show, the SRK taxonomy provides a set of conceptual "eyeglasses" that can help keep us from mixing apples and oranges.

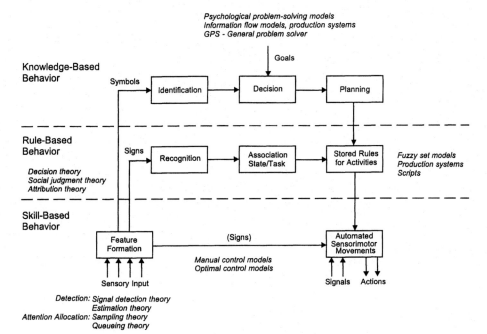

FIG. 11.6. Mapping of human performance models onto the SRK taxonomy (Rasmussen, 1986a). Reprinted with permission of the author.

Stages of Skill Acquisition. The next example we discuss is centered on the stages that people go through as they gain more expertise. H. L. Dreyfus and S. E. Dreyfus (1988) proposed an influential model that identifies the following five stages of skill acquisition:

1. *Novice*—In the first stage of skill acquisition, students learn explicit facts, features, and rules that can be readily verbalized. Performance at this stage is essentially based on algorithmic processing. The knowledge that is used is context-free in the sense that it is not yet sensitive to situated factors that may mediate performance.

2. *Advanced Beginner*—As students gain more practical experience in concrete situations, they begin to take into account more contextual factors and thus develop more sophisticated rules for performing a task. New episodes are perceived as being similar to prior examples, thereby enabling rudimentary recognition processes.

3. *Competent Performer*—At this stage, students use specific goals to prioritize facts according to their relevance. They adopt a hierarchical process by which a plan is developed to organize the situation, and then use that plan to examine only those factors that are most relevant to the current goal and plan. This process is more efficient than those observed in previous stages because it involves more selective information processing.

4. *Proficient Performer*—Whereas previous stages involve deliberate, conscious choice, activity at this stage is the result of experience-based associations connecting context and current stimuli with plans that have proven to be successful. However, when unfamiliar and particularly important events are encountered, students may still revert from this recognitional process to a more deliberate, analytical approach to decision making.

5. *Expert*—At this stage, students are able to deal with task demands in an effortless and automatic fashion. They rely on learned, experience-based, context-sensitive associations that provide them with a deep understanding of the situation, and that allows them to engage in fluid, intuitive actions. In the words of H. L. Dreyfus and S. E. Dreyfus (1988), "When things are proceeding normally, experts don't solve problems and don't make decisions; they do what normally works" (pp. 30–31). Rather than following rules, experts exploit both their experience base and the information in the environment to guide action. Because expert processes are largely perceptual and automatic, they remain tacit and are thus not verbalizable.

Olsen and Rasmussen (1989) reinterpreted and elaborated on this five-stage model by using the constructs in the SRK taxonomy. Recall that the levels in the taxonomy are based on fundamentally different ways of representing goal-relevant constraints in the environment. From this perspective, the notion of expertise cannot be meaningfully addressed without referring to a particular set of constraints in the environment. Accordingly, the first general point that Olsen and Rasmussen made is that expertise is not a fixed property of a person, but rather a dynamically varying relationship between the demands imposed by the environment and the resources of a particular person. Thus, to say that someone is an expert is an incomplete statement without specifying the kinds of situations for which their expertise holds. This observation is consistent with H. L. Dreyfus and S. E. Dreyfus' (1988) context-specific view of expertise.

This explicitly relational view of expertise led Olsen and Rasmussen (1989) to make two initial clarifications to H. L. Dreyfus and S. E. Dreyfus' (1988) five-stage model. These clarifications are in the spirit of the original model, but draw more attention to certain aspects of expertise and skill acquisition. First, the novice stage already presupposes certain basic skills that are required to bootstrap the process of skill acquisition. For example, in complex sociotechnical systems, we would expect even novice workers to have rudimentary motor and perceptual skills. If workers do not have these basic skills, then the road to expertise may be blocked. Accordingly, Olsen and Rasmussen introduced the concept of a prenovice to capture this primitive level of competence. Second, the relational view of expertise raises the question of whether the expert stage provides a comprehensive characterization of expertise. Consider, for example, someone who, at the RBB on KBB level, can explain the basis for their expert performance and use that insight to advise and instruct others expertly. These competencies are not explicitly captured by Dreyfus and Dreyfus' expert stage. Nevertheless, it would not be unreasonable to characterize such a person as an expert, albeit of a different type. Accordingly, Olsen and Rasmussen introduced the concept of a reflective expert to capture this additional dimension of competence. It is one thing to be able to deal with situations via automatic, holistic associative processes, but it is another to be able to explain the rationale behind your actions (should there be one) and thereby teach others to improve their competence. Making these points explicit helps us realize that expertise is a diverse and complex relational phenomenon.

Olsen and Rasmussen (1989) went on to use the three levels of cognitive control in the SRK taxonomy to capture their elaborated view of expertise. The most important contribution of their work is the realization that people can be more or less skilled at each level of cognitive control. For example, H. L. Dreyfus and S. E. Dreyfus' (1988) expert can be reinterpreted as someone who is an expert at the SBB level for certain situations. Such a person can deal with a particular class of demands using fluid and automatic, perceptual-motor skills. In contrast, a reflective expert is someone who is competent at the RBB and KBB levels. Such a person can explain the rationale behind actions, whether it be in terms of rules or model-based reasoning. Note that these competencies can be observed either in two different people (e.g., a "doer" vs. a "knower") or preferably in the same person (e.g., a "reflective practitioner"; Schön, 1983).

Associating different kinds of expertise with each level of cognitive control has additional significant implications. First, learning can take place within each level, not just from one level to the next. As Sanderson and Harwood (1988) observed, there is a tendency to think that skill acquisition involves a transition from KBB to RBB to SBB. This is the process that is built into H. L. Dreyfus and S. E. Dreyfus' (1988) five stages. Although a great deal of skill acquisition may be captured by this view, some tasks may require learning completely within a level. For example, learning to talk or walk involves exploration and development within the SBB level. Children do not learn these activities by first reasoning from first principles. Other tasks may require a skill acquisition process that begins primarily at RBB and then transitions primarily to SBB. A task that we first learn by following instructions or directions (e.g., cooking or mechanical assembly) would be an example. Second, differentiating expertise at different levels in the SRK taxonomy also highlights the issue of meta-expertise. Do workers know when they should be operating at each level of cognitive control? We could argue that a meta-expert is someone who knows when he is no longer an expert, and thus pops up from a SBB or RBB level to a KBB level. Workers who lack such meta-expertise might instead persist with the SBB or RBB levels, despite the fact that the situation has changed and their expertise is no longer valid.

In summary, this example shows how the constructs in the SRK taxonomy provide a differentiated, relational view of human expertise that captures, and adds to, the influential, five-stage model proposed by H. L. Dreyfus and S. E. Dreyfus (1988). The outcome is a set of insights that reflects the richness and diversity of expertise and skill acquisition.

Theories of Expertise Effects in Memory Recall. The next example that we present to illustrate the value of the SRK taxonomy is centered on a body of psychological research investigating the relationship between expertise and memory recall performance. In his seminal research on chess expertise, de Groot (1946/1965) found that chess masters were able to reconstruct almost perfectly a chess position after having viewed it for only a few seconds. In contrast, lesser skilled chess players exhibited a far less impressive level of memory recall performance. This basic finding has been replicated in many domains other than chess (see Vicente & Wang, 1998, for a recent review).

Vicente (1988) reviewed the theories that had been developed to identify the psychological mechanisms that are responsible for the observed relationship between expertise and memory recall performance. Two competing theories were identified, a perceptual chunking theory and a conceptual chunking theory. According to the *perceptual* account (Chase & Simon, 1973), chess experts have a vast number of "chunks" stored in long-term memory. These chunks are perceptual structures that contain information about patterns of pieces on a chessboard. Thus, the superior memory recall ability of experts is attributed to the fact that they can recognize and code familiar chunks of pieces. The chunks of pieces are easily and quickly perceived because they are associated with familiar perceptual forms. Novices do not have enough experience to recognize familiar chunks, and so they can encode pieces only on an individual basis. As a result, their memory recall performance is well below that of experts.

The *conceptual* chunking theory (Egan & Schwartz, 1979) provides a seemingly competing psychological explanation for experts' superior recall performance. According to this view, the chunking phenomenon is based on the meaningfulness of the material to be recalled, not its perceptual form. Thus, experts' skill is a result of the organization of functional concepts in long-term memory. Given this knowledge base, they can identify the appropriate conceptual category for coding the material to be recalled (e.g., the chess strategy known as Queen's Gambit). They can then use knowledge of this conceptual category to reconstruct low-level details (i.e., the location of individual chess pieces) so that those details are consistent with the category in question. Novices have not developed a differentiated knowledge base so they cannot recognize meaningful conceptual categories; nor do they know how those categories constrain low-level details. As a result, their memory recall performance is much weaker. The key point to keep in mind is that, according to the conceptual account, it is the meaningfulness of the material, not its perceptual familiarity, that provides the psychological basis for expert recall performance.

As mentioned earlier, these two theories were originally proposed as alternative explanations for the observed relationship between expertise and memory recall performance. Vicente (1988) showed that the theories are actually complementary explanations by using the SRK taxonomy to develop an integrative theoretical account of the chunking phenomenon. From this perspective, the perceptual and conceptual accounts represent two alternative ways in which chunking can occur. The perceptual account draws on the RBB level of cognitive control. The environment (i.e., the set of pieces to be recalled) is interpreted as signs (i.e., familiar perceptual cues) that correspond to groups of chess pieces that have been frequently encountered in the past. Thus, chunking is possible because of a recognition process. The conceptual account, on the other hand, draws on the KBB level of cognitive control. In this case, the environment is interpreted as symbols (i.e., knowledge structures) that correspond to meaningful relationships between chess pieces. Thus, chunking is possible because of conceptual inferences. The SRK taxonomy also makes clear that these processes can interact. Thus, chunking can be driven by a combination of the two levels of cognitive control. As far as we know, this was the first time that the perceptual and conceptual accounts had been viewed as complementary rather than competing explanations, and the first time that they had been integrated into an overarching framework.

This unification allowed us to make predictions that could be tested empirically. For example, relatively pure conceptual chunking should be possible in situations where the material to be recalled is meaningful to people but not in a familiar visual form. Vicente (1992a) conducted a memory recall study in the domain of process control using DURESS as an experimental test-bed. The recall performance of novices was compared with that of experts who were familiar with thermal-hydraulics theory but who had never before seen either DURESS or the displays used in the experiment. The findings showed that the experts clearly outperformed the novices, thereby confirming the prediction that expertise effects in memory recall can be obtained under extreme conditions that induce KBB.

Similarly, relatively pure perceptual chunking should be possible in rote situations where the material to be recalled is presented in a familiar perceptual form but is not meaningful to people. Ericsson and Harris (1990) conducted an unusual memory recall study in the domain of chess. They gave a novice approximately 50 hours of practice at performing the memory recall task. By the end of the experiment, the novice's recall performance was just as accurate as that of a chess master. This finding supports the prediction that expertise effects in memory recall can be obtained under extreme conditions that induce RBB as well. Note that both of these empirical tests were conducted after Vicente (1988) made these predictions based on the SRK taxonomy. These results confirm Olsen and Rasmussen's (1989) claim that expertise can develop independently at each level of cognitive control. Typically, however, we would expect both types of processes to be active. For example, chess masters not only can recognize familiar patterns of pieces, but they can explain their semantic significance as well.

In summary, this example shows how the structure of the SRK taxonomy provides an umbrella framework under which purportedly competing psychological theories of expertise effects in memory recall performance can be integrated in a coherent manner. The distinctions provided by the taxonomy allow us to see structure where others see contradictions.

A Framework for Interface Design for Complex Sociotechnical Systems.
So far, we have focused primarily on examples that show how the SRK taxonomy can be used to add some conceptual clarity and coherence to research issues in

psychology and cognitive science. Now we show how the taxonomy can also be used to develop design principles. Given our earlier description emphasizing qualitative concepts, we might expect that the SRK taxonomy cannot be used in a formative manner. However, in the remainder of this section, we show that it is in fact possible to base the design of an interface on categories of human action and their limiting properties. In chapter 12, we provide a concrete example showing how the abstract principles described in this subsection can be applied in practice.

The example we present is based on the work that we have conducted on ecological interface design (EID), a theoretical framework for interface design for complex sociotechnical systems (Vicente, Christoffersen, & Hunter, 1996; Vicente & Rasmussen, 1990, 1992). The SRK taxonomy played a critical role in the theoretical development of the EID framework. By revealing some of this conceptual development, we can show the kind of role that SRK can play in systems design.

We followed a three-step process in laying out the theoretical foundations of the EID framework. First, we conducted a literature review to identify empirical findings that might be pertinent to the aspects of interface design in which we were most interested. Second, we used the SRK taxonomy as an "umbrella" framework for integrating, under a common language, the variety of research results that were uncovered. Third, we then used the theoretical constructs of SRK to deduce from these findings a set of three principles for interface design.

The results of the first step are described in detail in Vicente and Rasmussen (1992). The path of reasoning followed in the second step is depicted in Fig. 11.7. The literature review revealed that (a) lower levels of cognitive control tend to be executed more quickly, more effectively, and with less effort than higher levels, and (b) people have a definite preference for carrying out tasks by relying on lower levels of cognitive control, even when an interface is not designed to support this type of behavior. These two points suggest that information should be presented in a way that allows workers to effectively rely on lower levels of cognitive control.

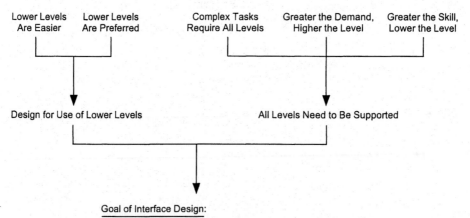

FIG. 11.7. Theoretical rationale behind the development of the EID framework. Adapted from Vicente, K. J., & Rasmussen, J. (1992). Ecological interface design: Theoretical foundations. *IEEE Transactions on Systems, Man, and Cybernetics, SMC-22,* 589–606. Copyright © 1992 IEEE. Reprinted with permission.

Intuitively, it would be nice to allow workers to meet task demands by relying on their powerful perceptual-motor abilities that have been honed over time, both by evolution and by experience with everyday skills. However, even if information is presented in such a way that a task can be accomplished using lower levels of cognitive control, higher levels may nonetheless be triggered because the level of cognitive control that is activated is determined not just by how information is presented but also by task demands and the worker's level of skill. No matter what we do, there will always be some situations in complex sociotechnical systems that require workers to engage in laborious problem-solving activities. This, and the fact that any reasonably complex task will require a complex interaction between all three levels of cognitive control (see Fig. 11.5), suggests that an interface should provide the appropriate support for all three levels. Therefore, a framework should strive to achieve the following twofold goal: to design interfaces in such a way as not to force cognitive control to a higher level than the demands of the task require (thereby inducing automated skills whenever possible), while at the same time providing the appropriate support for all three levels (thereby supporting analytical reasoning whenever necessary).

The third step in the development of the EID framework was the derivation of a set of formative principles that would allow designers to develop interfaces that satisfy this twofold goal. The SRK taxonomy provides a general indication of how to do this, because there are constraints on inducing each level of cognitive control. For instance, SBB can be activated only when information is presented in the form of time-space signals. RBB, on the other hand, is triggered by familiar perceptual forms (signs). And finally, KBB is activated by meaningful relational structures (symbols). By using this knowledge of the categories of human action and their limiting properties, we developed the following three design principles (see chap. 12 for an application example):

1. *SBB—To support interaction via time-space signals, workers should be able to act directly on the display.* This is the familiar idea of direct manipulation (Hutchins, Hollan, & Norman, 1986; Shneiderman, 1983). To take advantage of the fluency and cognitive economy of SBB, we should allow workers to directly manipulate the display. Furthermore, to provide a visualization of the dynamics of the work domain, we should present information in terms of time-space signals.

2. *RBB—Provide a consistent one-to-one mapping between the work domain constraints and the cues or signs provided by the interface.* People have a natural tendency to pick out cues in the environment to guide action rather than having to engage in analytical reasoning. If we are to support this level of cognitive control, then the cues provided by the interface should not just be probabilistically correlated with the actual state of the work domain. Otherwise, the rules that workers develop will work only some of the time (i.e., the rules would be only locally valid). The way around this problem is to create an interface that presents workers with signs that have a one-to-one mapping with the underlying structure of the work domain. In this way, the signs are not merely correlated with, but are actually diagnostic of, the state of the work domain. This approach should allow workers to enjoy the cognitive economy of RBB in a way that leads to more robust performance.

3. *KBB—Represent the work domain in the form of an abstraction hierarchy to serve as an externalized, faithful model that will support knowledge-based problem solving.* As we mentioned previously, there are several reasons why workers will need to engage in KBB. It is important to support workers under these situations because analytical reasoning is an effortful and error-prone cognitive activity. Because KBB involves reasoning based on a mental model, it would be useful for the interface to provide an externalized model so that workers do not have to encode, store, and retrieve all of this information mentally. The AH provides a useful basis for achieving this goal because it is a psychologically useful way of representing how the work domain actually functions (see chap. 7). Thus, building a visualization of the AH representation of the work domain into the interface should provide an effective basis for supporting KBB.

In conclusion, without an overarching framework, it would have been difficult to go from the mass of results in the literature to practical design insights. For one, the relevant findings are spread across several disciplines including human factors, HCI, management science, psychology, and systems engineering. Consequently, the studies use different terms and concepts. This makes it difficult to find regularities in the literature. Furthermore, the pertinent studies were not all originally designed with the purpose of interface design in mind. As a result, it is not very easy to extract reliable design implications. Our example shows how the SRK taxonomy can help us overcome these obstacles. The concepts in the taxonomy can be used to integrate a diverse set of research findings in a way that allows us to derive practical implications for design.

Summary

The examples in this section should have shown you the unique value added by the SRK taxonomy to both basic research and systems design. Although it provides only a qualitative framework, its conceptual distinctions provide a very powerful set of "glasses" with which to view the world. This frame of reference can allow us to clarify and enrich existing theoretical claims, to reconcile seemingly competing theoretical accounts, to integrate a diverse set of empirical research findings, and to derive implications for design. Reason (1990) provided another example showing how the SRK taxonomy can be used in these ways in the particular area of human error. Even though the contributions of the taxonomy may seem modest initially, it provides us with a coherent conceptual basis for distinguishing between qualitatively different categories of human performance, thereby helping us gain many valuable insights. As Rasmussen (1983) pointed out, "we, like Eddington's [1958] ichthyologist . . . , will be able to obtain some of the results needed more readily by conceptual analysis before experiments than by data analysis afterwards" (p. 266).

WORKER COMPETENCIES ANALYSIS FOR PROCESS CONTROL MICROWORLD

In this section, we show how the SRK taxonomy can be used to conduct a worker competencies analysis, as a part of CWA. As in chapters 7 to 9, we use the DURESS II process control microworld as a case study.

Objective

The objectives of a worker competencies analysis are to consolidate the demands that previous levels of CWA place on workers, and then to determine how those requirements can be satisfied in a psychologically realistic manner. The results of the first step can be summarized in the form of an "inventory," broken down by level of cognitive control. Using this list and more detailed knowledge of human information-processing capabilities (e.g., Wickens, 1992), we can then derive implications for training, interface design, and perhaps selection as well.[4]

Skill-Based Behavior

Most of the demands associated with industrial process control are cognitive rather than perceptual-motor in nature. Consequently, there are comparatively few demands in DURESS II that can be dealt with using SBB. Nevertheless, there are a few important competencies that are required at this level of cognitive control. First, workers should have basic perceptual skills. In particular, it would be useful for them to either recognize, or easily keep track of, the labels for the various components in DURESS II (see Table 6.2). This requirement has straightforward implications for training or interface design.

Second, workers should be able to interact directly with the display rather than have to communicate their intentions using some arbitrary command language. The former will allow workers to rely on the fluency and power of their perceptual-motor skills, whereas the latter would force workers to engage in slower and more effortful cognitive processes. This requirement has implications for the form of HCI.

Third, as we pointed out in chapter 8, timing is a very important aspect of proficient control in DURESS II. The various components have time constants, so it is important for workers to be sensitive to the dynamics of the work domain. The very same action can be perfectly appropriate or completely disastrous, depending on when it is taken. Thus, it would be useful to provide a visualization of the work domain dynamics (e.g., time constants, rates of change), so that workers could develop an intuitive appreciation for these important parameters. Providing such transparent feedback should help workers develop appropriate expectations regarding the evolution of the state of the work domain. In addition, this feedback should allow workers to use time-space signals in the environment to guide their actions, rather than having to rely on rules or algorithms stored in memory.

Rule-Based Behavior

The worker competencies at the RBB level can be divided into two categories. On the one hand, it would be useful to create salient perceptual cues that could be used to support robust RBB. On the other hand, it would also be useful for workers to be aware of the various shortcuts and strategies that they can use to effectively control the process. Each of these categories is described in turn.

---

[4]We do not address implications for selection in this case study. For some preliminary thoughts on this issue, see Torenvliet et al. (1998).

The goal of creating salient and diagnostic signs can be accomplished by following the logic behind EID, described earlier. More specifically, the idea is to create a one-to-one mapping between the deep properties of the work domain and the perceptual cues provided by the interface. Several specific examples can be offered for DURESS II. For instance, in chapters 7 and 8, we determined that actors should be sensitive to the mass balance constraint governing each reservoir. This constraint provides a first principles description of the work domain that is usually represented in the form of a differential equation (see Table 6.3). It would be useful to present this constraint in a way that would allow workers to use the relationship in guiding their actions, but without having to solve a differential equation or engage in any other abstract, symbol manipulation. One way to accomplish this goal would be to create a graphic visualization of the mass balance constraint (see chap. 12 for a design example). In this way, workers could take this important constraint into account when controlling the work domain, but they could do so by relying on perceptual processes rather than analytical processes. In other words, workers could interpret the information in the display as a sign rather than as a symbol. The result should be cognitively more efficient performance.

The same basic idea can be applied to another example, namely the relationship between reservoir temperature, energy inventory, and mass inventory (see chaps. 7 and 8). This constraint also needs to be taken into account if proficient control is to be achieved. However, rather than presenting the relationship in abstract form as an algebraic equation, we can develop a graphical visualization that shows the very same constraint in a more concrete, perceptual form (see chap. 12 for a design example). In doing so, we make it easier for workers to control the process by using the associative recognition processes comprising RBB rather than the analytical problem-solving processes comprising KBB.

In addition to developing such visualizations of goal-relevant constraints, it is also possible to support RBB by ensuring that workers are aware of procedural constraints that can be exploited to control the process effectively. The control task and strategies analyses presented in chapters 8 and 9 provide some valuable requirements that we can use for this purpose. For example, in chapter 8, we determined that pumps should not be turned on before valves, otherwise there will be no downstream path for the water to follow and the pumps may be damaged. We also determined that the "configure heater step" should not be initiated before the "configure FWS step," otherwise an empty reservoir will be heated and the reservoir vessel may be damaged. It is important for workers to be sensitive to these sequential constraints. This can be accomplished either through interface design (e.g., by making the sequential constraints perceptually obvious) or through training (e.g., by instructing workers to follow rules that do not violate the sequential constraints). Analogous tactics could be used to deal with other requirements identified in the control task analysis. For example, a shortcut linking the *observe* box and the *task* circle was identified in the decision ladder for start-up (see Fig. 8.9). A proficient worker should be able to pursue this shunt whenever possible. Thus, the training and interface design requirements should ensure that this behavioral objective is indeed satisfied during systems design.

The strategies analysis conducted in chapter 9 also identified a number of procedural constraints that have implications for supporting RBB. For example, we developed information flow maps for three different strategies that can be used to perform the FWS configuration task in DURESS II (see Fig. 9.12). Each of these flow maps identifies a different category of processes that can be used to accomplish the same control task. A proficient worker should be able to use any of these strategies effectively, and should know when each strategy is applicable (see Table 9.4). As with the other examples presented earlier, these behavioral objectives can be achieved either through interface design or through training. For example, a display could be tailored to the sequential flow constraints of each strategy, and workers could be instructed on how to follow each strategy effectively. In either case, we can use our knowledge about how RBB can be triggered to ensure that these requirements could be dealt with using this level of cognitive control, rather than forcing workers to engage unnecessarily in more effortful and error-prone KBB.

Knowledge-Based Behavior

Even if we design training programs and interfaces to encourage the use of lower levels of cognitive control, workers will still need to engage in KBB. Because analytical problem solving is a slow, resource demanding, and error prone activity, it is important that we provide the appropriate support for it. One way to accomplish this objective is to follow the EID framework and build the relationships identified by a work domain analysis of DURESS II (see chap. 7) into an interface in the form of an AH. Such a design would allow workers to see the degrees of freedom that they have available for action. For example, workers would be able to see that changing the mass flow to affect demand may have the unintended side-effect of changing temperature, whereas the reverse need not be true. Similarly, workers would be able to see that changing the state of one of the components in one feedwater stream to control the flow to one reservoir could unintentionally change the flow to the other reservoir. By presenting these relationships in the interface, we would be creating a faithful, external model of the process that workers could use during problem solving.

Similarly, the set of relationships identified in the work domain analysis could also serve as the basis for a training program (e.g., Hunter, Vicente, & Tanabe, 1996). We could teach workers how to think about DURESS II at each level of the AH. We could also instruct them about the linkages between levels. Then, we could give them instructional exercises to allow them to become fluent in using this type of representation to support effective problem solving. The idea is analogous to that described in the previous paragraph, the difference being that we would be inducing deep knowledge in a top-down manner through instruction rather than in a bottom-up manner through interface design.

Summary

These examples should have given you a more concrete feel for how the SRK taxonomy can be used as a conceptual guide when conducting a worker competencies analysis. The outcome of such an analysis should be a number of detailed implications for interface design and training. Using knowledge of the signals, signs, and symbols distinctions, we can present information in a form that will make it more likely that particular levels of cognitive control will be activated. Furthermore, using the requirements that were identified in previous phases of CWA, we can identify the behavioral objectives that a training program should seek to satisfy.

To be sure, the SRK taxonomy does not—and in fact, cannot—take the place of existing knowledge about human performance. For example, the taxonomy will not tell us how to design an effective training program. To do that, we will need a great deal of additional knowledge about training (e.g., Patrick, 1992). Having said that, the SRK taxonomy does play a useful role by helping us organize the insights we gain in previous layers of CWA and the findings we uncover in the human performance literature. Integrating all of this knowledge under a common framework makes it easier to derive implications for systems design. In the following chapter, we provide several design examples to support this claim.

SUMMARY AND IMPLICATIONS

In chapter 2, we pointed out that a framework for work analysis should be based on an ecological approach, beginning by uncovering the constraints in the environment and gradually moving to uncovering the constraints in the human cognitive system (see Fig. 11.1). This chapter has described worker competencies analysis, the final phase of CWA that is the culmination of this transition from the environmental to the psychological. It is in this phase that we most clearly see how the requirements identified in previous phases build up and constrain the competencies that workers need to have to function effectively.

As shown in Fig. 11.8, the SRK taxonomy was introduced as a modeling tool that can be used to conduct a worker competencies analysis. The taxonomy itself is not a model, but rather a qualitative framework within which detailed, quantitative models can be subsumed. As we mentioned, the SRK taxonomy is not a replacement for everything that is known about human cognition. Instead, it is a coherent framework for integrating requirements and psychological knowledge so as to make the implications for systems design more obvious.

We illustrated this final phase of CWA in the context of the DURESS II case study. This example shows how worker competencies analysis provides us with implications for systems design, particularly interface design and training. At each level of cognitive control, we can identify the competencies that an ideal worker should exhibit.[5] These competencies can be turned into behavioral objectives that must be satisfied by a training program. In addition, we can use the concepts in the SRK taxonomy to determine in what form information should be presented to support competence required at each level of cognitive control.

> Now that we have finished explaining the last of CWA's phases, this is a good time for you to revisit the definitions that we provided in the Glossary at the beginning of the book. If we have accomplished our objectives, then by now you should have a much better appreciation for the distinctions that are described there. Consult the Glossary to ensure that you have an intuitive appreciation for each concept. If something doesn't make sense to you, then revisit the sections of the book that describe that concept. Although this may seem like a tedious academic exercise, it is of great practical importance. To apply CWA in practice is tantamount to looking at the world using the concepts defined in the Glossary. Therefore, it is essential to ensure that you have a firm and comprehensive grasp of the definitions presented there.

---

[5]Because DURESS II can be controlled by a single worker, the worker competencies analysis we presented in this chapter does not reveal the fact that a different set of competencies will usually be required for different workers, depending on the role they play in the organization (see chap. 10).

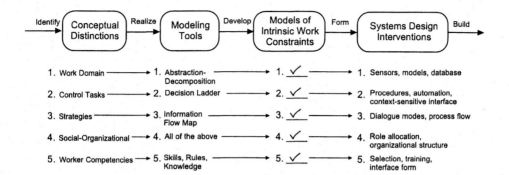

FIG. 11.8. An outline of the CWA framework with all five phases (i.e., rows) filled in.

Transition

In each of these last five chapters, we have tried to show how each layer of CWA has important implications for systems design (see Fig. 11.8). As we mentioned at the outset of the book, it is important to demonstrate this connection because work analysis is not a end in itself, but rather a means to achieving the end of better systems design. Therefore, CWA should be judged by the value it adds to systems design. The proof is in the pudding, if you will. To demonstrate this value added in a more convincing fashion, the next chapter provides several examples of how CWA can be applied, both in design and in research.

Implications for Design and Research 12

If human factors participates in [systems] development, the discipline is obliged to ensure that, whatever other utility its research has, that research must be performed in such a way that its outputs are relevant and useful to development.
—Meister (1989, pp. 57–58)

How much impact do science and research actually have on products? Less than you might think, less than you might hope.
—Norman (1998, p. 2)

PURPOSE

The objective of this chapter is to give you a better feel for how the products of a CWA can be used in both systems design and research. As we mentioned before, this book is primarily about analysis, but work analysis is a means, not an end in itself. During the last five chapters, we have pointed out how CWA can be used to derive requirements for the design of sensors, databases, constraint-based procedures, automation, context-sensitive interface mechanisms, human-computer dialogue modes, dialogue process flows, organizational structures, training programs, selection criteria, and interface forms. It is important to illustrate some of these connections to systems design in concrete detail so that the payoffs that arise from the effort that is invested in conducting a CWA become more tangible. The first part of the chapter deals with implications for design by showing how the concepts in CWA have been used to design innovative interfaces for library information retrieval and for the DURESS II microworld. The second part of the chapter shows that CWA has important implications for cognitive engineering research as well. We illustrate this additional source of value by explaining how the concepts in CWA can help solve difficulties that typically arise in microworld research.

To avoid misunderstandings, it is important to realize that this chapter is not intended to be exhaustive. Completely detailing the relevance of CWA to systems design and research would require much more space than we have here. Accordingly, the examples we present are intended to provide a flavor of the implications, not an exhaustive or definitive account. If we are successful, by the end of the chapter, you will see that CWA has a great deal to offer to both designers and researchers.

IMPLICATIONS FOR DESIGN

In this section, we provide two examples showing how CWA can be used in systems design. The first example comes from library information retrieval and shows how a CWA can lead to the design of an innovative and successful interface. The second example comes from DURESS II and shows how just the work domain and worker competencies phases of CWA can be used to design a novel and effective interface.

Example #1: Full-Scope Document Retrieval System for Public Libraries

The first example we use to illustrate CWA derives from a "humanistic" application domain. More specifically, we draw on the research program conducted by Pejtersen (1973, 1979, 1984, 1988, 1992; Pejtersen & Goodstein, 1990) in the domain of library information retrieval. There are several reasons why this is a very important example to review. First, it is one of the most detailed applications of CWA to date. Thus, it provides a thorough example to illustrate the design implications of CWA. Second, this CWA resulted in the design of an innovative full-scope document retrieval system that has been made available as a commercially available product, and is currently being used in several public libraries in Scandinavia. That information system has been recognized by one international design award and one national design award and has been nominated for several others. Thus, the case study is a CWA "success story." Third, this example shows how CWA can be used to identify novel and powerful functionality for a computer-based system. The document retrieval system that arose from this CWA consists, not just of an interface (i.e., displays and controls), but of a database and search engine as well. Fourth, the example shows the breadth of CWA. The choice of DURESS II as a case study in the previous chapters may have created the impression that CWA can be used only in application domains where workers are controlling a physical system whose behavior is governed by laws of physics that can be described by equations. Pejtersen's work shows us that this impression is mistaken. As we see, the library information retrieval example is very different in nature from the DURESS II case study. There are no physical components, no conservation laws, and no equations. Nevertheless, there are qualitative behavior-shaping constraints that can be modeled. As a result, it is still possible to apply the conceptual distinctions in CWA to great effect, despite the enormous difference in application domains.

First, we briefly describe the results of each phase of CWA for this example.[1] Second, we describe how these findings were used to design the BookHouse (Pejtersen, 1992; Pejtersen & Goodstein, 1990), an innovative full-scope document retrieval system that can be used by library patrons to retrieve fiction books from public libraries.[2] Third, we discuss some of the broader conclusions arising from this example.

Work Domain Analysis. When we go to a public library to find a book, the work domain is the set of books that are available to be retrieved. Ideally, the choice of a book should be based on the degree of fit between the book contents and the needs of the library patron. To facilitate this process, books should be indexed according to attributes that are meaningful to library patrons. But as anyone who has ever been in a library knows, this is rarely the case. Most libraries use an indexing

---

[1]Just as with the DURESS II case study, the library information retrieval, as modeled here, does not have a significant social-organizational component. Thus, the social organization and cooperation phase of CWA is omitted.

[2]A Danish version of the BookHouse interface can be found on the www at www.isl.ruc.dk/boghus/. The software can currently be obtained by writing to erling.johannsen@risoe.dk.

scheme that is based on the Library of Congress or Dewey Decimal System codes. But does the book code KF299.I5N33 mean anything to you? Probably not. The reason for this is that the indexing schemes used by most libraries are somewhat arbitrary from the viewpoint of most library patrons.

Pejtersen (1973, 1979, 1988) conducted a number of field studies to overcome this problem for the particular case of fiction books. She observed approximately 450 interactions between library patrons and librarians in public libraries in Denmark as the latter tried to help the former find a book that they would enjoy reading. Based on the results of these field studies, Pejtersen created the AMP classification scheme, which defines a set of indexing attributes for cataloguing fiction books. As shown in the two columns on the left in Fig. 12.1, these attributes can be structured in the form of an AH, with higher levels representing functional properties (e.g., author's intention) and lower levels representing physical properties (e.g., typography). The important feature of the AMP classification scheme is that it describes books according to the attributes that people usually use when searching for books, not according to some arbitrary scheme. This symbiotic relationship is shown in the right-hand column of Fig. 12.1, which identifies the user needs that correspond to each level of the AH. Note that, although there are no physical components, conservation laws, or equations, the work domain of fiction books can be described using the AH as a modeling tool.

Abstraction Level	Document Content	User Needs
Functional Purpose	Why: Author Intention; Information; Education; Emotional Experience.	Goals: Why ? User's Ultimate Task and Goal
Abstract Function	Why: Professional Paradigm; Style Literary or Professional Quality; or School.	Value Criteria: Why? Value Criteria Related to Reading Process and/or Product
Generalized Function	What: General Frame of Content; Cultural Environment, Historical Period, Professional Context.	Content Frame: What? General Topical Interest of Historical, Geographical or Social Setting.
Physical Function	What: Specific, Factual Content. Episodic, Course of Events; Factual Descriptions.	Subject Matter: What? Topical Interest in Specific Content
Physical Form	How: Level of Readability. Physical Characteristics of Document; Form, Size, Color, Typography, Source, Year of Print.	Accessibility: How? User's Reading Ability, Perceptual and Cognitive Capability

FIG. 12.1. The attributes in the AMP classification scheme can be represented in the form of an abstraction hierarchy with each level defining information relevant to different types of user needs. Adapted from J. Rasmussen, A. M. Pejtersen, & L. P. Goodstein, *Cognitive systems engineering*, Copyright © (1994, John Wiley & Sons, Inc). Reprinted by permission of John Wiley & Sons, Inc.

Figure 12.2 shows an example of how the AMP classification scheme can be used to catalogue a book. The underlined labels on the left correspond to the dimensions of the classification scheme and the descriptions on the right represent the entries for a particular book, *Kinflicks* by Lisa Alther. This multidimensional classification scheme has important implications for systems design. More specifically, it provides a rich basis for creating a computer database structure that indexes books in a user-relevant fashion.

Control Task Analysis. Once the field of action has been modeled, the next step in CWA is to identify the control tasks that need to be performed. In contrast with the DURESS II case study, there is not much structure in the control task analysis for the library information retrieval example. As shown in Fig. 12.3, this layer of CWA is relatively straightforward and does not add much design leverage. The first control task is to acquire information about the user's needs (i.e., what kind of

---

LISA ALTHER: KINFLICKS 1977

Front Page Colours:	White, red and black
Front Page Pictures:	Faces
Controlled Browsing Terms:	Age groups, attitude to life, female roles, development, identity, illness, live together, love, parents.
Controlled Subject Terms:	America, death, daughters, identity problems, marriage, mothers, love affairs, personal development, relationship between, revival, sickbeds, student days, youth.
Subject Matter:	A woman's visit to her mother's sickbed and her revival of her youth, student days and marriage. Her experience of her mother's death.
Place:	USA. Tennesee.
Setting:	Southern states. Middleclass. High school, Feminism.
Time:	1960's
Cognition/Information:	Realistic description of the American society and of a woman's love affairs, her development and identity problems. The relationship between mothers and daughters.
Emotional Experience:	Humoristic.
Literary Form:	Novel. Related in first and third person. Feminist novel. Developmental novel.
Readability:	Average.
Typography:	Normal.

FIG. 12.2. An example of how a book is indexed using the AMP classification scheme. Adapted from A. M. Pejtersen (1988), Search strategies and database design for information retrieval from libraries. In L. P. Goodstein, H. B. Andersen, & S. E. Olsen (Eds.), *Tasks, errors, and mental models: A festschrift to celebrate the 60th birthday of Professor Jens Rasmussen* (pp. 171–190), published by Taylor & Francis. Copyright © 1988 Taylor & Francis Ltd. Reprinted by permission.

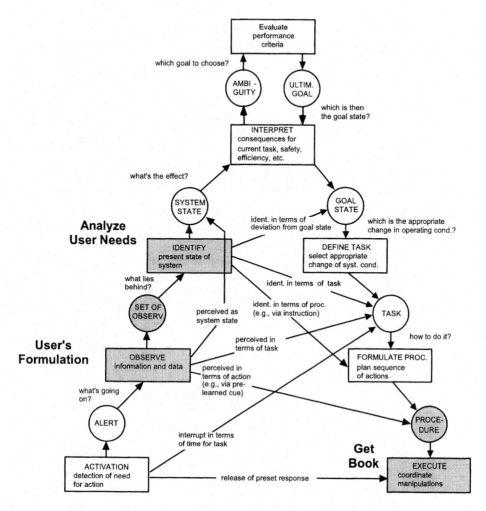

FIG. 12.3. A summary of the control task analysis for library information retrieval using the decision ladder template.

attributes they are looking for). In Pejtersen's field studies, this information was communicated verbally by the library patron to the librarian. However, if a computer-based system were designed, then the same kind of information could be input into the computer by the patron.

The second control task in Fig. 12.3 is to analyze the user's needs, based on the given information. This step involves mapping the user's formulation onto the existing database of books to identify one or more books of interest. In Pejtersen's field studies, this step was performed by the librarian, who served as an intermediary between the patron and the books. However, if a computer-based system were designed, then this step could be performed by the computer instead. As we see later, this approach was adopted in the BookHouse interface.

The final control task is a relatively trivial one. The user must retrieve the books of interest. This would usually be done manually, rather than automatically, by going to the shelves where the desired book is located.

Strategies Analysis. Although there is not much structure in the control task analysis for the library example, there is a rich set of constraints in the strategies layer. We focus on alternative strategies that could be used to perform the control task box identified as "analyze user needs" in Fig. 12.3. Table 12.1 lists some of the strategies that Pejtersen identified when she observed interactions between library patrons and librarians.

The first strategy, *bibliographical search,* is straightforward. The user communicates their needs in library terms (e.g., title, author, etc.), so the transformation that has to be performed by the intermediary (i.e., librarian or computer) is relatively trivial. This strategy was not observed very frequently in Pejtersen's field studies. When looking for fiction, users frequently know what kind of book they are looking for but usually not which particular book.

The second strategy, *analytical search*, is more intricate. In this case, the user communicates their needs in terms of the dimensions of the AMP classification scheme. For example, a user might express a need for a book that takes place in Canada (place) in the 20th century (time) and that is about the escapades of two brothers named Bob and Doug (subject matter). The intermediary must then map this input information onto the attributes of the books in the library. Consequently, this strategy requires a model-based search in a multidimensional space. As we might expect, Pejtersen found that this strategy was not observed very frequently nor was it found to be very successful on the rare occasions when it was used. The reason why seems clear. The analytical search strategy puts tremendous demand on intermediaries. They must know the attributes of each book in the library along each of the dimensions of the AMP classification scheme (e.g., Fig. 12.2), and then they must be able to search efficiently through this multidimensional knowledge base. The knowledge and computational power that are required to carry out this strategy effectively and reliably are beyond the capabilities of any one person unaided.

The third strategy, *search by analogy,* is simple, at least conceptually. In this case, the user says that they would like a book that is similar to a book that they have already read (e.g., *A Clockwork Orange*). The intermediary then tries to find a book that is similar to this referent. The key point, however, is that similarity is defined in a user-relevant, nonarbitrary manner using the various dimensions in the AMP classification scheme. This strategy is also very demanding on intermediaries because it requires that they have knowledge of all of the books in the library according to the dimensions of the AMP classification scheme.

The fourth strategy identified by Pejtersen is referred to as a *browsing strategy.* As the name indicates, this strategy is adopted when users have vague intentions

---

TABLE 12.1   A Subset of the Library Information Retrieval Strategies Identified by Pejtersen (1973, 1979, 1988) in Her Field Studies of User-Librarian Interactions

1. Bibliographical Search

2. Analytical Search

3. Search by Analogy

4. Browsing Strategy

Note.   See text for description of each strategy.

about the kind of book that they are interested in retrieving. As a result, the intermediary engages in an exploratory process until the user detects a book or attribute of interest. The objective is to match the ill-defined need to the contents of a particular book.

Interestingly, the relationships between these strategies are very similar to those between the troubleshooting strategies identified by Rasmussen and Jensen (1973, 1974). First, the various strategies have different degrees of resource requirements. For example, the analytical search strategy requires a great deal of knowledge about the book contents and poses substantial memory and computational loads. In contrast, the bibliographical search strategy imposes less of a demand on knowledge, memory, and computation. Second, Pejtersen found that strategy shifts tended to occur during impasses in the interaction between library patrons and librarians. These shifts provided a mechanism for dealing with resource-demand mismatches, thereby confirming the findings we presented in chapter 9. Third, it should also be clear from the description provided earlier that the various strategies require different types of information. Thus, knowledge of the strategies allows us to begin to think about designing human-computer dialogue modes and process flows that are tailored to the unique requirements of each strategy.

Worker Competencies. Despite the various layers already described, there are still many remaining degrees of freedom in the library information retrieval. Pejtersen conducted a worker competencies analysis to resolve these degrees of freedom in a way that is compatible with human cognitive capabilities and limitations (see Rasmussen et al., 1994, for a detailed account). The three principles of the EID framework (see previous chapter) were adopted to design the form of the interface.[3] Thus, the overarching goal for interface design was to encourage the use of SBB and RBB whenever possible, while at the same time, supporting KBB. Particular attention was paid to the fact that users of a library information retrieval system can be quite diverse in terms of age, interests, and knowledge. The challenge, then, was to present the required information in a form that would allow such a broad population to interact efficiently and effectively with the resulting device.

Three general techniques were adopted to take advantage of the competencies that a general population could already be expected to exhibit. First, users were divided into two general categories: children (age 6 to 16) and adults (age 16 to 70). This step was necessary because the analysis revealed that each of these groups had a different set of competencies. Second, KBB was supported by creating a conceptual metaphor that would allow users to leverage their existing knowledge structures. More specifically, the library information retrieval system was portrayed as a book house comprised of different rooms (see later discussion). Because users would already be familiar with navigation, this spatial metaphor would allow them to deal with the demands of information retrieval without having to learn an entirely new set of competencies. Third, RBB was supported by making extensive use of iconic representations. This decision has two benefits. It allows users to take advantage of

---

[3]The relationship between the BookHouse and the EID framework is somewhat subtle. As currently formulated, the EID framework is based solely on the work domain and worker competencies phases of CWA, as evidenced by the roles played by the abstraction-decomposition space and the SRK taxonomy (see chap. 11). The BookHouse was based on these phases of CWA as well as the control task analysis and strategies analysis phases.

FIG. 12.4. The introductory, welcoming display in the BookHouse. From J. Rasmussen, A. M. Pejtersen, & L. P. Goodstein, *Cognitive systems engineering*, Copyright © (1994, John Wiley & Sons, Inc). Reprinted by permission of John Wiley & Sons, Inc.

their powerful pattern recognition capabilities, thereby reducing the burden on resource intensive analytical processing. In addition, this decision also takes advantage of preexisting competencies because the icons were based on images that users would already be familiar with (see later discussion). Collectively, these techniques were intended to allow users to cope with the demands identified in the previous three phases of CWA in a way that is compatible with their cognitive capabilities and limitations.

BookHouse Document Retrieval System. The findings obtained from this CWA were used to create the BookHouse, a novel and innovative full-scope document retrieval system for retrieving fiction books from a public library.[4] In the remainder of this subsection, we provide a summary description of the BookHouse interface (see Pejtersen, 1992; Pejtersen & Goodstein, 1990; Rasmussen et al., 1994, for more details). You may wish to follow along interactively by accessing the BookHouse interface on the WWW (www.isl.ruc.dk/boghus/). Although that version is in Danish, it may nevertheless help you get a better feel for the BookHouse features.

Figure 12.4 shows the first-level, introductory display welcoming users to the interface. Note that the text shown is in Danish. The user enters the BookHouse through the open doors in the middle by clicking with a mouse. Note also that there is a one-line text explanation at the bottom of the screen. This text line is used in subsequent displays as well. It explains to first time users what commands are possible at each level. After users gain a little experience, they can readily recognize the graphical icons in the displays, and no longer need to refer to the text line.

Figure 12.5 illustrates the second-level display in the BookHouse interface. Here we can begin to see the icon-based nature of the interface. There are three databases

---

[4]Of course, the results of the CWA do not completely specify the design. Many of the detailed attributes of the BookHouse interface (e.g., colors, navigation mechanisms, font size and type) were determined by using standard HCI design guidelines and knowledge.

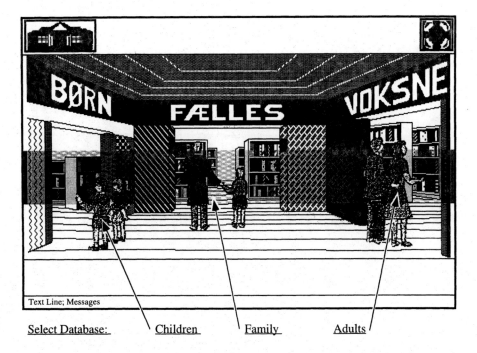

Select Database: Children   Family   Adults

FIG. 12.5. The second-level display in the BookHouse showing the three databases that are available. From J. Rasmussen, A. M. Pejtersen, & L. P. Goodstein, *Cognitive systems engineering*, Copyright © (1994, John Wiley & Sons, Inc). Reprinted by permission of John Wiley & Sons, Inc.

of books to choose from, one for children (represented by two icons of two children on the left), one for families (represented by one icon of a child and another of an adult in the middle), and one for adults (represented by two icons of two adults on the right). The user selects a database by clicking on one of these three areas. The life preserver in the upper right corner of the display is an icon that, if selected, brings up a help screen. The upper left corner of the display contains a miniature version of the first-level display (Fig. 12.4). By clicking in that area, users can return to the first level. This navigation technique is used in lower level displays as well. Users are always one click away from any of the levels that are above the one in which they are currently located (see Fig. 12.6).

Figure 12.6 shows one of the third-level displays, known as the strategies room. Here, there are four clickable areas, each corresponding to one of the four strategies that has been implemented in the interface: search by analogy, browsing with pictures, analytical search, and browsing by book descriptions. Clicking on any one of these areas takes us to one of the fourth-level displays, each of which is tailored to the information requirements of a particular search strategy.

For example, Fig. 12.7 depicts the fourth-level display for the analytical search strategy. Each of the objects in the room represents one of the dimensions in the AMP classification scheme. For instance, the clock on the wall specifies the time (i.e., the historical period), and the globe on the desk represents the geographical location. If we click on one of these objects, we get a number of keywords associated with the corresponding dimension in the classification scheme. Figure 12.8 provides

Start Again Select Other Database   Help

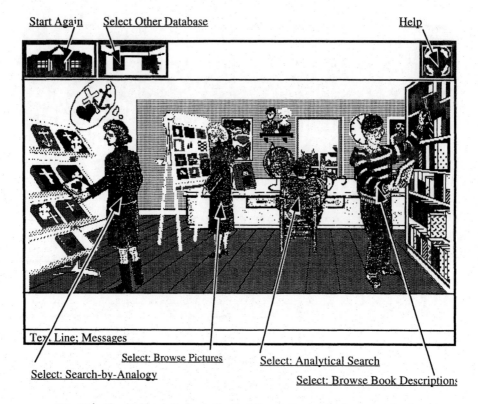

Text Line; Messages

Select: Browse Pictures   Select: Analytical Search
Select: Search-by-Analogy
                                        Select: Browse Book Descriptions

FIG. 12.6. The third-level display in the BookHouse showing the four strategies that are available. From J. Rasmussen, A. M. Pejtersen, & L. P. Goodstein, *Cognitive systems engineering*, Copyright © (1994, John Wiley & Sons, Inc). Reprinted by permission of John Wiley & Sons, Inc.

an example for the entries that are available in the database along the plot dimension. Again, the text is in Danish. Users can search for a book by specifying a value along one or more dimensions. For instance, it may be possible to search for a book that takes place in France (geographical location) during the 17th century (time dimension) and that has a happy ending (emotional experience). This user-centered, multidimensional search is possible because each book in the database has been classified according to the dimensions specified by the AMP classification scheme. The contrast with the Library of Congress or Dewey Decimal System indexing schemes is quite noticeable.

Returning momentarily to the strategies room in Fig. 12.6, if we choose the browsing-by-book-descriptions strategy, we are taken to a fourth-level display like that shown in Fig. 12.9. A book description is presented at random. We can see that the entry is organized according to each of the dimensions in the AMP classification scheme (see the capitalized labels on the left of Fig. 12.9). If the user is not interested in the book, then she can continue to browse and a second random book description can be called up, and so on. Some words in the book descriptions are presented in red (shown in italics in Fig. 12.9). These are keywords that the user can select to find other books.

Choose Other Strategy

Specify Plot Specify Actors Specify Environment Specify Author Intention
(Social or Work Place) Specify Time

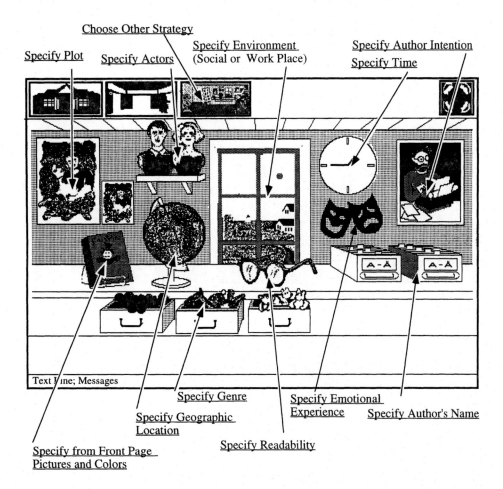

Text Line; Messages

Specify Genre Specify Emotional
Experience Specify Author's Name
Specify Geographic
Location

Specify Readability

Specify from Front Page
Pictures and Colors

FIG. 12.7. A fourth-level display in the BookHouse showing the support provided for the analytical search strategy. From J. Rasmussen, A. M. Pejtersen, & L. P. Goodstein, *Cognitive systems engineering*, Copyright © (1994, John Wiley & Sons, Inc). Reprinted by permission of John Wiley & Sons, Inc.

If we pick the search-by-analogy strategy from the strategies room in Fig. 12.6, then we are taken to the fourth-level display shown in Fig. 12.10. For this strategy, the user is required to identify the author and title of a book to be used as a referent for the search. The BookHouse will then try to find books that are similar to that referent. Note, however, that similarity is defined in a user-centered, nonarbitrary, fashion, as defined by the dimensions in the AMP classification scheme.

Finally, if we pick the browsing-by-pictures strategy from the strategies room in Fig. 12.6, then we are taken to the fourth-level display in Fig. 12.11. This display presents users with a number of icons, each representing one or more keywords. If the user clicks on a particular icon, the books that are linked with these keywords are identified in the one line of text at the bottom of the screen. A text-based description of each of these books (e.g., Fig. 12.9) can then be viewed. There are many pages of icons available to choose from. Thus, this strategy also allows users

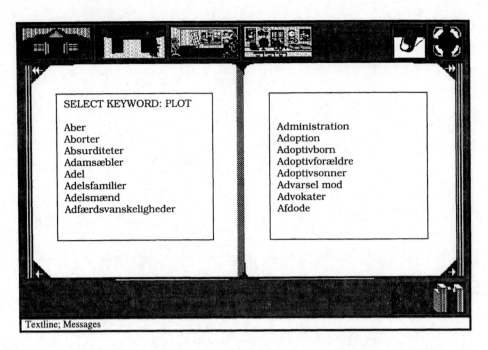

FIG. 12.8. Some keywords that are available under the plot dimension of the AMP classification scheme. From J. Rasmussen, A. M. Pejtersen, & L. P. Goodstein, *Cognitive systems engineering*, Copyright © (1994, John Wiley & Sons, Inc). Reprinted by permission of John Wiley & Sons, Inc.

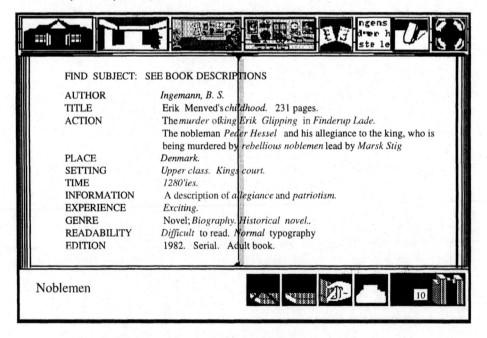

FIG. 12.9. A fourth-level display in the BookHouse showing the support provided for the browsing search strategy. From J. Rasmussen, A. M. Pejtersen, & L. P. Goodstein, *Cognitive systems engineering*, Copyright © (1994, John Wiley & Sons, Inc). Reprinted by permission of John Wiley & Sons, Inc.

FIG. 12.10. A fourth-level display in the BookHouse showing the support provided for the search by analogy strategy. From J. Rasmussen, A. M. Pejtersen, & L. P. Goodstein, *Cognitive systems engineering,* Copyright © (1994, John Wiley & Sons, Inc). Reprinted by permission of John Wiley & Sons, Inc.

FIG. 12.11. A fourth-level display in the BookHouse showing the support provided for the browsing by pictures search strategy. From J. Rasmussen, A. M. Pejtersen, & L. P. Goodstein, *Cognitive systems engineering,* Copyright © (1994, John Wiley & Sons, Inc). Reprinted by permission of John Wiley & Sons, Inc.

to browse for books but in a more intuitive, perceptual manner than the text-based browsing screen depicted in Fig. 12.9.

The BookHouse has many other features that we have not described here (see Pejtersen, 1992; Pejtersen & Goodstein, 1990; Rasmussen et al., 1994, for a more detailed account). We hope this account has given you a general feel for how the insights derived from the CWA were used to design an interface that is qualitatively different from those that are currently in use in most libraries.

Discussion. There are two general points to emerge from the BookHouse example. First, and most important, the example shows how the CWA framework can be applied to a "humanistic" rather than a "technical" application domain. Earlier chapters may have created the impression that CWA can be applied only to physical systems, like DURESS II, that are governed by physical laws described in terms of equations. The BookHouse clearly shows that CWA is much broader in scope. There are no differential equations or conservation laws in the library information retrieval domain. Nevertheless, there are qualitative behavior-shaping constraints (e.g., the dimensions in the AMP classification scheme, the search strategies, human cognitive limitations and capabilities) that can be identified by a CWA and that can then be built into a device. The important point to realize is that constraints need not be restricted to quantitative equations. Of course, the content of each layer of analysis in this example is quite different from that in DURESS II, but the same layers of analysis apply in both cases.

Second, the BookHouse interface also allows us to illustrate a point that we first made in chapter 4, namely that a work analysis framework should not just describe current practice. One of the reasons that we gave in support of this claim was that there may be ways of doing work that are potentially very effective, but currently unexplored, because workers do not have the requisite information support. The BookHouse provides a very clear-cut example of this claim. When Pejtersen observed user-librarian interactions, she found that the analytical search strategy was the least frequently used. Moreover, it was also the most difficult strategy to use, rarely leading to effective searches, because of the great demands it imposed in terms of knowledge and computation. If we followed a purely descriptive approach to work analysis, we might be very tempted to conclude that this strategy is not very good—perhaps because it is too "analytical" or because it does not match the properties of human decision making—and thus, not worthwhile supporting. This decision would have been a mistake.

When the BookHouse was introduced, the pattern of strategy usage changed drastically (Pejtersen, 1992; Rasmussen et al., 1994). The analytical search strategy became the most frequently used of all. Why such a big change? The simple answer is that the strategy is no longer analytical for the users because of all of the work done by the designer. The more resource-demanding aspects of the strategies were off-loaded to the computer. Before, librarians had to know the contents of each book in the library according to the dimensions that users adopted when searching for a book. Now, all of this knowledge has been built into the database. Moreover, the complex information-processing activity associated with searching through a high-dimensional state space has also been off-loaded from the librarian to the computer. Thus, all the user has to do now is specify the attributes of interest along the corresponding indexing dimensions. The computer then integrates the attributes and searches the database. This example shows the unique value of a formative approach. Work analysis should try to discover unexplored possibilities for accomplishing work rather than getting too stuck in the details of current practice. The example also shows another underappreciated point: Whether or not a strategy imposes analytical demands on human psychological resources depends on the interface that we make available to workers. In the unaided case, the analytical strategy requires extensive human analytical problem solving, whereas in the aided case with the BookHouse, *the very same strategy* requires minimal human analytical problem solving.

Summary. The BookHouse information system provides an example showing that CWA has practical implications for systems design (e.g., database design, human-computer dialogue modes and process flows, and interface form). The interface was very favorably received by users, was made available as a commercially available software product, was recognized by two design awards, and is currently being used in several public libraries in Scandinavia. It is perhaps the most notable success story associated with CWA. The example also allows us to make another very important point: CWA and formative models for system design can be applied to a wide range of application domains, not just those that involve the control of physical processes.

Example #2: Interface Design for DURESS II

The second example illustrates how the work domain and worker competencies phases of CWA have been used to create an innovative interface for DURESS II (Pawlak & Vicente, 1996; Vicente & Rasmussen, 1990). First, the abstraction-decomposition space was used to conduct a work domain analysis. The resulting representation (see chap. 7) identified the information content and structure of the interface. Second, the SRK taxonomy was used to conduct a worker competencies analysis. The resulting insights (see chap. 11) were used to guide the design of the visual form of the interface (i.e., how the information in the abstraction-decomposition space should be displayed in terms of signals, signs, and symbols). The final product is an interface that is consistent with the principles of the EID framework, as described in the previous chapter.[5] We refer to this interface as the P+F interface because it provides both physical and functional representations of DURESS II.

Supporting Knowledge-Based Behavior. The first step that was taken in designing the P+F interface was to represent DURESS II in the form of an abstraction-decomposition space to identify the content and structure of the interface. Such representations provide an externalized mental model that can support effective problem-solving activities. Figure 12.12 summarizes the work domain representation that we developed in chapter 7 for DURESS II. There are three levels of resolution along the part–whole hierarchy: system, subsystem, and component.

For the AH, there are five levels. At the top level of *functional purpose,* the higher level purposes of the domain are specified: to keep the water in each reservoir at the criterion temperature, and to satisfy the respective output demand for each reservoir. At the second level, *abstract function,* the structure of the system is described according to first principles, conservation of mass and energy. For the mass balance, there are three variables: the rate at which mass is flowing into the reservoir (source), the rate at which mass is flowing out of the reservoir (sink), and the current mass inventory. The energy topology is similar, the only difference being that there are two energy sources, the incoming water and the heater. At the level of *generalized function,* DURESS II is described in terms of several standard heating and liquid flow functions that instantiate the mass and energy topology specified previously. In the energy flow, there is a heating source, a cooling source, a means for removing heat, and a means for storing "heat" (or more accurately in engineering terms, thermal energy). In the mass flow, there is a means for adding water, a means for removing water, and a means for storing water. At the level of *physical function,* the system is described in terms of the components that realize the functions specified at the level above. These are the variables over which the worker has control. The

---

[5]As with the BookHouse, the results of the analyses do not completely specify the design (see Footnote 4). HCI design knowledge was used to determine the detailed attributes of the interface.

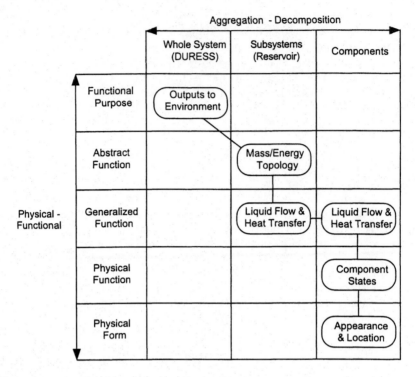

FIG. 12.12. Abstraction-decomposition space for DURESS II (adapted from Vicente, 1991).

final level of *physical form* is similar to the level of physical function. The difference is that whereas physical function specifies the settings and causal relationships between components, physical form specifies the spatial layout and appearance of those very same components.

To summarize, the abstraction-decomposition representation of DURESS II illustrates the various layers of functional structure that are inherent in the domain. Because it describes the way the process actually functions, this representation can be viewed as a faithful model of the work domain (i.e., it specifies the way workers should think about the work domain during problem solving), thereby providing a systematic basis for interface design. The way in which this representation was built into a particular visual form is described next.

Supporting Rule-Based Behavior. The next step in building the P+F interface was to embed the work domain representation into the interface. We can think of this phase as making visible the otherwise unobservable goal-relevant properties of the domain. According to the principles of EID, this step is accomplished by mapping the relational structures of the domain, represented in this case by the process' dynamic equations, onto the visible geometric properties of the interface objects. The goal is to allow workers to deal effectively with many task demands by relying on their powerful pattern recognition capabilities. Figure 12.13 illustrates how this has been accomplished in DURESS II for two pieces of higher level functional information that are frequently overlooked in traditional design approaches.

a) Mass Balance:

1) State Equation

$$\frac{dV(t)}{dt} = \frac{W_i(t) - W_o(t)}{\rho}$$

2) Geometry

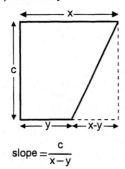

$$slope = \frac{c}{x-y}$$

3) Mapping

$$\frac{dV(t)}{dt} = \frac{1}{slope}$$

$$W_i(t) = x$$

$$W_o(t) = y$$

$$\rho = c$$

b) Temperature:

1) Algebraic Equation

$$T_w(t) = \frac{E_{tot}(t)}{V(t)\rho\, C_p}$$

2) Geometry

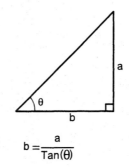

$$b = \frac{a}{Tan(\theta)}$$

3) Mapping

$$E_{tot}(t) = a$$

$$T_w(t) = b$$

$$V(t)\,\rho\, C_p = Tan(\theta)$$

Legend: $V(t)$ = reservoir volume ρ = density

$W_i(t)$ = input flowrate C_p = specific heat   } Constants

$W_o(t)$ = output flowrate   capacity

$T_w(t)$ = reservoir temperature

$E_{tot}(t)$ = total reservoir energy

FIG. 12.13. Making visible the invisible in DURESS II. The 1:1 mappings between abstract work domain constraints to concrete visual forms. Reprinted with permission from Vicente and Rasmussen (1990). Copyright © 1990 by Lawrence Erlbaum Associates, Inc.

In Fig. 12.13a, the state equation describing the mass balance is given. This equation describes a relationship between a set of variables, and thus can be considered a form of domain knowledge. Following the goal of making visible the invisible, a geometric figure was developed to represent perceptually the relational structures specified by the state equation. As a result, there is an isomorphic mapping between the geometry of the trapezoid and the relationships between the variables describing the mass balance. Because the general form of the energy balance equation is very similar to that of the mass balance equation, a mapping similar to that shown in Fig. 12.13a can also be developed to represent the energy balance.

Figure 12.14 provides a close-up view of the mass balance display derived from the mapping in Fig. 12.13a. The two views show the display under alternate conditions. The input is shown at the top of the graphics (MI1). The volume level is indicated by the scale on the left side of each graphic (V1) and by the shaded area of the container. The output, MO1, is shown at the bottom. The sloped line connecting the input and output relies on a funnel metaphor. For example, if the top is wider than the bottom (i.e., input > output, as in Fig. 12.14a), then it is easy to visualize the consequence, namely, that volume should be increasing. Conversely, if the bottom is wider than the top (i.e., output > input, as in Fig. 12.14b), then volume should be decreasing. Thus, the slope of the line represents the rate at which the volume should be changing.

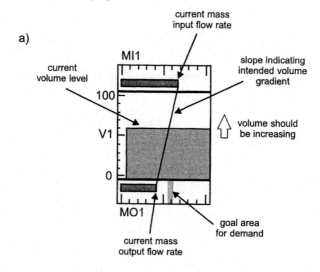

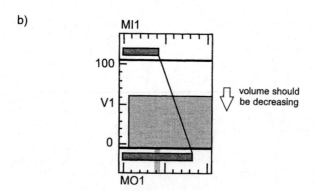

FIG. 12.14. Close-up view of the mass balance display in the P+F interface (Pawlak & Vicente, 1996). When input is greater than output (a), the slope is positive indicating that the volume should be increasing. When input is less than output (b), the slope is negative indicating that the volume should be decreasing. Reprinted from *International Journal of Human-Computer Studies, 44,* W. S. Pawlak & K. J. Vicente, "Inducing effective operator control through ecological interface design," pp. 653–688, Copyright © 1996. Reprinted by permission of the publisher, Academic Press.

The next display, illustrated in Fig. 12.15, shows the energy balance under two different conditions. Because it is based on essentially the same mapping function, the energy balance display functions in a very similar manner to the mass balance display. As shown in Fig. 12.15a, the primary difference is that the energy input to the reservoir (EI1) is partialed out according to its two contributors. The rate at which energy is added by the feedwater stream (FWS) is shown as a lightly shaded bar, and the rate at which energy is added by the heater is shown as a dark bar. The energy inventory is represented by the area, E1. Again, a funnel metaphor is used, so that, for example, if the total energy input flow rate is greater than the output (as in Fig. 12.15a), then energy inventory should be increasing. Figure 12.15b shows the same display in a different state. Now, the incoming water is adding energy at a greater rate than the heater. Another difference is that the total energy input flow rate equals the energy output flow rate, so the sloped line is perpendicular, indicating that the energy inventory should remain stable.

Returning to Fig. 12.13, the algebraic equation relating the temperature of the water in the reservoir, the energy contained therein, and the reservoir volume is shown in Fig. 12.13b. Again, a geometric figure was developed to illustrate the relationships specified by the equation. The result, in this case, is an isomorphic mapping between the trigonometric relationships describing the geometry of the triangle and the algebraic equation describing the behavior of the work domain.

A close-up of the display based on this mapping is provided in Fig. 12.16. The horizontal line with a ball on the end emanates from the current volume level (V1). Change in the height of this line always accompanies any change in mass inventory (i.e., the bar will always be at the same height as the water level, V1). The diagonal line in the center display rotates about its leftmost endpoint (connected to the top left of the T1 box) and is always tangent to the ball on the end of the horizontal line. The slope of the diagonal line represents the function that maps the relationship between volume and energy onto temperature. This mapping is indicated by the line emanating from the current energy inventory level (E1) that comes across and reflects off of the diagonal line at a right angle down onto the current value on the temperature scale (T1). The goal temperature is indicated by the thin shaded area on the temperature scale. This goal area reflects up from the temperature scale, off of the diagonal line, and onto the energy inventory scale. As a result, the energy level required to achieve the goal temperature is directly visible.

Figure 12.17 shows how this display reveals changes in work domain state. In Fig. 12.17a, the energy inventory is being held constant while the volume is increased slightly. The increase in volume causes the horizontal line emanating from the volume level to go up. This change in vertical position, in turn, forces the diagonal line in the center display to rotate counterclockwise, increasing its slope. The energy inventory level is constant and thus remains at the same height, but it now reflects off of the diagonal line at a different point. As a result, the new temperature is lower than the old temperature, which makes intuitive sense because the energy per unit volume is lower than before, due to the increase in volume. Note also that the energy level required to achieve the same goal temperature is now greater than before because of the increase in volume.

Figure 12.17b shows how this display changes when volume is held constant and energy inventory level is decreased. In this case, the change is much simpler. Because volume is constant, the height of the horizontal bar is fixed, which in turn, fixes the slope of the diagonal line in the center. The only change that occurs when energy

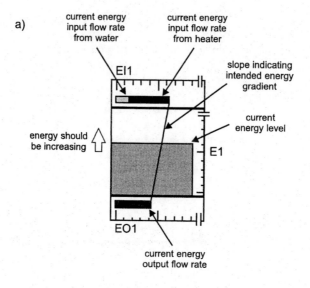

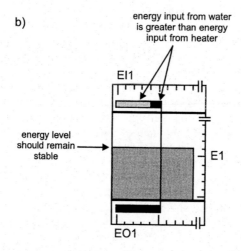

FIG. 12.15. Close-up view of the energy balance display in the P+F interface (Pawlak & Vicente, 1996). This display functions in the same manner as the mass balance display, except that there are two energy input sources, the water coming into the reservoir (lightly shaded bar) and the heater (darkly shaded bar). Figure (a) shows the case where the heater is adding energy at a greater rate than the incoming water. Also, in this case, the energy inventory should be increasing because the energy input is greater than the output, as indicated by the positive slope. Figure (b) shows the case where the incoming water is adding energy at a greater rate than the heater. Also, in this case, the energy inventory should be stable because the energy input equals the energy output, indicated by the perpendicular line. Reprinted from *International Journal of Human-Computer Studies, 44,* W. S. Pawlak & K. J. Vicente, "Inducing effective operator control through ecological interface design," pp. 653–688, Copyright © 1996. Reprinted by permission of the publisher, Academic Press.

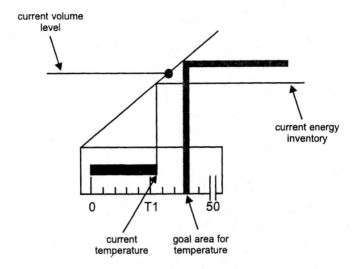

current volume
level

current energy
inventory

0   T1   50

current
temperature

goal area for
temperature

FIG. 12.16. Close-up view of the portion of the P+F interface showing the relation between volume level, energy inventory level, and reservoir temperature (Pawlak & Vicente, 1996). Reprinted from *International Journal of Human-Computer Studies, 44,* W. S. Pawlak & K. J. Vicente, "Inducing effective operator control through ecological interface design," pp. 653–688, Copyright © 1996. Reprinted by permission of the publisher, Academic Press.

inventory is decreased is that the horizontal line emanating from the far right goes down, which means that it intersects the diagonal at a different point, indicated by the dashed line in Fig. 12.17b. Again, the new temperature is lower than the previous one because there is now less energy in the same amount of water.

The key behind the mappings described in Fig. 12.13, and the resulting displays illustrated in Figs. 12.14 to 12.17, is that they translate the constraints inherent in the work domain into constraints on the interface geometry. This is equivalent to embedding knowledge about the domain in the interface. The great advantage of providing an externalized representation of the domain constraints in the interface is that it unburdens the worker's memory. For example, rather than having to remember, retrieve, and calculate the relationship describing the mass balance from memory, the worker can rely on the interface to provide this information directly. Making visible the invisible, then, allows workers to control the work domain by relying more on the powerful capabilities of perception and action than on comparatively more effortful and error-prone inferential processes.

The product of the design process up until this point is shown in Fig. 12.18. Information at every level of the AH is included. Beginning at the top level of functional purpose, the demand and temperature set points are represented as thin goal areas on the mass output (MO1, MO2) and temperature scales (T1, T2) in the interface (all of the variable definitions were provided in Table 6.2). The level of abstract function is represented by the boxed group of graphics on the right. This is the most innovative portion of the P+F interface, providing additional higher order functional information in the form of first principles (i.e., mass and energy conservation laws). The rectangular graphic on the left represents the mass balance for the reservoir (depicted in Fig. 12.14), and the rectangular graphic on the right represents the energy balance (depicted in Fig. 12.15). The display in the center

a)

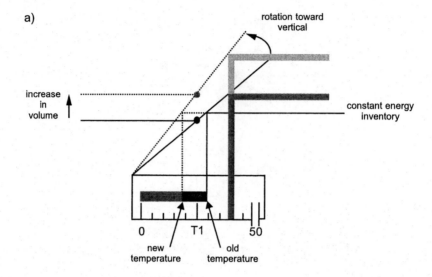

b)

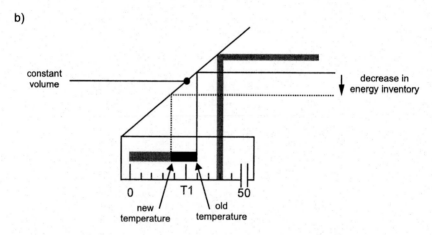

FIG. 12.17. The dynamics of the display in Fig. 12.16 (Pawlak & Vicente, 1996). Figure (a) reveals how the display changes when energy inventory is held constant while volume is increased, leading to a decrease in temperature. Figure (b) reveals how the display changes when volume is held constant while energy inventory is decreased, leading to a decrease in temperature. Reprinted from *International Journal of Human-Computer Studies, 44*, W. S. Pawlak & K. J. Vicente, "Inducing effective operator control through ecological interface design," pp. 653–688, Copyright © 1996. Reprinted by permission of the publisher, Academic Press.

with the diagonal line (depicted in Figs. 12.16 and 12.17) shows the relationship between volume, energy inventory, and temperature. Note that off-scale markers are represented in the output temperature scales and the energy input, inventory, and output scales as well. These were added to the interface by creating a gap in the scale at the off-scale point, thereby allowing workers to discriminate the maximum value from off-scale. At the generalized function level, the flow rates in each FWS (e.g., FVA, FPA, FA1, FA2) and the heat transfer rates (e.g., FH1) are displayed as

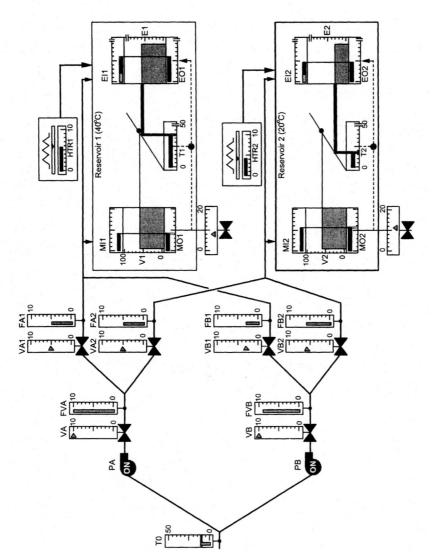

FIG. 12.18. The complete P+F interface for DURESS II. This figure shows the process in steady state, and therefore provides a baseline against which to compare Figs. 12.19 and 12.20. Reprinted with permission from Vicente and Rasmussen (1990). Copyright © 1990 by Lawrence Erlbaum Associates, Inc.

bar scales. At the level of physical function, the valve settings (e.g., VB) and heater settings (e.g., HTR2) are indicated by the small triangular pointers on the respective scales. Because the pump settings (e.g., PB) are discrete (either ON or OFF), they are directly labeled on the pumps themselves. At the final level of physical form, the relative spatial layout of the components and the connections between them are also represented.

To summarize, the P+F interface in Fig. 12.18 has exploited the mappings described in Fig. 12.13 to make the process dynamics visible to the worker. Unlike traditional process control interfaces that concentrate on presenting physical information, the P+F interface in Fig. 12.18 includes both physical and functional information (i.e., the levels of generalized function and abstract function), thereby making visible the abstraction-decomposition space.

Supporting Skill-Based Behavior. The final step in designing the P+F interface was concerned with action. EID suggests that, whenever possible, commands should be communicated by directly acting on the display—the familiar idea of direct manipulation. The intent here is to close the perception-action loop, thereby taking advantage of skilled motor control. As an example, to change the setting of the components in the P+F interface in Fig. 12.18 (e.g., VA, or HTR1), it is sufficient to "drag" the triangular set points to the new value using a mouse. For the pumps, a change in state is achieved by merely clicking on the pump icon itself. This direct manipulation design is preferable to abstract, and often awkward, command languages (Hutchins et al., 1986; Shneiderman, 1983).

The P+F Interface in Action. An impression of the dynamics of the P+F interface can be obtained by viewing Figs. 12.18 to 12.20, which show the progression of events before and after a fault, namely a block in valve VA1. In Fig. 12.18, the process is at steady state. The remaining figures presuppose this state as an initial condition. Figure 12.19 illustrates the state of the process 5 seconds after the onset of the fault, and shows a number of significant changes relative to Fig. 12.18. FA1, the flow through the blocked valve, has started to decrease, as has FVA. This reduction in inflow, in turn, causes the mass and energy inputs to Reservoir 1 (MI1 and EI1, respectively) to decrease. As a result, the input flow rates for both mass and energy are now less than their respective output flow rates, thereby causing the two lines connecting the input and output flows to deviate from vertical. Using the funnel metaphor on which these balance displays are based, the bottom is now wider than the top, so the two slanted lines indicate an expected negative gradient, which should cause the levels to decrease. The decrease in FA1 also causes a temporary surge in FA2, which propagates to a slight increase in MI2 and EI2. In this case, the balance lines are slightly tilted in the opposite direction (i.e., inputs slightly greater than outputs), indicating an expected small positive gradient. Figure 12.20 shows the state of the work domain 20 seconds after the fault, by which point VA1 has become completely blocked. This fact is clearly shown by the fact that VA1 is open but there is no flow through the valve (i.e., FA1 = 0). As a result, FVA has decreased as well, now supplying only the flow being drawn through VA2. Because FWS A is no longer supplying water to Reservoir 1, MI1 and EI1 have decreased considerably. Examining EI1, the fact that the lightly shaded bar (representing the contribution of the incoming water) is much smaller than before and the dark bar (representing the contribution of the heater) is the same length as before indicates that the reduction

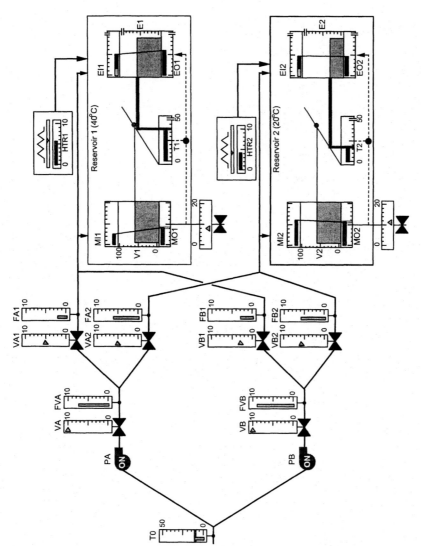

FIG. 12.19. P+F interface during a transient showing the dynamic content of the interface (Pawlak & Vicente, 1996). This snapshot was taken 5 s after the beginning of a block of valve VA1. Reprinted from *International Journal of Human-Computer Studies*, *44*, W. S. Pawlak & K. J. Vicente, "Inducing effective operator control through ecological interface design," pp. 653–688, Copyright © 1996. Reprinted by permission of the publisher, Academic Press.

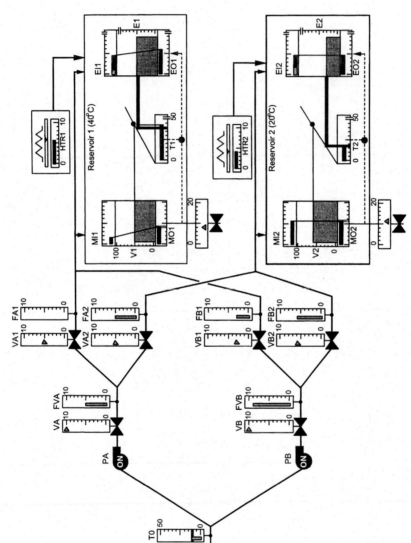

FIG. 12.20. P+F interface during a transient showing the dynamic content of the interface (Pawlak & Vicente, 1996). This snapshot was taken 20 s after the start of the fault, at which point valve VA1 had become completely blocked. Reprinted from *International Journal of Human-Computer Studies, 44*, W. S. Pawlak & K. J. Vicente, "Inducing effective operator control through ecological interface design," pp. 653–688, Copyright © 1996. Reprinted by permission of the publisher, Academic Press.

in EI1 is caused by a reduction in the flow of water, not a reduction in the transfer of heat. This reduction in inflow causes the sloped lines for the mass and energy balances to be even more negative than in Fig. 12.19. As expected, the volume (V1) and the energy inventory (E1) have decreased as a result. Furthermore, because there is no longer any cold water coming into Reservoir 1 and the volume is decreasing, the temperature (T1) has increased and is now slightly outside of the goal region.

Discussion. Like the BookHouse example presented earlier, the P+F interface for DURESS II shows how some of the phases of CWA can uncover important insights that have practical implications for systems design. The result in this case was a novel and innovative interface for a process control microworld. This interface has been compared against a more traditional interface in a number of experiments, and has been consistently found to lead to better performance (see Vicente, 1996, for a review). In addition, parts of the P+F interface have been adapted by Toshiba in Japan to create a novel interface for a full-scope nuclear power plant simulator (Itoh et al., 1995). Thus, this example is a success story of technology transfer from research to industry for CWA. It is important to point out, however, that the Toshiba control room remains to be evaluated empirically.

Summary

The goal of this section has been to show that CWA provides value added to systems design. We showed how parts of the CWA framework have been applied to two very different application domains, thereby illustrating the framework's broad range of applicability. In either case, the analysis led to an interface that was innovative in that it was markedly different from interfaces that had been developed in the past for each domain. Each interface was also successful, as evidenced by data from empirical evaluations. Each interface was also influential in the sense that it led to technology transfer. In addition, the BookHouse example is particularly notable because (a) it was turned into a commercially available product, (b) it is currently being used in several public libraries, and (c) it has been recognized by design awards.

As encouraging as they are, these facts do not prove that CWA will always lead to very effective designs (see chap. 13). Nevertheless, they show that CWA generates important implications for systems design, and that it can be used to create interfaces that are both innovative and effective.

IMPLICATIONS FOR RESEARCH

As mentioned earlier in this chapter, the distinctions comprising CWA have implications not just for systems design but for research as well (Vicente, 1995a). In this section, we focus specifically on typical problems that are encountered in empirical cognitive engineering research with microworlds—small-scale computer simulations that are intended to be representative of industrial-scale complex human machine systems (Brehmer & Dörner, 1993). Three different types of research problems are addressed: (a) design of experiments, (b) analysis of data, and (c) classification of dependent variables. We provide examples of how CWA can be applied to each of these issues by drawing on our research with the DURESS II microworld.

Design of Experiments

Typical Issues. One of the challenges that arises in the conduct of microworld research is how to compare performance on different trials in a meaningful way. Because of the richness of microworlds, task demands are broader in scope than in traditional experiments. As a result, we often find that there can be a significant degree of uncontrolled variability in the difficulty associated with different trials in a microworld. This variability presents a problem for the sensitive comparison of performance across trials.

Another issue that arises frequently in microworld research is the need to gradually increase the complexity of trials during the practice phase of an experiment. Microworld tasks can be very complex and, in fact, overwhelming to a novice. Consequently, it is important to present subjects with simpler trials first, so that they can gradually become familiar with the domain, and only then introduce trials of greater complexity.

To deal with these typical issues, it would be useful to have an objective way of identifying the complexity of trials in a particular microworld.

Benefits of CWA. The CWA of DURESS II presented earlier in this book led to a solution to these two problems. As part of the strategy analysis, a taxonomy of trials based on the range of feasible strategies was identified.

Recall from chapter 9 that one of the important structural properties of DURESS II is that each FWS can supply water only at a maximum rate of 10 units/second of flow. This means that different demand set points can be satisfied using three different configurations of the FWSs (see Table 9.4). First, if the sum of the output flow demands for the two reservoirs does not exceed 10 units/second, then a single FWS can be used to feed both reservoirs. This configuration allows workers to use a smaller number of components to control the process. Second, if the sum of the demands exceeds 10 units/second but neither of the demands exceeds 10 units/second, then a decoupling strategy can be adopted. That is, subjects can use one FWS to feed one reservoir and the other FWS to feed the other reservoir, keeping the crossover valves VA2 and VB1 closed. This configuration facilitates the control of the process because there are no interactions between FWSs, thereby eliminating the possibility that manipulating a control valve will have an undesired effect on the other reservoir. Third and finally, if one of the output flow demand set points is greater than 10 units/second, then that reservoir cannot be supplied by just one FWS. As a result, a many-to-many strategy must be adopted by opening at least one of the crossover valves (VA2 or VB1). This many-to-many mapping between FWSs and reservoirs complicates the control of the process because adjusting the flow to one reservoir can affect the flow to the other.

This taxonomy shows the relationship between task demands (i.e., demand set points) and feasible strategies (and therefore, in this case, task complexity). This insight provided by CWA was put to use in several experiments on DURESS II to address the issues mentioned at the beginning of this subsection (e.g., Christoffersen, Hunter, & Vicente, 1996, 1997, in press; Pawlak & Vicente, 1996). When it was important to compare performance across trials, an effort was made to match those trials by having the demands fall in the same category of the taxonomy just described, thereby equating the complexity of the two trials in question. Similarly, when it was important to gradually increase the complexity of trials during practice, demand pairs that could be satisfied by simpler strategies were used for the initial practice trials. After subjects became familiar with the work domain, subsequent trials were introduced that were more complex in that they could not be solved using simplifying strategies.

Therefore, in both instances, the application of CWA to the experimental design of microworld research provides the important benefit of giving the experimenter better control over the trials that are presented to subjects.

Analysis of Data

Typical Issues. Microworld research also leads to difficulties in terms of data analysis. For example, because of the richness of the tasks and the extensive level of practice that subjects typically receive in such studies, coarse measures of performance rarely reveal differences between treatment groups (Moray, Lootsteen, & Pajak, 1986; Sanderson, Verhage, & Fuld, 1989). Also, because so many data are typically generated by microworld experiments, an exhaustive analysis is usually unrealistic because of the inordinate amount of time involved. Both of these issues create a need for focusing the data analysis effort in a direction that is economic, yet promising in terms of revealing differences between subject groups.

Benefits of CWA. The CWA for DURESS II led to an important insight for data analysis (Pawlak & Vicente, 1996).[6] The analysis revealed that the sensitivity of the temperature to the heaters depends on both the demand and volume. The former is a goal set point, and so subjects have no choice but to satisfy it if they are to perform the task. However, the choice of volume is discretionary. The CWA indicates that if subjects decide to stabilize the system at a low volume, then it will be more difficult to stabilize the temperature within the goal region because the temperature will be very sensitive to changes in heater setting. Consequently, stabilizing the process at low steady state volumes can be considered to be dysfunctional because it makes the process more difficult to control.

This insight was applied to analyze the data from an experiment comparing the performance of two interfaces for DURESS II (Pawlak & Vicente, 1996). One of these interfaces represented the work domain relationships more thoroughly than the other, so it was hypothesized that it would lead to improved performance. However, an analysis of typical measures of performance (e.g., task completion time) showed no difference between the two interfaces on normal (i.e., nonfault) trials. The CWA suggested the hypothesis that the two interface groups might differ in terms of their steady state volumes, the rationale being that subjects with the interface showing the work domain relationships should realize that low volumes should be avoided. An analysis of the steady state volumes for the two interface groups confirmed this hypothesis. Furthermore, it was also found that steady state volume was significantly negatively correlated with the number of heater actions for low demands. This result was also motivated by the CWA, because one would expect that low volumes would

---

[6]This portion of the CWA was not described in the portions that were presented earlier in this book.

lead to difficulties in stabilizing temperature, and therefore, an increased number of heater actions.

It is extremely unlikely that we would have had the idea of conducting these analyses were it not for the CWA conducted before the experiment. Therefore, one of the important benefits of CWA for microworld research is that it can lead to the identification of sensitive data analyses that can show differences between treatment groups that would otherwise not be observed.

Classification of Dependent Variables

Typical Issues. Because there are typically many more degrees of freedom in microworld research than in traditional laboratory experiments, there are many different types of dependent variables that can potentially be adopted. The question then becomes which measures should we adopt in a particular study? Furthermore, if we decide to adopt a number of different measures, how can we organize these dependent variables in a meaningful and concise way?

Benefits of CWA. CWA can provide some help with this aspect of microworld research as well. Each layer of constraint defines a distinct class of potential dependent variables. The first layer, work domain analysis, defines variables that identify the state of the work domain. The obvious candidates for measurements here are state variables. However, sometimes other variables are also useful. For example, higher order constructs that are derivable from state variables can frequently provide information about the state of a work domain that is perhaps more meaningful than that provided by state variables alone. An example would be a measure of how close the work domain is to exceeding one or more of its safety boundaries.

The second layer, control task analysis, defines product measures of performance that describe *what* subjects do. Examples would include: task completion time, number of errors, number of control actions, number of worker verbalizations, and nonverbal behaviors. All of these data can be used to determine what is being done, but by themselves (i.e., without further analysis) they do not indicate *how* the tasks are being done because the exact same sequence of overt behavior could have been generated by a number of different strategies. Note that this level of measurement is quite distinct from the previous one; measures of work domain state and product measures of subject performance should not be confused because they are conceptually different.

The third layer, strategies analysis, defines process measures of performance that describe *how* subjects do what they do. Such measures go beyond merely describing what workers are doing, and try to capture and describe the strategies that are generating the patterns of behavior identified by the product measures described previously. Examples of such process-tracing measures include: eye movement analysis, verbal protocol analysis, state-space diagrams, and action transition graphs (Howie & Vicente, 1998; Moray et al., 1986; Sanderson et al., 1989).

Note that, in some cases, the same raw data (e.g., worker verbalizations) can be relevant to both this level and the previous one. The difference depends on the extent to which the data are interpreted by the experimenter. For example, if we merely classify verbalizations according to which control tasks they are referring to, then we are in the realm of product measures because all we can reliably infer is what tasks are being done, not how they are being done. On the other hand, if we take those same raw data and process them more extensively (e.g., by looking at sequential dependencies, and cross-correlating multiple measures), we would be in the realm of process measures because we would be developing an understanding of the strategies that are behind the verbalizations. Nevertheless, it is important to keep in mind that the question of what tasks are done and how tasks are done are conceptually distinct because the very same task can be done in multiple ways and the very same sequence of overt behavior could have been generated by different strategies. Therefore, the two types of dependent variables should not be confused because the inferences we can make from each are different.

The fourth layer of CWA, social organization and cooperation analysis, defines measures of team or group communication and cooperation that can be used to understand the degree and quality of coordination across multiple actors. Examples include: the direction and frequency of communication between different actors, measures of nonverbal cooperative behavior, speech act analyses, and the form of communications (e.g., orders, advice, instructions or neutral facts). This level of measurement is important because it investigates the way in which teams/groups/organizations work within the constraints defined by the demands of the work domain, the tasks that are required in that domain, and the different strategies that each actor might use. Thus, although building on previous levels, this level of measurement contributes toward an understanding of a conceptually distinct set of issues, namely the form and content of team coordination.

Finally, the fifth level, worker competencies analysis, defines measures that describe subjects' level of expertise. It is useful to distinguish between two different subclasses at this level, one situation-dependent and the other situation-independent. Regarding the former, several different types of online measures of competency can be adopted including measures of situation awareness (Endsley, 1995) and mental work load (O'Donnell & Eggemeier, 1986). These dependent variables assess operators' competencies in specific situations. In contrast, situation-independent measures assess competence in a broader fashion. For example, knowledge elicitation measures developed in the psychology and cognitive science literatures (e.g., categorization tasks, transfer of training, control recipes, memory recall tasks) can be used to assess a subject's knowledge of a particular work domain (Christoffersen, Hunter, & Vicente, 1994; Cooke, 1994).

Again, note that this level is distinct from the previous ones because it is possible for workers to perform the exact same task in the same way, or to perform a task using the same strategy, but still exhibit different competencies. As an example of the former, two workers could exhibit the same overt sequence of behaviors, but one could be using spatial resources whereas another could be using verbal resources (cf. Wickens, 1992). Similarly, two workers could be performing fault diagnosis using a hypothesis-and-test strategy, but they could be using very different psychological resources (e.g., if the workers were of different levels of expertise, or if one of them was using an interface that off-loaded part of the demands of that strategy to a computer aid).

This taxonomy of dependent variables was used to identify a set of measures for a longitudinal experiment with DURESS II (Christoffersen et al., in press). The taxonomy provided several benefits. First, it allowed us to ensure that we had collected the dependent variables that were required to answer all of the questions of interest. For example, sometimes experimenters make claims about subjects' mental models without having any measures at the level of worker competencies. Such claims are indefensible. Second, the taxonomy allowed us to associate a given set of measures with each specific experimental question of interest (e.g., knowledge elicitation tests with the question of what impact interface design had on worker competencies). These linkages made the data analysis effort much more manageable because the numerous dependent variables could be segregated into meaningful groups.

These examples show that CWA also has important benefits for the classification of dependent variables in microworld research. More specifically, CWA can help experimenters generate a comprehensive and coherent measurement plan that can, in turn, lead to deep insights and defensible inferences.

Summary

In this section, we have shown that the value of CWA extends beyond systems design. The distinctions and concepts in the framework can also be used productively in cognitive engineering research. We provided several examples of how CWA can help us deal with the difficulties associated with experimental design, analysis of data, and classification of dependent variables in microworld research. These examples just scratch the surface of the potential relevance of CWA to research, however. Many other applications are foreseeable (e.g., Yu et al., 1997).

THE BOTTOM LINE

There is no intrinsic practical value in conducting a work analysis because it is a means, not an end. Thus, it is important to demonstrate that a framework for work analysis can be used to generate valuable, and preferably unique, implications. In this chapter, we have illustrated the value that CWA can add to both systems design and research. In the first part of the chapter, we showed that CWA can be used to design effective and innovative interfaces for both humanistic and technical application domains. The BookHouse information system and the P+F interface for DURESS II illustrate that CWA has a wide range of applicability in systems design, much broader than the DURESS II case study that we have been using would suggest. In the second part of the chapter, we showed that CWA can also be used to address challenging issues in microworld research. These examples should convince you that CWA has an important and unique contribution to make to both systems design and research. Perhaps even more important, we hope that these examples give you ideas for new ways in which CWA can be applied in your own design and research activities. Indeed, this is one of the primary purposes of this book—to make these concepts more readily available to a wider audience so that the framework can be applied in new ways.

FINAL WORDS IV

Designing for Adaptation: Safety, Productivity, and Health and the Global Knowledge-Based Economy

13

Unpredictability is a source of disturbances throughout the lifetime of the system. . . . attempts to fight against them are also continuous.
—Norros (1996, p. 175)

Change is the invariant, not knowledge.
—Blum (1996, p. 160)

PURPOSE

In this capstone chapter, we have three final objectives. The first is to discuss how the CWA framework is intended to improve safety, productivity, and worker health. Our second objective is to address a number of caveats that we hope put our claims in perspective and thereby minimize any misunderstandings. Finally, our third objective is to briefly discuss the trend toward a knowledge-based global economy and how this trend is increasing the requirement for adaptation on the part of workers, managers, organizations, and technology. Given the characteristics of the knowledge-based global economy, the need to design computer-based systems that support adaptation, and thus the need for CWA, will only increase in the future.

HOW DOES COGNITIVE WORK ANALYSIS ADDRESS SAFETY, PRODUCTIVITY, AND HEALTH?

Three Problems: Briefly Revisiting the Evidence

In chapter 1, we introduced three perspectives that can be used to evaluate the success of a complex sociotechnical system, namely safety, productivity, and worker health. An effective sociotechnical system should have an acceptable level of safety, should lead to a high level of productivity, and should not jeopardize worker health. We also cited a body of evidence to indicate that corporations are currently having difficulty satisfying each of these three criteria. To give just one example from each criterion, it has been estimated that:

- Abnormal situations in the petrochemical industry have an annual impact of $20 billion on the U.S. economy alone (Bullemer & Nimmo, 1994).

- Companies lose $5,590 per year per computer because of the time that employees waste by "futzing" with their computers instead of performing productive work with them (Gibbs, 1997).
- Up to $80 billion per year in direct health care costs may be preventable in the United States alone by designing healthier work (Karasek & Theorell, 1990).

Given the magnitudes involved, any progress that we can make along any one of these dimensions can lead to very substantial gains.

However, in chapter 1, we did not choose one of these three criteria as the focus for the remainder of the book. We made a much stronger claim, namely that progress on all three criteria can be achieved by a single, comprehensive approach to the analysis of work in complex, sociotechnical systems. At the time, you may have found it difficult to fathom how progress could possibly be made on all three fronts without having to trade off one criterion for another. Now that we have described the rationale behind CWA and its conceptual structure, we are in a position to revisit our claim.

A Potential Three-Fold Remedy: Designing for Adaptation

Safety. As systems become increasingly complex, there is a commensurate need to manage risk. System safety has been a traditional topic of concern in technical systems, such as nuclear and chemical process plants. Names such as Three Mile Island, Bhopal, and Chernobyl quickly remind us of the stakes involved. However, the recent, well-publicized disruptions to global financial markets have made it clear that, in contemporary society, threats to safety can take many other forms as well. The management of risk is more broadly relevant than we might first suspect.

Virtually all of the empirical evidence we know of (e.g., Halbach, 1994; Hirschhorn, 1984; Kaufmann et al., 1992; Leveson, 1995; Norros, 1996; Perrow, 1984; Pool, 1997; Rasmussen, 1969; Reason, 1990; Waldrop, 1987) tells us that the most significant threat to the safety of complex sociotechnical systems is posed by events that are both unfamiliar to workers and that have not been anticipated by designers (see chap. 1). Procedural approaches cannot, by definition, cope with this problem because a detailed procedure cannot be devised if the instigating event is unanticipated (see chap. 3). This point was clearly made by our story of "malicious procedural compliance" that opened the book. The only viable approach to radically improving system safety is to figure out some way to help workers adapt to the unique, unanticipated contingencies with which they may be faced. That is, *we must design for the unanticipated.* After all, this is the primary reason why people are included at all in complex sociotechnical systems: They are supposed to cope with the situations that are beyond the reach of automation and designers' foresight. Otherwise, workers would be automated out of the system and autonomous (or "lights-out") systems would be commonplace. But despite the enormous investments made in achieving that objective and the vast improvements that have been made in the capabilities of automation technology, the completely autonomous operation of these systems is still a pipe dream, and will remain so for the foreseeable future.

The most promising path for improving safety in complex sociotechnical systems is to *deliberately* design to support adaptation. We know through painful experience that unanticipated events will occur, so we need to provide workers with the support that they need to play the role of adaptive problem solvers reliably and effectively. We cannot expect workers to fulfill this onerous responsibility unless we provide them with the appropriate set of tools. As we pointed out in chapter 5, this is precisely the philosophy behind CWA—uncover the information requirements that can lead to the creation of computer-based information systems that help workers finish the design (Rasmussen & Goodstein, 1987). CWA instantiates this philosophy by adopting a constraint-based approach throughout. Rather than designing to support the mythical one best way or designing a "band-aid" that will fix the latest "human error," the objective is to identify the constraints that define the boundaries of system safety and reliable performance. We need to make these boundaries visible to workers and provide them with the tools that they need to navigate and explore the remaining space of functional possibilities for action (see Fig. 13.1). By creating such a flexible design, competent groups of workers should be able to self-organize and adapt to meet the unique demands imposed by unanticipated events. CWA provides a systematic framework for developing such a design. It identifies the various layers of behavior-shaping constraints in a complex sociotechnical system, thereby delimiting a dynamic space of bounded flexibility within which functional adaptation is possible.

In summary, systematically supporting workers in finishing the design seems to be the only viable path to creating flexible, adaptive sociotechnical systems that will be able to cope with the unanticipated, and thus lead to marked improvements in safety.

Productivity. The negative relationship between investment in information technology and productivity is so notorious that it has a label, the *productivity paradox* (Landauer, 1995). The greater the investment in computers, the lower the growth in productivity. This outcome is not the one that was expected by the corporations that invested, and continue to invest, so heavily in information technology. With a few exceptions, the expected improvements in productivity have not materialized. This is a cause for great concern, given the relationship between corporate productivity gains and a nation's economic prosperity.

According to Landauer (1995), part of the problem is that those activities that can be reduced to a set of rules or an algorithm have largely already been completely

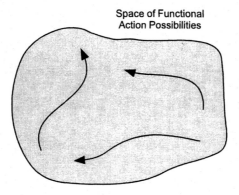

**Space of Functional
Action Possibilities**

FIG. 13.1. The behavior-shaping constraints in CWA define a dynamic space of functional action possibilities. It is up to workers to choose a trajectory in this space to meet the demands of the moment.

automated with computer technology (see chap. 1). This we know how to do, and we have done it well. But as a consequence, what remains are intellectual activities that we do not know how to automate fully. These activities will generally require workers to engage in discretionary decision making. Thus, if we are going to make large strides in productivity growth, we need to design computer systems that provide workers with new types of functionality that are required to perform these discretionary activities effectively and reliably.

How can we design computer systems that have such useful functionality? CWA addresses this issue by taking a formative approach to work analysis (see chap. 5). Rather than prescribing how work *should* be done or only describing how it *is* currently being done, CWA also seeks to identify how work *could* be done if the appropriate tools were made available. The key to this approach is the modeling of intrinsic work constraints. To the extent possible, the characteristics of the application domain should shape the design of the device. For example, rather than assuming up front what is to be automated and what is not, CWA lets decisions about automation emerge as outputs based on the requirements that are identified during the analysis. Similarly, rather than assuming up front what information is to be presented in an interface, CWA lets decisions about information content emerge as outputs of the analysis. This is the fundamental idea behind the formative approach—let the intrinsic constraints of the application domain shape the functionality of the device.

By adopting this formative approach, we should expect to design innovative computer-based systems that provide useful, not just usable, features. The Book-House interface is a very good example (see chap. 12). It provides a qualitatively different kind of information support than that offered by traditional library information retrieval systems. But this innovative functionality would not have been possible were it not for the formative nature of the CWA on which the design was based. The characteristics of the application (e.g., the attributes that library patrons can use to think about books, and the strategies that patrons and librarians can adopt to search for books) largely determined what functionality was to be built into the BookHouse interface. Similarly, the P+F interface for DURESS II (see chap. 12) provides a qualitatively different kind of information support than that offered by traditional process control interfaces. Rather than figuring out how to present whatever information is already made available by a predetermined set of sensors, the work domain analysis that was used to create the interface explicitly sought to identify what information was required to get the job done, and thus, what sensors are needed in the first place. The functionality of the interface was not assumed up front but emerged as an output of the analysis. This formative approach is very different from that typically used to design interfaces for process control systems (Reising & Sanderson, 1996; Vicente, Christoffersen, et al., 1996).

If Landauer (1995) was correct, then we would expect that information systems like the BookHouse and the P+F interface should lead to quantum improvements in productivity. Providing workers with the functionality that is required to support intellectual activities involving discretionary decision making should allow us to make a noticeable dent in the productivity paradox.

As this book goes to press, there is new empirical evidence to indicate that the philosophy behind CWA can indeed lead to a positive impact on productivity. Table 13.1 summarizes the data from Brynjolffson and Hitt's (1998) study of the interaction between organizational structure (specifically, degree of decentralization) and investment in information technology. The first row of the table shows that, when a

TABLE 13.1 The Impact of Decentralization in Organizational Structure and Investment in Information Technology (IT) on Corporate Productivity, as Identified by Brynjolffson and Hitt's (1998) Study of 232 Corporations

Decentralization	Investment in IT	
	Low	High
Low	0	−0.0366
	(n = 69)	(n = 47)
High	0.0161	0.0455
	(n = 47)	(n = 69)

Productivity was measured as output divided by input costs in 1990 US dollars. All values are relative to the low-low cell in the upper left, which has been normalized to zero.

corporation has a more centralized organizational structure, then increased investment in information technology actually reduces productivity. In contrast, the second row shows that, when a corporation has a more decentralized organizational structure, then increased investment in information technology improves productivity. This pattern of results is consistent with the claim that using information technology to provide workers with the autonomy to finish the design, as CWA recommends, can lead to comparatively large improvements in corporate productivity.

Health. Although it is rarely explicitly considered in the design of computer-based systems, worker health is at least as important as safety and productivity. People spend a large portion of their lives at work, and the extent to which their workplace is designed to induce health has an impact on all aspects of their lives, not just on the quality of working life (Reed, 1996). The evidence available shows that impacts on worker health manifest themselves, not just in direct physiological terms (e.g., systolic blood pressure, incidence of heart disease, number of cardio-vascular deaths), but in less direct terms as well (e.g., absenteeism, job turnover, involvement in leisure, involvement in political activity). Aside from the obvious welfare impact, these effects also have enormous economic costs. For example, it has been estimated that the productivity losses due to preventable job stress amount to $300 billion per year in the United States alone (Matteson & Ivancevich, 1987), not including the preventable direct health care costs that are estimated to be $80 billion per year in the United States alone (Karasek & Theorell, 1990). Therefore, whether we are interested in human welfare or improved economic return, a focus on improving worker health is easily justified by the available data.

According to Karasek and Theorell's (1990) demand-control model (see chap. 1), the most important factor contributing to worker health is the degree of job autonomy. Workers who lack control over how they perform their job are the ones who suffer the most. For some reason, having to follow orders or procedures that define the one right way of doing the job, and that thereby take all of the thinking and innovation out of a job, exacts its toll on a workers' health. The following quote from Kanigel's (1997) biography of Frederick Taylor gives us an intuitive feel for the problem:

> The work itself might be no more physically demanding, but somehow, by day's end,
> it felt as if it were. Going strictly by somebody else's say-so, rigidly following directions,

doing it by the clock, made Taylor's brand of work distasteful. You had to do it in the one best way prescribed for you and not in your old, idiosyncratic, if perhaps less efficient way. (pp. 209–210)

Of course, the correlation between autonomy and health can just as well be viewed in a positive light as an opportunity to improve worker health.

CWA capitalizes on this insight. Rather than basing the design of a device on detailed proceduralized descriptions of tasks, CWA identifies the constraints that should be respected and then gives workers the autonomy to deal with the demands as they see fit, within the aforementioned constraints. This approach represents a radical shift in philosophy. Rather than forcing workers to follow the one best way, CWA gives workers the responsibility of finishing the design. If Karasek and Theorell's (1990) demand-control model is correct, then we should expect to obtain significant improvements in worker health as a result. Such an improvement should, in turn, benefit both the quality of life and corporate productivity.

Summary: All Roads Lead to Rome

Surprising as it may seem, the roads to improved safety, improved productivity, and improved worker health all appear to overlap. By adopting the philosophy of designing for adaptation, CWA is in a position to make contributions on all three fronts. As far as we can tell, there are no trade-offs. And given the costs associated with the difficulties that corporations are currently experiencing along each front, the potential payoff is very large indeed. This does not mean, however, that the path toward safe, productive, and healthy computer-based work is an easy one with a guaranteed payoff, as the caveats described next make clear.

CAVEATS

In an effort to describe and advocate a relatively new approach to any research problem, it is very easy (and commonplace) for authors to overstate their case and thereby give a distorted impression of the uniqueness and value of their approach. There is a very fine line between advocacy and exaggeration. Moreover, the more challenging the research problem, the greater the chance that authors may be accused of overselling their approach. We have tried to be sensitive to this concern throughout the book. Nevertheless, it is possible that we may have unwittingly caused you to believe that we have made claims that we do not hold. In a final attempt to rectify any such misunderstandings, this section provides a number of caveats concerning the limitations and boundary conditions of CWA. We have organized the points around questions, much like the FAQ (frequently asked question) format frequently found on the WWW.

Are There Other Approaches to Cognitive Engineering?

Yes. As with any comparatively new area of study, several different approaches have been developed. Norman (1981, 1986), Woods and colleagues (Hollnagel & Woods, 1983; Roth & Woods, 1988; Woods & Hollnagel, 1987; Woods & Roth, 1988a, 1988b),

Andriole and Adelman (1995), and Dowell and Long (1998; see also the peer commentary following the main article) all put forth alternative visions that have some similarities to the one we have described, but many significant differences as well. The approach described in this book is based on a view of cognitive engineering originating in the 1960s with the work conducted in the Electronics Department at Risø National Laboratory in Roskilde, Denmark (see Appendix). A comparison between this approach and the others would be very valuable but is premature at this point. Cognitive engineering is an emerging research area, so there are very few comprehensive expositions of alternative approaches. Consequently, the relationship between the ideas in this book and those cited earlier is unclear. You are encouraged to consult the references just cited to draw your own conclusions about the relative strengths and weaknesses of the various approaches.

Why Is the Framework Called Cognitive Work Analysis?

The honest answer to this question is that the label is a historical accident (Vicente, 1995c). As we mentioned in chapter 3, human factors researchers developed traditional task analysis in the 1950s (R. B. Miller, 1953). More recently, researchers have proposed cognitive task analysis methods to deal with jobs whose demands are more cognitive in nature (see Schraagen et al., 1997, for a recent review). The label *cognitive* was thereby added to the traditional label *task analysis* as a differentiator. CWA was developed to deal with open systems that require adaptation on the part of workers (Rasmussen et al., 1994). Thus, the word *task* was changed to *work* to broaden the scope of analysis beyond particular tasks and to reflect the importance of work domain representations in providing support for the unanticipated (see chap. 7).

This evolutionary naming process is less than ideal because the label that we are left with is not particularly apt for our framework. For instance, there is nothing particularly privileged about the cognitive level of analysis. Following its ecological orientation, CWA pays at least as much attention to the environment as it does to human cognition. Moreover, experience has taught us—sometimes painfully—that cognitive factors are nested within, and thus strongly shaped by, social-organizational factors. Perhaps *sociotechnical work analysis* would have been a more apt label for the framework. However, the choice was made, and it is probably too late to change it. The bottom line is that we should pay more attention to the distinctions and concepts comprising CWA than to its name.

Is Cognitive Engineering Trying to Take Over the "Turf" of Other Disciplines That Have Been Concerned With How to Introduce Information Technology Into Complex Sociotechnical Systems?

No, that would be silly. No one discipline can possibly lay sole claim to the problems with which this book is concerned. These problems are so complex and multifaceted that they far outstrip the contributions of any one discipline. Rather than fighting over turf, we need the help and expertise of researchers from a wide variety of disciplines, including (but not limited to) systems engineering, computer science, industrial/organizational psychology, cognitive science, anthropology, human factors, information science, sociology, and management science. To take but one example, it would be ludicrous to suggest that the concepts presented in chapter 10, no matter how useful they may be, could possibly take the place of the life works of all of the scientists conducting research on organizational behavior. Cognitive engineering must draw on, not replace, the insights developed in disciplines that are concerned with the introduction of information technology into the workplace.

Accordingly, the primary value added by CWA is to provide a single unified framework within which the contributions of all of these disciplines—spanning both engineering and human sciences—can be pulled together into a coherent whole (Rasmussen et al., 1994). In other words, cognitive engineering is better viewed, not as a new discipline, but as a "conceptual market place" where the complementary contributions of these disciplines can be integrated in a systematic way that is useful for systems design (Rasmussen, 1988a).

Is Cognitive Work Analysis a Brand New Way of Approaching Work Analysis?

Again, the answer is no. Although the specifics of CWA (e.g., the modeling tools for each layer) are relatively unique as a group, the basic idea behind CWA owes much to the discipline of control engineering. The standard control theoretic approach to analyzing and designing systems involves (a) understanding the plant (i.e., environmental constraints), (b) identifying potential sources of disturbances, (c) defining the control objectives (i.e., goals), and (d) making sure that the means are available to control effectively. Moreover, changes in strategies can be implemented via adaptive control to achieve robust performance over a variety of conditions. All of these concepts are similar to those in CWA, which should not be surprising given that Rasmussen's training is in control engineering. The primary difference is that the quantitative models and analysis techniques that have been developed in control engineering are not sophisticated enough to handle comprehensively the all-important human and social-organizational factors with which we have been concerned in this book. Thus, CWA can be viewed as the qualitative application of control engineering concepts to the unique demands imposed by complex sociotechnical systems.

Is Cognitive Work Analysis Meant to Replace Other Methods of Work Analysis?

No. We agree with Weinberg (1982), who stated: "There's no solution that applies to every problem, and what may be the best approach in one circumstance may be precisely the worst in another" (p. 22). To give an extreme example, it would be inappropriate to use CWA to design a remote control for a programmable VCR. That application domain is essentially closed, not open, and does not satisfy the dimensions of complexity that we listed in chapter 1. CWA is intended to be useful for the unique challenges imposed by complex sociotechnical systems. If you are concerned with a different type of problem, then it is likely that you should look for a different method of work analysis.

Why Haven't Methods for Knowledge Elicitation Been Discussed in the Book?

The answer to this question can be found back in chapter 1, where we introduced the distinction between the conceptual and methodological perspectives on work analysis. The conceptual perspective is primarily concerned with models, because it describes the types of concepts to be used for modeling and the types of representations that should be developed by the time the work analysis is complete. In contrast, the methodological perspective is primarily concerned with data collection, because it describes the process and the methods that an analyst should adopt in actually performing a work analysis. The methods that are chosen to collect data for work analysis are eclectic and the activities that analysts actually engage in while conducting an analysis are very opportunistic, chaotic, nonlinear, and iterative (cf. Card, 1996). Thus, we chose to write the book from the conceptual perspective. As a result, the focus has been on modeling and representation.

This does not mean that knowledge elicitation methods (e.g., naturalistic observations, interviews, questionnaires, verbal protocols, storyboards, mock-ups, usability trials, card-sorting tasks, simulator experiments) are not important. On the contrary, they are essential because they provide a means of collecting data (see chap. 4). And in fact, if anything, CWA requires a broader knowledge elicitation effort than most approaches to work analysis. Instead of focusing primarily on current or prospective users, analysts must also elicit knowledge from engineers, computer scientists, and design documents. The insights that are needed to conduct a CWA, especially in earlier phases, must be obtained from system designers, not just users.

Figure 13.2 shows the complementary relationship between knowledge elicitation methods and CWA. Knowledge elicitation methods provide different ways of obtaining data for a work analysis, whereas CWA provides a modeling framework for integrating those data into a coherent and useful set of models.

Are All Phases of Cognitive Work Analysis Equally Important for All Application Domains?

This issue has not been studied empirically in as much detail as it deserves, but it seems that the answer is no. CWA identifies five layers of constraint for analysis. However, it is likely that application domains will differ considerably in terms of how much structure they have in each layer. The library information retrieval and the DURESS II examples discussed in the previous chapter provide an illustrative contrast. Figure 13.3 illustrates a speculative representation of the relative amount of constraint in each application domain, broken down by each phase of

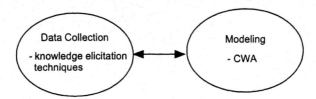

FIG. 13.2. The complementary relation between knowledge elicitation techniques and CWA.

a) <u>DURESS II:</u>

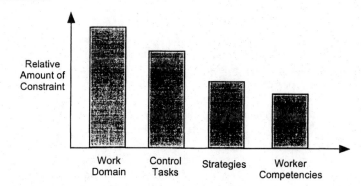

b) <u>Library Information Retrieval:</u>

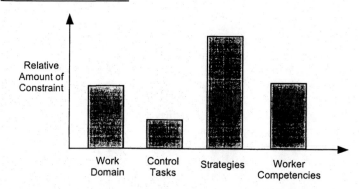

FIG. 13.3. A speculative representation of the relative amount of constraint in the library information retrieval and DURESS II examples, as a function of CWA phase.

CWA.[1] Note that the amount of constraint depicted is the independent contribution from each CWA phase. DURESS II, being a process control microworld, has a great deal of work domain and control task structure. The number of components, their type, their functions, and the connections between them are all prespecified by design. And although there are many ways to accomplish most tasks, the relationships between variables, represented by equations, strongly bound the possibilities for action. In contrast, the fiction information retrieval has very little control task structure. The task to be performed is relatively open-ended—find a book that you like. When we move to strategies, we find a different picture. Here, the fiction retrieval application domain is very rich in structure, as evidenced by the strategies described in the previous chapter.

If these differences are any indication, then we would expect to find that different layers of CWA will lead to more or less insights for different application domains. Figure 13.4 illustrates this point by comparing the amount of constraint in a hypo-

---

[1]The social-organizational layer of constraint is omitted because neither example has a substantial social component, at least as described here.

a) Technical Application Domain:

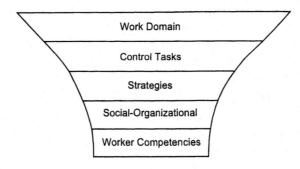

b) Social Application Domain:

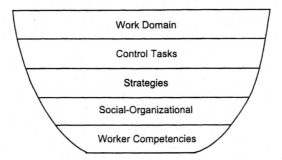

FIG. 13.4. A speculative representation of the cumulative amount of constraint in a hypothetical technical domain and hypothetical social domain, as a function of CWA phase.

thetical, prototypical technical application domain and a hypothetical, prototypical humanistic application domain as a function of CWA phase. The width of the funneling in the figure represents the *cumulative* degrees of freedom for action. In both cases, we start off with many degrees of freedom, but as each layer of constraint is introduced, the number of degrees of freedom is reduced. For applications whose demands are primarily technical in nature, we would probably find that the work domain analysis provides a great deal of leverage for design because it is tightly coupled. As a result, cognitive and social constraints may play a comparatively lesser role because there would be fewer remaining degrees of freedom. This pattern is represented in Fig. 13.4a by a relatively rapid narrowing of degrees of freedom in the first few phases of CWA. In contrast, for applications whose demands are primarily social in nature, we would probably find that the work domain analysis does not provide much leverage for design because it is loosely coupled. As a result, there may be many degrees of freedom available for action that must be resolved by social and cognitive constraints. This pattern is represented in Fig. 13.4b by a relatively slow narrowing of degrees of freedom in the first few phases of CWA. Of course, different "constraint profiles" are possible, depending on the structure of the application domain. Rasmussen et al. (1994) described a taxonomy of domain characteristics that could be used to study these issues empirically in a systematic fashion.

Therefore, although all phases of CWA are relevant to all complex sociotechnical systems, different phases will be more important for different application domains. These differences will probably manifest themselves in systems design as well. After all, the greater the structure that is uncovered, the more insights are provided for systems design. For example, work domain analysis may not provide much design leverage for loosely coupled domains whose demands are primarily cognitive or social in nature. Conversely, worker competencies analysis may not provide as much design leverage for tightly coupled domains whose demands are primarily technical in nature. If these speculations turn out to be correct, they would not mean that CWA is not useful. Rather, they would mean that different phases are more or less useful for different application domains.

Can Subsets of Cognitive Work Analysis Be Used to Solve Particular Problems, or Does the Entire Framework Have to Be Used to Obtain Satisfactory Results?

Although the various phases of CWA form a cohesive whole, it is possible and meaningful to use a subset of the phases to tackle some problems. For example, in the previous chapter, we showed how work domain analysis and worker competencies analysis could serve as a basis for interface design. Other phases could be used for other purposes. For example, in chapters 7 and 8, we showed how the abstraction-decomposition space and the decision ladder, respectively, can both be used to analyze verbal protocols. In chapter 11, we showed how the SRK taxonomy can be used to derive, or resolve, theoretical claims in psychology and cognitive engineering. Also, the abstraction-decomposition space can be used to conduct an analytical evaluation of the content of an existing or a proposed interface (Dinadis & Vicente, in press; Hajdukiewicz, 1998; Reising & Sanderson, 1996; Sharp, 1996; Vicente, 1990b).

Does the Application of Cognitive Work Analysis Guarantee the Design of a Better Computer-Based Information System?

No. Be suspicious of anyone who offers you guarantees. There is no silver bullet. Having said that, CWA does provide a systematic and conceptually coherent basis for uncovering design requirements. Therefore, although success cannot be guaranteed, CWA does provide a firm foundation for systems design.

Will the Application of Cognitive Work Analysis Lead Directly to the Design of a New Device and Organizational Structure, Thereby Removing Intuition and Creativity From the Design Process?

Definitely not. As we pointed out at the beginning of the book, our intent is not to provide a cookbook for systems design. CWA does try to reduce the distance between analysis and design by linking layers of analysis with classes of systems design interventions (see Fig. 5.8). The objective is to make systems design less of a black art, and more of a systematic process. Nevertheless, there will always be a gap that must be filled by the creativity of the designers. Although this gap makes the systems design process subject to the idiosyncratic biases and errors of individual designers, we believe that these elements can never be completely removed. Engineering design is an inherently creative process (Ferguson, 1977). Having said that, we do believe that it would be useful to formalize the concepts comprising CWA so that this distance can be made considerably shorter than it is now.

Can Cognitive Work Analysis Be Used for Evolutionary Design Problems Where a "Clean Slate" Approach Is Not Feasible Because Some Attributes of the Design Are Frozen a Priori?

Yes. For example, the abstraction-decomposition space can be used as a descriptive frame of reference for organizing the data that have been made available by a previously defined sensor set, rather than as a formative frame of reference for identifying which data could be made available through improved sensor placement or design (e.g., Dinadis & Vicente, in press; Hajdukiewicz, 1998; Reising & Sanderson, 1996; Sharp, 1996). The other modeling tools in CWA can be used in an analogous fashion. For example, in chapter 8 we showed how the decision ladder can also be used in a descriptive, rather than a formative, manner. Having said that, it is important to reemphasize the point that we made in chapters 4 and 5. Using CWA in an evolutionary manner will result in a design that is not nearly as effective as it could be, because the design is inheriting whatever deficiencies or limitations existed in the previous design. Rather than letting the design solution be shaped by the characteristics of the problem context, an evolutionary approach lets the design solution be shaped by historical factors that are frequently idiosyncratic, or in some cases, no longer even valid. Although these compromises are frequently necessary in practice, they will generally not result in qualitative changes in design solutions or quantum leaps in performance (Blum, 1996).

It is important to point out that the focus on a "clean slate" approach to design is an advantage in some cases. A number of industries are engaged in "first of a kind engineering" (FOAKE) projects, whose objective is to design a sociotechnical system that is qualitatively different from any that have been designed in the past (e.g., Cannon-Bowers et al., 1997; Roth & Mumaw, 1995). Conventional approaches to work analysis are not very well suited to FOAKE design projects, because the usual knowledge elicitation sources are not available. For example, analysts cannot interview experienced workers to get design insights, because such workers do not exist until after the system has been designed and is in operation. Similarly, analysts cannot examine the previous-generation design to see how it can be improved because there is no comparable previous design to work with. In contrast, with its focus on modeling intrinsic work constraints, CWA provides an analytical basis for identifying requirements for FOAKE projects. Although these projects are certainly more challenging, analysts can still try to identify behavior-shaping constraints to uncover requirements for systems design, even though there are no previous designs or expert workers.

Is Cognitive Work Analysis a Definitive, Mature Framework That Has Repeatedly Proven Itself in Industrial Applications?

This question is difficult to answer in a clear-cut way. Approaches to systems design that are used in industry are rarely evaluated. For example, it is exceedingly difficult to find an empirical study showing that approach X improved a particular company's productivity level by Y%. And even when such data are available, we cannot know that the approach is actually responsible for the improvement unless the appropriate control conditions are in place. Thus, to claim that a technique has proven itself would require a comparative study of at least two alternative approaches under comparable conditions. Such a study would be very difficult to conduct under representative conditions because of the difficulty in controlling for all of the relevant factors (T. D. Cook & Campbell, 1979). Even if such a comparative experiment could be properly designed, it would be very difficult to find someone to pay for it. Corporate customers are not interested in paying you to prove that your approach is a good one. If they believe you have something to offer them, they will want you to solve their problems, not conduct an experiment. As a result, there are no convincing data that we know of to show that CWA or any other approach to the design of computer-based systems for complex sociotechnical systems has been "proven" in this sense. This does not mean that none of these approaches is effective, but merely that their value in industry has not been demonstrated in a scientifically acceptable way.

One alternative to a representative, scientific evaluation is a practical proof of principle. How many companies have adopted this approach in the way they do design? CWA has some limited success using this criterion. For example, parts of the CWA framework were used by Westinghouse to design an advanced control room for nuclear power plants (Easter, 1987). Similarly, most of the CWA framework was used by Toshiba in Japan to design an advanced control room for a future-generation nuclear power plant (Itoh et al., 1990, 1995). This implementation is particularly noteworthy because it has been constructed on a very large scale, namely a fully functional, prototype control room connected to a full-scope simulator. Also, as we described in the previous chapter, the CWA framework has been used to create an award-winning library information retrieval system that was turned into a commercially available product currently being used in public libraries in Scandinavia. Thus, parts of CWA have been used in industry to tackle complex design problems in both engineering and humanistic application domains. To be fair, however, it is important to point out that the examples are few in number, particularly compared to other design approaches being used in industry. For example, human-centered systems design (Rouse, 1991) and contextual design (Beyer & Holtzblatt, 1998) have both been adopted in numerous organizations and appear to have changed, for the better, the way in which companies design computer-based information systems to support human work.

A third criterion that can be used to assess the maturity of CWA is to examine its research base. How many studies have been conducted to develop the theory or test the predictions of the framework? To how many different application domains has CWA been applied? If we do not include the studies that were involved in the original development of these ideas (see Appendix), then parts of CWA have been explored in the following application domains: aviation (Dinadis & Vicente, in press), command and control (Chalmers, 1997; Moray, Sanderson, & Vicente, 1992; C. L. Smith, 1992; C. L. Smith & Sage, 1991), computer programming (Leveson, in press), engineering design (Burns & Vicente, 1995; Pejtersen et al., 1997; Rasmussen, 1990b; Rouse & Cody, 1988; Rouse, Cody, Boff, & Frey, 1990), information retrieval (Pejtersen, 1992; Rasmussen et al., 1994; Xu, 1996; Xu et al., 1996), medicine (Hajdukiewicz, 1998; Rasmussen et al., 1994; Sharp, 1996; Sharp & Helmicki, 1996; Xiao, 1994; Xiao et al., 1997), process control (Burns & Vicente, 1997; Christoffersen et al., 1996, 1997, in press; Dinadis & Vicente, 1996; Goodstein, 1985; Goodstein, Hedegård, Højberg, & Lind, 1984; Howie & Vicente, 1997; Jamieson & Vicente, 1998; Janzen & Vicente, in press; Meshkati, Buller, & Azadeh, 1994; Olsson & P. L. Lee, 1994; Pawlak & Vicente, 1996; Reising & Sanderson, 1996; Torenvliet et al., 1998; van Passen, 1995; Vicente, 1992a, 1992b; Vicente & Rasmussen, 1990, 1992; Vicente et al., 1995; Yamaguchi, Furukawa, & Tanabe, 1997; Yu et al., 1997), and workplace design (Dainoff & Mark, 1995).[2] Many of the articles just cited represent preliminary research efforts, and many important issues remain to be addressed. Nevertheless, there is a substantial number of theoretical and empirical studies that are beginning to explore different aspects of CWA.

In summary, the value of CWA has not been scientifically demonstrated in industry. It has been applied to several very complex design projects, but the number of cases is small. As a result, we can hardly claim that CWA has repeatedly proven itself in the corporate design world. There is, however, a growing research base underlying CWA. Most of these studies have been conducted in the context of process control, but a wide variety of other application domains have also been investigated. Although far from definitive, the empirical results that have been obtained so far are very encouraging. Thus, the prospects for CWA look good, but many challenges remain to be addressed.

What Are the Primary Limitations of Cognitive Work Analysis?

Because the framework has not yet been extensively used in industry, it is difficult to be sure. Nevertheless, there seem to be at least three significant points of concern. First, to conduct a CWA for a complex sociotechnical system could take an incredible amount of effort, more effort than many companies are willing to invest. This is a significant obstacle that must be overcome if CWA is going to be used in industry. We see two potential ways of overcoming this issue. One approach is to develop computer-based tools that can be used to assist in conducting the analysis. Such support for analysts could greatly reduce the level of effort involved. To do this, however, the various phases of CWA would need to be formalized so that they could be represented in a computational form. Throughout the book, we have tried to identify areas where we believe such formalization is necessary. Another approach is to develop "prototypical template analyses" for certain classes of application domains (e.g., aviation, finance, power plants, manufacturing, hospitals). These templates would capture the general constraints that are common to each class of domain, while leaving room for the idiosyncratic details of particular applications.

---

[2]Despite its length, this list is not intended to be exhaustive. Rather, it is intended to illustrate the breadth of application domains that have been tackled to date, and the relative number of studies in each application domain.

Although a great deal of effort would be required to develop the initial prototypes, they would greatly reduce the effort involved in subsequent analyses. Analysts would merely have to fill in an existing structure rather than developing one from scratch. Both of these approaches to reducing the level of effort associated with CWA should be pursued, and their viability evaluated.

A second point of significant concern is that CWA is based on a formative approach that takes a clean slate perspective on design, whereas most corporate design projects are evolutionary in nature. We do not know how big an obstacle this is. On the positive side, we have demonstrated that CWA can be used descriptively as well, so there is no reason why it cannot be used in evolutionary design projects (see earlier discussion). Furthermore, the clean slate approach is an advantage for FOAKE projects (see previous discussion). On the negative side, we do not know how much value will be lost by using the framework in a more descriptive manner. It could be that, by using CWA in this way, innovative designs would be more difficult to come by. Only more experience in applying the framework in industry can tell us whether this concern is significant or not.

A third point of potential concern is that by giving workers the responsibility of finishing the design, we are putting a large burden on their shoulders, perhaps too large. There is no doubt that CWA will lead to designs that require a high degree of competency on the part of workers. This requirement may translate into greater training needs. It may even be possible that not all workers would be comfortable with the enhanced level of responsibility that comes with being required to finish the design. If so, then this would translate into new types of selection requirements. Clearly, the change in philosophy that comes with CWA has widespread implications that are not easy to anticipate or cope with.

Having said that, there are several reasons to believe that these changes are worthwhile, or even more strongly, necessary. In chapter 5, we argued that workers should finish the design because complex sociotechnical systems are open systems that introduce unanticipated disturbances. These disturbances do not go away if we create a more traditional procedural role for workers. In fact, we could argue that, by adopting such a traditional role, we are actually putting more of a burden on workers. When disturbances do occur, workers will have to take on a role for which they have not been properly trained, and for which they do not have the required information support. From this perspective, the philosophy of designing for adaptation is viewed as a natural solution to dealing with the demands imposed by very complex systems. Either we recognize and confront those demands and design accordingly, or we start building less complex systems so that the demands on workers will be lessened. Given the rapid pace of technological development (see later discussion), the latter scenario seems very unlikely.

LOOKING AHEAD: THE RELATIONSHIP BETWEEN A KNOWLEDGE-BASED GLOBAL ECONOMY AND THE DEMAND FOR ADAPTATION

Everywhere we look in the media, we see a number of buzzwords appearing with increasing frequency, such as *innovation, knowledge worker, wealth creation, flexible manufacturing, free trade,* and *global village.* What do these terms have in common, and what is their significance for the need to support worker adaptation? We believe that these terms are symptomatic of a fundamental shift toward a knowledge-based global economy that will lead to an increasing demand for workers, managers, organizations, and technology to adapt in the future.

The Knowledge-Based Global Economy

There are a number of trends that have acted together to transform qualitatively the nature of contemporary industry and economics. These changes, in turn, pose a new set of requirements for corporate success.

Brzustowski (1998), the president of the National Sciences and Engineering Research Council of Canada, identified the following contributors:

- The market is full of products based on recent advances in science and technology.
- These products are made all over the world, regardless of brand name.
- High quality is expected in the market, and is being achieved everywhere.
- Today's new high-tech products become tomorrow's commodity products.
- The market for medium and low-tech goods is fiercely competitive.
- Commodity prices are low and distance is not a factor even for bulk materials.
- Everybody has to compete on price, so productivity improvement is key.
- Changes in demand and market conditions can be rapid and unpredictable.
- New knowledge appears in new goods and services fast, and it's getting faster.

According to Brzustowski, these factors have led to a knowledge-based global economy that is qualitatively different from the economy of the past.

The fact that we are now living in a qualitatively different economic world has been pointed out, not just by researchers like Brzustowski, but by many in the corporate world as well. To take just one very recent example, Davis and Meyer (1998), who are affiliated with the Ernst & Young Center for Business Innovation, identified essentially the same set of characteristics as Brzustowski. These changes have increased the speed and connectedness of change in the economic world, a phenomenon Davis and Meyer referred to as "BLUR." Under these turbulent conditions, knowledge, information, learning, and speed of response are the valuable commodities for corporate survival and competitiveness. What implications does this set of changes have for success in the work place?

The Future Demand for Worker Adaptation

We believe that the trend toward a knowledge-based global economy will only increase the need for adaptation in the future. Workers, managers, organizations, and even technology will all have to become more flexible and adaptive than in the past. There are a number of arguments from a diverse set of sources that we can put forth in support of this claim.

Workers. The U.S. National Research Council Committee on Human Factors issued a report a few years ago, documenting what it believed were the most fertile areas for human factors research in the next few decades (Nickerson, 1995). The report was written by a diverse group of experts, spanning the entire range of human factors, from physical to psychological to social-organizational aspects. One of the important points made in the report is that the nature of work will continue to change:

> Technological change means changes in job requirements. The ability to satisfy changing, and not entirely predictable, job requirements in a complex, culturally diverse, and constantly evolving environment will require a literate workforce that has good problem solving skills and learning skills. (p. 22)

Consequently, "a critical aspect of industrial competitiveness will be the ability to adapt quickly to rapid technological developments and constantly changing market conditions" (p. 42). Therefore, in the future, the workforce will have to be more versatile and adaptive than in the past.

A similar conclusion was reached by the participants in a workshop on cognitive task analysis sponsored by NATO and the U.S. Office of Naval Research (Schraagen, 1998). The goal of the workshop was to assess the state of the art in the area by bringing together a total of 23 experts from six countries. One of the main points raised in the workshop was that, as technological capabilities improve, proceduralized tasks will become increasingly automated. As a result, the primary role played by workers in the system is to deal with those situations that cannot be readily anticipated by designers, and thus, automated. Therefore, the need to support worker flexibility and adaptation will only increase in the future.

The Organization for Economic Cooperation and Development (OECD) has also noted the relationship between the global, knowledge-based economy and worker adaptation:

> As we move into 'knowledge-based' economies the importance of human capital becomes even more significant than ever. . . . There is an increased awareness of the importance of lifelong learning in a society where economic, social and technological change call for flexibility, adaptation and learning throughout life. (OECD, 1998, p. 3)

The OECD is devoting considerable effort to understanding these increasingly important issues.

In his influential monograph on automation in industrial plants, Hirschhorn (1984) also highlighted the importance of worker adaptation. His thesis was that workers must be able to learn, adapt, and actively control the automation if the full potential of automation is to be realized. His justification for this conclusion is that automated systems inevitably fail. As a result, human ingenuity is continually required to deal with unanticipated situations for which there are no predetermined sequence of actions. To achieve this vision, Hirschhorn believed that organizations should become more distributed, and that more autonomy and greater responsibility should be provided to workers. Only with such a flexible coordinative structure will it be possible for workers to adapt and continue to learn. It should be clear from this brief description that Hirschhorn's conclusions are very similar to CWA's philosophy of having the workers finish the design.

In his recent monograph on the interaction between engineering and society, Pool (1997) reached essentially the same conclusion but from a somewhat different path. By reviewing the influence that society has had on technology, Pool noted that there is an increasing need to "do more with less." This trend will only increase with the demands imposed by the knowledge-based global economy. Because of the premium that is put on increasing efficiency and providing new functionality, engineering systems have become—and will continue to become—more and more complex. According to Pool, this increase in complexity has had an unintended result, namely an increase in unanticipated events. In the terms of chapter 3, as systems become more complex, they become more open too. Unanticipated disturbances are bound to occur, albeit infrequently. As a result, Pool claimed that there is a greater need for worker and organizational adaptation than in the past. By inference, we can expect that the requirement to support adaptation, and thus the need for CWA will increase in the future.

Cannon-Bowers et al. (1997) described an excellent example of the trend identified by Pool (1997). Because of a tremendous reduction in operating budget, the U.S. Navy is under pressure to greatly reduce costs. As a result, the navy has set the goal of reducing the manning level on a future generation of ships from the current level of 350 workers to an envisioned level of 95 workers. Clearly, this is a very ambitious target, but the magnitude of the design problem is compounded by the fact that the missions that such ships are expected to play in the future will increase in complexity. There will be more missions, they will be more varied in nature, and their nature will be highly unpredictable. To meet this challenge to be able to do more with less, the Navy has initiated a FOAKE project (see earlier discussion) to achieve affordability through improved human-systems integration.

Managers. Adaptation is relevant to managers as well, as evidenced by Morone's (1993) fascinating field studies. He studied corporate general managers that have been successful in turning their companies (Motorola, Corning, and General Electric) into global leaders in high-technology markets. Such markets are particularly affected by the symptoms accompanying the trend toward a knowledge-based global economy, exhibiting a high degree of volatility and uncertainty. Accordingly, managers must be able to adapt to all of this turbulent change if their companies are going to be competitive and survive. As one manager put it, "You need a strategic intent—but within that context, you have to be totally opportunistic. . . . You can't know what's around the next corner, so construct an organization that is able to adapt" (p. 119). In another part of his book, Morone observed that:

> [Some of the businesses studied] followed the general course that had been hoped for, but the specific form they took, the specific market and technological developments to which they had to respond as they followed that general course, were not, and could not have been, anticipated. (p. 190)

Thus, yet again we see the need to adapt to disturbances. Finally, another manager pointed out: "A lot of people think of product development as involving a lot of planning, but . . . the key is learning, and an organization's ability to learn" (p. 224). The high-tech markets studied by Morone have been particularly volatile in the past, and show every evidence of continuing to be so. However, as the trends identified by Brzustowski (1998) affect different sectors of the economy, we can expect to see the same strong need for adaptation to uncertainty in other industries too.

Organizations. Adaptation is relevant at an organizational level as well. In his national bestseller, Senge (1990) discussed the importance of organizational learning in the knowledge-based global economy. The problem is that traditional ways of governing and managing are largely outdated and are breaking down rapidly because of the changes documented by Brzustowski (1998). The static, hierarchical organizational structures that have dominated in the past are no longer appropriate given the current pace of change. According to Senge, to succeed and survive, modern organizations must learn to accept, embrace, and seek change. This, in turn, requires a decentralized organizational structure whose individuals are committed to learning and have a shared mental model of the world.

More recently, Davis and Meyer (1998) identified a similar set of corporate requirements to cope with the phenomenon that they referred to as BLUR (see previous discussion). Because the pace of change has accelerated, the rules that guided corporate decision making in the past are no longer reliable. As a result, companies have to give up on the idea of stable solutions to business problems. Instead, they have to move from relying on prediction, planning, and foresight to building in flexibility, speed, and self-organization. In short, "strategy in the age of BLUR means that an enterprise must adapt to its environment. . . . It needs to be every bit as adaptable as the economy in which it participates" (Davis & Meyer, 1998, p. 114).

Technology. It is interesting to note that the need to cope with disturbances has impacted technology as well. For example, Blum (1996) argued for a paradigm shift in software engineering. He started off his argument with a bold claim, namely that software is being developed within a framework whose concepts are now obsolete. Blum referred to this paradigm as *technological design* because it is based on a techno-centric view. The goal is to follow a two-step process: First, a complete, formal statement of specifications is identified, and second, this statement is used to create a computer program that meets the formal specifications. The basic idea is to define what is to be built (i.e., the *solution* to the design problem), and then build it. In this view, the primary focus of the design effort is on the software in the computer. Although this process may seem intuitively appealing, the fact that the first step is to define the solution to the problem, rather than the problem itself, should give us pause.

Technological design seemed appropriate decades ago during software engineering's formative years. At that time, large projects were perceived to have stable, closed requirements. However, the context for computing has changed drastically since then. According to Blum (1996), projects have become more and more complex, so the requirements are now more open than before (cf. Pool, 1997). That is, the environment in which complex computer programs are to function is dynamic and generally unpredictable; change is inevitable. In this type of dynamic environment, requirements cannot be comprehensively identified once and for all before the programming begins. On the contrary, there is a strong need for a continuous response to changing requirements. In Blum's words: "Because the learning, responding, and expanding do not cease once a product is delivered, design must be recognized as a continuing activity that is never completed" (p. 198). Note the similarity to our notion of continually finishing the design. Technological design cannot accommodate this need for adaptation because the formal specifications are supposed to be frozen before the programming begins. Thus, there is a mismatch between current software design practices and the nature of complex software design projects in the dynamic knowledge-based global economy.

According to Blum (1996), a new set of concepts—indeed, a new paradigm—is required for contemporary software design projects. Blum referred to this paradigm as *ecological (or adaptive) design* because it is based on a contextual, evolutionary view. The goal is to follow a one-step process: Represent the problem to be solved in the language of the application domain using formal conceptual models that are automatically transformed into implementation code. The basic idea is to focus on the problem in the world to be solved, rather than the software itself. After all, software is merely the means by which that problem is solved, so it should play a subordinate role in design (thus the title of Blum's book: *Beyond Programming*). As Blum put it: "Programming is to developing systems as grammar is to writing a novel. A necessary tool to be mastered and used automatically" (p. x). Thus, the design effort should focus on modeling the problem to be solved, and the power of computer technology can then be used to automate the transformation from conceptual models to software implementation.

The result of adaptive design is an evolutionary system that evolves over time as the external requirements change. Designers can modify, update, and add to the conceptual models describing the problem to be solved as the needs of the application domain change. Each time a change is made in the conceptual model, implementation code is automatically generated because the conceptual models are formalized. This may seem like a magical process, but Blum (1996) described a complex software project, the Oncology Clinical Information System (OCIS), that was constructed using his adaptive design paradigm. This information system has been in use for over a decade, and empirical evaluations have indicated that the system is of high quality, high reliability, and high adaptability, despite being highly complex. Thus, OCIS is an existence proof showing that adaptive design is indeed a viable paradigm.

Blum's (1996) arguments can be interpreted in light of the knowledge-based global economy. Change is becoming ever-present, and thus computer programs must adapt accordingly. This means that software engineers must be provided with tools that allow them to evolve a design as required. Blum went on to identify a number of trends that will only increase the need for adaptive design in the future. His conclusion was that we need to go beyond programming to a new paradigm of software design.[3]

Summary. The knowledge-based global economy will continue to transform the landscape of modern work. Changing and unpredictable circumstances will be the norm. As a result, there is an increasing need for workers, managers, organizations, and technology to become more flexible and adaptive than they have been

---

[3]It would be interesting to see if the concepts in CWA could be used to identify the conceptual models that are required to support Blum's (1996) paradigm of adaptive design. Such an integration might result in a twofold contribution by formalizing CWA and by providing a systematic framework for developing conceptual models to support adaptive design.

in the past. These trends suggest that the requirement to design for adaptation, and thus the need for CWA, will only increase in the future.

The Relationship Between Adaptation and Learning

Along with the buzzwords identified at the beginning of this section, we also frequently find another set, such as life-long learning, continuous improvement, and learning organization. How does continuous learning fit into the picture we have been drawing?

So far, we have established that the knowledge-based global economy has created a strong need for adaptation to change. In essence, this is a call for *learning to learn*. By adapting to disturbances, workers are, in effect, engaging in opportunities for learning. We briefly alluded to this relationship in chapter 3 when we pointed out that the constraint-based approach provides opportunities for learning, while also satisfying the demands of a disturbance, by accommodating variability in workers' actions. Hirschhorn (1984) made this point in the context of process control: "Each time operators diagnose a novel situation, they become learners, reconstructing and reconfiguring their knowledge" (p. 95). The connection between adaptation and learning goes well beyond process control, however. Related conclusions have been reached in other research areas, such as psychology, control theory, and cognitive science (e.g., E. J. Gibson, 1991; Johannson, 1993; Narendra, 1986; Norros, 1996).

Nowhere is the robust relationship between action variability and learning opportunities as well thought out as in the study of human motor control in ecological psychology, thanks to the seminal work of Nicholai Bernstein. This relationship was clearly expressed in a book written approximately 50 years ago but that has only been published much more recently (Bernstein, 1996). Bernstein was concerned with a very different problem than we are, so his terminology differs from ours. For instance, he used the term *dexterity* to refer to the capability to find "a motor solution for any situation and in any condition" (p. 21). In our terms, he was referring to the capability to adapt to unanticipated demands.

According to Bernstein (1996), the need for dexterity increases under the following conditions: (a) The problem to be solved becomes more complex; (b) the problem to be solved becomes more variable; (c) the number of unique and unexpected problems that need to be solved in real time increases. Although the content is obviously different for motor control, these generic characteristics are surprisingly similar to those that are associated with the knowledge-based global economy and complex engineering design projects. Demands on workers are becoming more complex; those problems take different forms, rarely repeating themselves; and, there is an increasing need to deal with novel, unanticipated situations in a timely fashion. Just as these characteristics lead to an increase in the need for dexterity in motor control, they also lead to an increase in the need for adaptation in complex sociotechnical systems.

These similarities establish the connection between motor dexterity and worker adaptation. How does the relationship with learning fit in? The answer to this question can be found in the rationale behind Bernstein's (1996) beautiful phrase, *"repetition without repetition"* (p. 204, emphasis in original). Having workers finish the design means that they will repeatedly have to generate a solution to a problem, rather than following a prepackaged procedural solution. As Bernstein pointed out in the context of motor control, "during a correctly organized exercise, a student is repeating many times, *not the means for solving* a given motor problem, but *the process of its solution*, the changing and improving of the means" (p. 205, emphasis in original).[4]

But because workers are solving the problem anew each time, opportunities for learning are created. This process is explained by the following extended quotation from Bernstein (1996):

> *Repetitions* of a movement or action are necessary in order to *solve a motor problem* many times (better and better) and *to find the best ways* of solving it. Repetitive solutions of a problem are also necessary because, in natural conditions, external conditions never repeat themselves and the course of the movement is never ideally reproduced. Consequently, it is necessary *to gain experience relevant to all various modifications* of a task, primarily, to all the impressions that underlie the sensory corrections of a movement. This experience is necessary for the animal not to be confused by future modifications of the task and external conditions, no matter how small they are, and to be able to adapt rapidly. (p. 176, emphasis in original)

Bernstein's statements can be generalized to the case of workers in complex sociotechnical systems. Here too, conditions rarely repeat themselves precisely, so workers should gain experience with solving problems under a wide variety of initial conditions. Supporting workers to be adaptive problem solvers accommodates this need for learning because it promotes and supports situated action. By accommodating context-conditioned variability in action, we can help workers gain valuable experience so that they are not confused if they have to perform the same task in a different way in the future because of a change in context. In short, designing for adaptation is equivalent to designing for continuous learning.

The Future: What Can We Be Sure Of?

The particular framework for CWA that we have described in this book may or may not help us obtain the answers we need to tackle the challenges imposed by a knowledge-based global economy. Only through more research and application will we be able to tell for sure. Having said that, there are a few things that appear to be indisputable. As sociotechnical systems become more and more complex, change will become the norm, not the exception. Therefore, to be competitive in this knowledge-based global economy, there will be an increasing demand for workers, managers, organizations, and technology to be flexible and adaptive. At the same time, there will be an accompanying need for learning to learn. Accordingly, com-

---

[4]Bernstein's (1996) notion of "repetition without repetition" is pictorially illustrated on the cover of this book. The photograph shown there was taken by Frank Gilbreth who, along with Frederick Taylor, was one of the developers of the "one best way" approach to work analysis described in chapter 3. Gilbreth developed a technique for time-lapsed photography that enabled him to record the motions that workers engaged in to perform certain tasks. For example, by putting light sources on workers' wrists, Gilbreth was able to record the trajectory that workers' hands took to complete a task. The photograph on the cover of this book documents the motions necessary to move and file 16 boxes full of glass. The worker's trajectories are notably variable and "organic" in nature, particularly compared to the rigid, unerringly repetitive trajectories produced by industrial robots (cf. Mandel, 1989). Thus, despite the highly procedural nature of this physical task, it is clear that the worker is not mechanically repeating the means for solving the motor problem.

puter-based information systems should be deliberately and systematically designed to help workers effectively fulfill these challenging roles. In short, we should systematically design for adaptation.

This objective may seem daunting, and in fact, it is. However, the good news is that, based on the evidence we have available, we believe that pursuing this objective will allow us to kill three birds with one stone. Designing for adaptation will lead to improvements in safety, productivity, and worker health. There is no need to trade off one criterion for another. Thus, we should strive to integrate the insights of humanistic and technical disciplines into a single systematic and coherent framework for work analysis. Doing so should help us uncover the insights that we require to design safe, productive, and healthy computer-based work in complex sociotechnical systems.

Appendix:
Historical Addendum[1]

PURPOSE

Sometimes, when a set of ideas is presented in a rational, logical form, as we have tried to do throughout this book, readers are left with uneasy, lingering questions about the origins of those ideas. Where did all these concepts come from? Are they just idle, armchair speculations? Are they merely based on one person's private intuitions? Are they based on any empirical foundation? If so, were the data collected under conditions that are representative of the settings and situations to which we are interested in applying these concepts (i.e., complex sociotechnical systems requiring worker adaptation)? The purpose of this Appendix is to answer these questions by providing a very brief historical overview of the studies and insights conducted between 1962 and 1979 that eventually led to the framework that we have described in this book (cf. Sanderson & Harwood, 1988). Putting CWA into its historical context should help us better appreciate, not just the origins of the framework, but also the research infrastructure that may be necessary for cognitive engineering to flourish.

RISØ NATIONAL LABORATORY

The ideas presented in this book grew out of a research program conducted in the Electronics Department of Risø National Laboratory in Roskilde, Denmark. Risø National Laboratory (or Research Establishment Risø, as it was first known) was created in 1956 and Niels Bohr, the Danish Nobel laureate in physics, served as its first chairman of the board. Risø was given the charge of conducting research so that Denmark could effectively implement nuclear power within 5 years. Remarkably, this 5-year window was maintained for over a quarter of a century until it was decided that Denmark would not have any commercial nuclear power plants!

During this quarter century, Risø fostered an exceptionally unique environment for conducting research. Originally, the laboratory's funding came from the Danish Ministry of Finance, providing a vast supply of financial support. There was no requirement at all to bring in large research contracts from external funding agencies. Furthermore, there was no requirement to publish research results in academic journals. Instead, much of the work described herein was published in an internal series of green technical reports. Although Risø had collaborations with Danish universities, its researchers were not required to teach classes, supervise graduate students, or take on extensive administrative responsibilities. What they were required to do was conduct research to address a practical problem of great social relevance—

---

[1]This chapter has been adapted with permission from *Proceedings of the Human Factors and Ergonomics Society 41st Annual Meeting*, pp. 210–214, 1997. Copyright © 1997 by the Human Factors and Ergonomics Society, Inc. All right reserved.

how to effectively support Danish government and industry when the decision to introduce nuclear power was going to be made, roughly within 5 years.

Anyone familiar with academic or industrial research will recognize that these were extraordinary circumstances. All of this has changed gradually over the last 15 years or so. Like almost all other research institutions around the world, Risø is now under extensive external scrutiny and intense economic pressure. Researchers are required to spend substantial amounts of time writing proposals to bring in large amounts of external funding. They are also required to engage in extensive project management activities that are associated with large research contracts. Furthermore, researchers often cannot afford to adopt long-term research goals, but instead are required to produce demonstrable short-term products. They are also required to document their productivity by publishing extensively in academic circles—the familiar doctrine of publish or perish. But during the period described in this Appendix (1962–1979), none of these constraints were present to the degree that they are now.

PREHISTORY (1962–1963)

The first body of research reviewed is more accurately described as prehistory in that it was not directly concerned with issues that we readily recognize today as pertaining to cognitive engineering. Nevertheless, the activities of this period are relevant because they explain the origins of CWA.

Three research reactors were installed on site to support Risø's mission. The Electronics Department was responsible for the commissioning and safety certification of the instrumentation of the research reactors. And because the head of the Electronics Department (Jens Rasmussen) became chairman of the Risø Reactor Safety Committee, it should not be surprising to find that the Electronics Department's early efforts focused on analyzing the reliability of reactor equipment and instrumentation (Jensen, Rasmussen, & Timmerman, 1963; Rasmussen & Timmerman, 1962). Note that the focus was strictly on examining hardware reliability, not human reliability, using probabilistic mathematical models. The topics of investigation included the probability of equipment failure and the degree of redundancy required in backup safety systems to achieve the desired level of reactor safety. Furthermore, considerable attention was paid to collecting failure data under representative conditions, a theme that would reemerge in later work. Interestingly, there is one passage in these reports that foreshadows the next phase of research: "Low [probability] figures must be used with great care, because the reliability in this case may be governed by factors not dealt with in this report" (Jensen et al., 1963, p. 31).

MOTIVATING PROBLEMS (1968–1969)

The next phase of research was conducted during the mid-1960s. Based on the work just described, researchers found that they could design redundant reactor safety systems with extremely high technical reliability, yet accidents still occurred. To solve this mystery, a review of 29 cases with major consequences to either plant or personnel in the nuclear domain and of 100 accidents in air transportation was conducted (Rasmussen, 1969). The results of this review revealed that the reliability of complex systems, such as nuclear reactors, cannot be viewed from a strictly technical viewpoint without considering the human element in the system (Rasmussen, 1968a, 1968c). It became apparent that workers played a key role in overall system reliability and safety. The review also revealed that accident-causing errors arose because workers were confronted with unfamiliar situations that had not been, and could not have been, anticipated by system designers (Rasmussen, 1969). In contrast, under normal circumstances, a trained and experienced worker would often be able to compensate for even gross deficiencies in the interface (Rasmussen, 1968b, 1969). Consequently, the single most important concern in improving system safety is to provide workers with the support required to adapt to unfamiliar and unanticipated abnormal situations.

Around this time, a prototype console was installed in a room adjacent to one of the research reactors' control rooms. This console was instrumented with a set of displays that were well before their time in that they included the use of digital computers, overview displays, emergent feature graphics, and display of higher order functional information (Goodstein, 1968). Operator interaction with this experimental prototype design was observed over a number of years in the field. In today's language, we would say that these research activities consisted of prototype building and usability analysis in a naturalistic setting. These rich field observations led to a number of insights that had a significant impact on the direction of subsequent research.

EMPIRICAL AND CONCEPTUAL DEVELOPMENTS (1972–1979)

One of the important empirical foundations for CWA is a field study of electronic trouble-shooting strategies conducted by Rasmussen and Jensen (1973) (see chap. 9). Methodologically, the study is interesting because most of the data were based on verbal protocols, despite the fact that such reports were considered unreliable by many psychologists at that time. Also, the study was conducted under highly representative conditions, with professional technicians diagnosing complex faults in commercially available electronic equipment. In this sense, that study was a precursor of subsequent research that also shared these methodological features. Empirically, this study was interesting because it showed that there can be several, very different strategies that can be used to perform the same task, and that people would switch between those strategies during their problem-solving activities. These findings replicated and extended the seminal, basic research results that had been obtained by Bruner et al. (1956). The study also showed that the performance criteria workers adopt and the way in which workers formulate the task (e.g., find the faulty component as quickly as possible vs. understand why the equipment is faulty in an elegant way with minimal observations) have a very strong effect on their strategies. Thus, to understand why workers tend to use certain strategies, we have to identify the performance criteria and subjective task formulation that workers choose to adopt. Otherwise, adaptive behavior can actually seem irrational from an analytical perspective. Thus, the electronic trouble-shooting field study was important because it showed that expert strategic behavior in a representative setting could be described systematically, despite its apparent complexity.

At about this time, a similar research program to study search behavior in libraries was initiated by Pejtersen (1973) at the Royal Danish School of Information and Library Science. A cooperation was established with Risø to explore the potential for a generalization in terms of a shared theoretical framework covering both technical and humanistic work domains. Pejtersen's comprehensive field results turned out to be a very promising basis for testing this framework by design of a full-scope, prototype retrieval system to be tested under actual work conditions (see chap. 12).

Another important contribution during this period was the development of the decision ladder, which represents information-processing activities (activation, observation, identification, interpretation, evaluation, task definition, procedure formulation, execution) in the shape of a ladder with shortcuts in between the two legs of the ladder (Rasmussen, 1974) (see chap. 8). This conceptual development originated from a field study of the cognitive activities of professional operators starting up a conventional power plant over a 1-week period, again using verbal protocols. The decision ladder was based on two insights. First, it is possible to parse human cognitive activities into a basic number of recurring decision tasks. Second, and more important, experts do not follow all of the information-processing steps that a novice would perform. Instead, experts rely on their knowledge and experience and thereby exhibit direct shortcuts (or associations) that allow them to bypass several cognitive activities. These shortcuts account for the increased speed and reduced effort that are hallmarks of expert performance.

The verbal protocols from the two field studies just described also gave some insights into how workers represent complex systems during problem solving. These findings eventually coalesced in the development of the AH (Rasmussen, 1979b), a framework that describes complex work domains at various levels of abstraction in a psychologically relevant fashion (see chap. 7). The AH is based on several important observations. First, when experienced operators are solving problems in the context of complex sociotechnical systems, they spontaneously adopt, and switch between, different models of the work domain in order to match the immediate task demands. Some of these models provide physical information whereas others provide functional information. Second, it is possible to represent technical systems in a way that makes reference to purpose, thereby spanning material form and functional meaning. This is accomplished through the structural means–ends links between levels of the AH. The result is a representation framework that bridges the gap between the technical and the psychological, thereby supporting goal-directed human problem solving.

Another important precursor to the ideas in this book was the analysis of 516 human error reports from the nuclear industry (Rasmussen, 1978, 1979a). Once again, the focus was on analyzing data on human behavior in representative conditions. These analyses shed light on the task characteristics and psychological mechanisms that were responsible for human error in complex sociotechnical systems. Implications were also derived for quantitative risk analysis.

More important, these analyses eventually led to the skills, rules, knowledge (SRK) framework (Rasmussen, 1979b), which categorizes three qualitatively different ways in which people can interact with the environment (see chap. 11). As such, it provides a taxonomy for models of human performance, and it can be used to derive useful implications for design as well (e.g., Vicente & Rasmussen, 1992). The utility of the taxonomy is evident by the influence it has had on the cognitive engineering community (Norman, 1993; Reason, 1990; Sanderson & Harwood, 1988).

SUMMARY AND IMPLICATIONS

From this brief, historical review we can identify a number of themes that characterized the Risø research program between 1962 and 1979:

- The research started with a practical problem of social and economic relevance, not particular theories, methods, or generic academic curiosity.
- The concepts developed were continually informed by, and focused by, very intensive analyses of data collected under representative conditions (field studies, operating experience, human error reports).
- Early empirical results changed the original focus of the research (hardware reliability) to a new set of issues (supporting operator adaptation to novelty) that were found to be of greater relevance to the problem of interest.

We believe that the emergence of these themes was facilitated by the relatively unique infrastructure and institutional support that Risø enjoyed during this period. Researchers were explicitly problem-driven rather than paradigm-driven; they had the luxury of being able to tackle research issues that were relevant to the problem at hand rather than research issues that were fashionable with funding agencies or journal editors and reviewers; they could afford to adopt a long-term rather than a short-term approach to their work; they could choose to adopt meaningful research methods that were very time consuming and laborious rather than having to resort to methods that generated any kind of results efficiently; they could let the research findings dictate their next step rather than having to stick to the deliverables defined in a research contract; they had the time to focus and think uninterruptedly about their research rather than having to time-share most of their attention among a number of other activities such as administration, teaching, and proposal writing.

A research setting with these characteristics is almost unheard of in contemporary society. Yet we could argue that these are ideal conditions for cognitive engineering research. If so, then perhaps the largest challenge facing us may not be to perform the appropriate research, but rather to create the conditions so that such research can be conducted.

References

Abraham, R. H., & Shaw, C. D. (1982). *Dynamics—The geometry of behavior: Part 1. Periodic behavior.* Santa Cruz, CA: Aerial Press.

Abraham, R. H., & Shaw, C. D. (1983). *Dynamics—The geometry of behavior: Part 2. Chaotic behavior.* Santa Cruz, CA: Aerial Press.

Abraham, R. H., & Shaw, C. D. (1984). *Dynamics—The geometry of behavior: Part 3. Global behavior.* Santa Cruz, CA: Aerial Press.

Abraham, R. H., & Shaw, C. D. (1988). *Dynamics—The geometry of behavior: Part 4. Bifurcation behavior.* Santa Cruz, CA: Aerial Press.

Ahl, V., & Allen, T. F. H. (1996). *Hierarchy theory: A vision, vocabulary, and epistemology.* New York: Columbia University Press.

Alexander, C. (1964). *Notes on the synthesis of form.* Cambridge, MA: Harvard University Press.

Allen, T. F. H., & Starr, T. B. (1982). *Hierarchy: Perspectives for ecological complexity.* Chicago: University of Chicago Press.

Anderson, J. G. (1997). Clearing the way for physicians' use of clinical information systems. *Communications of the ACM, 40*(8), 83–90.

Andognini, G. C. (1997). *Report to management: IIPA/SSFI evaluation findings and recommendations.* Toronto: Ontario Hydro.

Andriole, S., & Adelman, L. (1995). *Cognitive systems engineering for user-computer interface design, prototyping, and evaluation.* Hillsdale, NJ: Lawrence Erlbaum Associates.

Bainbridge, L. (1984). Diagnostic skill in process operation. In M. Matthews & R. D. G. Webb (Eds.), *Proceedings of the 1984 International Conference on Occupational Ergonomics: Vol. 2. Reviews* (pp. 1–10). Toronto: Human Factors Association of Canada.

Bainbridge, L. (1997). The change in concepts needed to account for human behavior in complex dynamic tasks. *IEEE Transactions on Systems, Man and Cybernetics, SMC-27,* 351–359.

Bannon, L. J., & Bødker, S. (1991). Beyond the interface: Encountering artifacts in use. In J. M. Carroll (Ed.), *Designing interaction: Psychology at the human-computer interface* (pp. 227–253). Cambridge, England: Cambridge University Press.

Bartlett, F. C. (1932). *Remembering: A study in experimental and social psychology.* Cambridge, England: Cambridge University Press.

Bartlett, F. (1958). *Thinking: An experimental and social study.* London: Allen & Unwin.

Bateson, G. (1979). *Mind and nature: A necessary unity.* New York: Dutton.

Bedny, G., & Meister, D. (1997). *The Russian theory of activity: Current applications to design and learning.* Mahwah, NJ: Lawrence Erlbaum Associates.

Bentley, A. F. (1954). *Inquiry into inquiries: Essays in social theory.* Boston: Beacon Press.

Benyon, D. (1992). The role of task analysis in systems design. *Interacting With Computers, 4,* 102–123.

Bernstein, N. A. (1996). On dexterity and its development. In M. L. Latash & M. T. Turvey (Eds.), *Dexterity and its development* (pp. 1–244). Mahwah, NJ: Lawrence Erlbaum Associates.

Beyer, H., & Holtzblatt, K. (1998). *Contextual design: Defining customer-centered systems.* San Francisco: Morgan Kaufmann.

Bisantz, A., Cohen, S. M., Gravelle, M., & Wilson, K. (1996). *To cook or not to cook: A case study of decision aiding in quick-service restaurant environments* (Report No. GIT-CS-96/03). Atlanta: Georgia Institute of Technology, College of Computing, Cognitive Science Program.

Bisantz, A. M., & Vicente, K. J. (1994). Making the abstraction hierarchy concrete. *International Journal of Human-Computer Studies, 40,* 83–117.

Black, A. (1990). Visible planning on paper and on screen: The impact of working medium on decision-making by novice graphic designers. *Behaviour and Information Technology, 9,* 283–296.

Blum, B. I. (1996). *Beyond programming: To a new era of design.* New York: Oxford University Press.

Bødker, S. (1991). *Through the interface: A human activity approach to user interface design.* Hillsdale, NJ: Lawrence Erlbaum Associates.

Bødker, S., Greenbaum, J., & Kyng, M. (1991). Setting the stage for design as action. In J. Greenbaum & M. Kyng (Eds.), *Design at work: Cooperative design of computer systems* (pp. 139–154). Hillsdale, NJ: Lawrence Erlbaum Associates.

Bødker, S., & Grønbæk, K. (1996). Users and designers in mutual activity: An analysis of cooperative activities in systems design. In Y. Engeström & D. Middleton (Eds.), *Cognition and communication at work* (pp. 130–158). Cambridge, England: Cambridge University Press.

Bogner, M. S. (1994). *Human error in medicine.* Hillsdale, NJ: Lawrence Erlbaum Associates.

Booch, G. (1994). *Object-oriented anlaysis and design with applications (2nd ed.).* Reading, MA: Addison-Wesley.

Bösser, T. (1987). *Learning in man-computer interaction.* Berlin: Springer-Verlag.

Braitenberg, V. (1984). *Vehicles: Experiments in synthetic psychology.* Cambridge, MA: MIT Press.

Brehmer, B., & Dörner, D. (1993). Experiments with computer-simulated microworlds: Escaping both the narrow straits of the laboratory and the deep blue sea of the field study. *Computers in Human Behaviour, 9,* 171–184.

Bruner, J. S., Goodnow, J. J., & Austin, G. A. (1956). *A study of thinking.* New York: Wiley.

Brunswik, E. (1956). *Perception and the representative design of psychological experiments* (2nd ed.). Berkeley: University of California Press.

Brynjolfsson, E., & Hitt, L. M. (1998). Beyond the productivity paradox. *Communications of the ACM, 41*(8), 49–55.

Brzustowski, T. (1998, March 5). *Engineering design and innovation, what's the connection?* Clarice Chalmers' Design Lecture, Department of Mechanical & Industrial Engineering, University of Toronto, Toronto.

Bucciarelli, L. L. (1994). *Designing engineers.* Cambridge, MA: MIT Press.

Bullemer, P. T., & Nimmo, I. (1994). Understanding and supporting abnormal situation management in industrial process control environments: A new approach to training. In *Proceedings of the 1994 International Conference on Systems, Man, and Cybernetics* (pp. 391–396). Piscataway, NJ: IEEE.

Burns, C. M., & Vicente, K. J. (1995). A framework for understanding interdisciplinary interactions in design. In *Proceedings of DIS '95: Symposium on Designing Interactive Systems* (pp. 97–103). New York: ACM.

Burns, C. M., & Vicente, K. J. (1997). *A comprehensive experimental evaluation of functional displays in process supervision and control* (ABB Final Report No. CEL 97-05). Toronto: University of Toronto, Cognitive Engineering Laboratory.

Button, G., & Dourish, P. (1996). Technomethodology: Paradoxes and possibilities. In *Proceedings of CHI 96* (pp. 19–26). New York: ACM.

Cacciabue, P. C. (1997). A methodology of human factors analysis for systems engineering: Theory and applications. *IEEE Transactions on Systems, Man and Cybernetics, SMC-27,* 325–339.

Cannon-Bowers, J. A., Bost, R., Hamburger, T., Crisp, H., Osga, G., & Perry, A. (1997, April). *Achieving affordability through human systems integration.* Paper presented at the Third Annual Naval Aviation Systems Engineering Supportability Symposium, Arlington, VA.

Card, S. K. (1996). Pioneers and settlers: Methods used in successful user interface design. In M. Rudisill, C. Lewis, P. G. Polson, & T. D. McKay (Eds.), *Human-computer interface design: Success stories, emerging methods, and real-world context* (pp. 122–169). San Francisco: Morgan-Kaufmann.

Card, S. K., Moran, T. P., & Newell, A. (1983). *The psychology of human-computer interaction.* Hillsdale, NJ: Lawrence Erlbaum Associates.

Carroll, J. M. (1991). Introduction: The Kittle House manifesto. In J. M. Carroll (Ed.), *Designing interaction: Psychology at the human-computer interface* (pp. 1–16). Cambridge, England: Cambridge University Press.

Carroll, J. M., Kellogg, W. A., & Rosson, M. B. (1991). The task-artifact cycle. In J. M. Carroll (Ed.), *Designing interaction: Psychology at the human-computer interface* (pp. 74–102). Cambridge, England: Cambridge University Press.

Carroll, J. M., & Rosson, M. B. (1992). Getting around the task-artifact cycle: How to make claims and design by scenario. *ACM Transactions on Information Systems, 10,* 181–212.

Casey, S. (1993). *Set phasers on stun: And other true tales of design, technology, and human error.* Santa Barbara, CA: Aegean.

Chalmers, B. A. (1997). Design issues for a decision support system for a modern frigate. In *Proceedings of the 2nd Annual Symposium and Exhibition on Situational Awareness in the Tactical Air Environment* (pp. 127–149). Patuxent River, MD: Naval Air Warfare Center.

Chase, W. G., & Simon, H. A. (1973). The mind's eye in chess. In W. G. Chase (Ed.), *Visual information processing* (pp. 215–281). New York: Academic Press.

Christiansen, E. (1996). Tamed by a rose: Computers as tools in human activity. In B. A. Nardi (Ed.), *Context and consciousness: Activity theory and human-computer interaction* (pp. 175–198). Cambridge, MA: MIT Press.

Christoffersen, K., Hunter, C. N., & Vicente, K. J. (1994). *Cognitive "dipsticks": Knowledge elicitation techniques for cognitive engineering research* (Report No. CEL 94-01). Toronto: University of Toronto, Cognitive Engineering Laboratory.

Christoffersen, K., Hunter, C. N., & Vicente, K. J. (1996). A longitudinal study of the effects of ecological interface design on skill acquisition. *Human Factors, 38,* 523–541.

Christoffersen, K., Hunter, C. N., & Vicente, K. J. (1997). A longitudinal study of the effects of ecological interface design on fault management performance. *International Journal of Cognitive Ergonomics, 1,* 1–24.

Christoffersen, K., Hunter, C. N., & Vicente, K. J. (in press). A longitudinal study of the impact of ecological interface design on deep knowledge. *International Journal of Human-Computer Studies.*

Clegg, C., Axtell, C., Damodaran, L., Farbey, B., Hull, R., Lloyd-Jones, R., Nicholls, J., Sell, R., Tomlinson, C., Ainger, A., & Stewart, T. (1996). *The performance of information technology and the role of human and organizational factors.* London: Economic and Social Research Council.

Cook, R. I., Potter, S. S., Woods, D. D., & McDonald, J. M. (1991). Evaluating the human engineering of microprocessor-controlled operating room devices. *Journal of Clinical Monitoring, 7,* 217–226.

Cook, R. I., & Woods, D. D. (1996). Adapting new technology in the operating room. *Human Factors, 38,* 593–613.

Cook, T. D., & Campbell, D. T. (1979). *Quasi-experimentation: Design and analysis issues for field settings.* Boston: Houghton Mifflin.

Cooke, N. (1994). Varieties of knowledge elicitation techniques. *International Journal of Human-Computer Studies, 41,* 801–849.

Dainoff, M. J., & Mark, L. S. (1995). Use of a means-end abstraction hierarchy to conceptualize the ergonomic design of workplaces. In J. Flach, P. Hancock, J. Caird, & K. Vicente (Eds.), *Global perspectives on the ecology of human-machine systems* (pp. 273–292). Hillsdale, NJ: Lawrence Erlbaum Associates.

Davis, S., & Meyer, C. (1998). *Blur: The speed of change in the connected economy.* Reading, MA: Addison-Wesley.

de Groot, A. D. (1965). *Thought and choice in chess.* The Hague, Netherlands: Mouton. (Original work published 1946)

de Groot, A. D. (1990). Unifying psychology: A European view. *New Ideas in Psychology, 3,* 309–320.

De Keyser, V. (1991). Work analysis in French language ergonomics: Origins and current research trends. *Ergonomics, 34,* 653–669.

De Keyser, V., Decortis, F., & Van Daele, A. (1988). The approach of Francophone ergonomy: Studying new technologies. In V. De Keyser, T. Qvale, & B. Wilpert (Eds.), *The meaning of work and technological options* (pp. 147–163). Chichester, England: Wiley.

Dessouky, M. I., Moray, N., & Kijowski, B. (1995). Taxonomy of scheduling systems as a basis for the study of strategic behavior. *Human Factors, 37,* 443–472.

Deutsch, H. W., Callahan, P., & Edwards, B. (1996). Everyone gets the net. *The Forrester Report, 13*(9), 1–16.

Dewey, J., & Bentley, A. F. (1949). *Knowing and the known.* Boston: Beacon Press.

Diaper, D. (1989). *Task analysis for human computer interaction.* Chichester, England: Ellis Horwood.

Diaper, D., & Addison, M. (1992). Task analysis and systems analysis for software development. *Interacting With Computers, 4,* 124–139.

Dinadis, N., & Vicente, K. J. (1996). Ecological interface design for a power plant feedwater subsystem. *IEEE Transactions on Nuclear Science, 43,* 266–277.

Dinadis, N., & Vicente, K. J. (in press). Designing functional visualizations for aircraft system status displays. *International Journal of Aviation Psychology.*

Dörner, D. (1996). *The logic of failure: Why things go wrong and what we can do to make them right.* New York: Henry Holt. (Original work published 1989)

Dougherty, J. W. D., & Keller, C. M. (1985). Taskonomy: A practical approach to knowledge structures. In J. W. D. Dougherty (Ed.), *Directions in cognitive anthropology* (pp. 161–174). Urbana: University of Illinois Press.

Dowell, J., & Long, J. (1998). Conception of the cognitive engineering design problem. *Ergonomics, 41,* 126–139.

Dreyfus, H. L., & Dreyfus, S. E. (1988). *Mind over machine: The power of human intuition and expertise in the era of the computer.* New York: The Free Press.

Duncker, K. (1945). On problem solving. *Psychological Monographs, 58*(5, Whole No. 270).

Easter, J. R. (1987). Engineering human factors into the Westinghouse advanced control room. *Nuclear Engineering International, 32*(May), 35–38.

Eddington, A. (1958). *The philosophy of physical science.* Ann Arbor: University of Michigan Press.

Edwards, E., & Lees, F. P. (1974). *The human operator in process control.* London: Taylor & Francis.

Egan, D. E., & Schwartz, B. J. (1979). Chunking in recall of symbolic drawings. *Memory & Cognition, 7,* 149–158.

Einstein, A., & Infeld, L. (1938). *The evolution of physics: From early concepts to relativity and quanta.* New York: Simon & Schuster.

Endsley, M. R. (1995). Measurement of situation awareness in dynamic systems. *Human Factors, 37,* 65–84.

Engeström, Y., & Middleton, D. (1996). *Cognition and communication at work.* Cambridge, England: Cambridge University Press.

Ericsson, K. A., & Harris, M. S. (1990). *Expert chess memory without chess knowledge: A training study.* Paper presented at the 31st annual meeting of the Psychonomic Society, New Orleans.

Ferguson, E. S. (1977). The mind's eye: Nonverbal thought in technology. *Science, 197,* 827–836.

Flach, J. M. (1990a). Control with an eye for perception: Precursors to an active psychophysics. *Ecological Psychology, 2,* 83–111.

Flach, J. M. (1990b). The ecology of human-machine systems: I. Introduction. *Ecological Psychology, 2,* 191–205.

Flach, J., Hancock, P., Caird, J., & Vicente, K. J. (1995). *Global perspectives on the ecology of human-machine systems.* Hillsdale, NJ: Lawrence Erlbaum Associates.

Fox, M. S. (1992). *Design in the large: A project description.* Toronto: University of Toronto, Department of Industrial Engineering.

Frank, P. M. (1990). Fault diagnosis in dynamic systems using analytical and knowledge-based redundancy—A survey and some new results. *Automatica, 26,* 459–474.

FRIEND '21. (1995). *Human interface architecture guidelines.* Tokyo: Institute for Personalized Information Environment.

Frijda, N. H., & de Groot, A. D. (1981). *Otto Selz: His contribution to psychology.* The Hague, Netherlands: Mouton.

Gerson, E. M., & Star, S. L. (1986). Analyzing due process in the workplace. *ACM Transactions on Office Information Systems, 4,* 257–270.

Gibbs, W. W. (1997). Taking computers to task. *Scientific American, 277*(1), 82–89.

Gibson, E. J. (1991). *An odyssey in learning and perception.* Cambridge, MA: MIT Press.

Gibson, J. J. (1979). *The ecological approach to visual perception.* Boston: Houghton Mifflin.

Gibson, J. J., & Crooks, L. E. (1938). A theoretical field-analysis of automobile-driving. *American Journal of Psychology, 51,* 453–471.

Glaser, R., & Chi, M. T. H. (1988). Overview. In M. T. H. Chi, R. Glaser, & M. J. Farr (Eds.), *The nature of expertise* (pp. xv–xxviii). Hillsdale, NJ: Lawrence Erlbaum Associates.

Golomb, S. W. (1968, January). Mathematical models—Uses and limitations. *Astronautics & Aeronautics,* pp. 57–59.

Goodstein, L. P. (1968). An experimental computer-controlled instrumentation system for the research reactor DR-2. In *Application of on-line computers to nuclear reactors* (pp. 549–566). Halden, Norway: OECD Halden Reactor Project.

Goodstein, L. P. (1985). *Studies of operator computer cooperation on a small-scale nuclear power plant simulator* (Report No. Risø-M-2522). Roskilde, Denmark: Risø National Laboratory, Electronics Department.

Goodstein, L. P., Hedegård, J., Højberg, K. S., & Lind, M. (1984). *The GNP testbed for operator support evaluation* (Report No. Risø-M-2460). Roskilde, Denmark: Risø National Laboratory, Electronics Department.

Greenbaum, J., & Kyng, M. (1991). *Design at work: Cooperative design of computer systems.* Hillsdale, NJ: Lawrence Erlbaum Associates.

Greif, S. (1991). The role of German work psychology in the design of artifacts. In J. M. Carroll (Ed.), *Designing interaction: Psychology at the human-computer interface* (pp. 203–226). Cambridge, England: Cambridge University Press.

Guerlain, S. (1995). Using the critiquing approach to cope with brittle expert systems. In *Proceedings of the Human Factors and Ergonomics Society 39th Annual Meeting* (pp. 233–237). Santa Monica, CA: HFES.

Guerlain, S., & Bullemer, P. (1996). User-initiated notification: A concept for aiding the monitoring activities of process control operators. In *Proceedings of the Human Factors and Ergonomics Society 40th Annual Meeting* (pp. 283–287). Santa Monica, CA: HFES.

Hajdukiewicz, J. R. (1998). *Development of a structured approach for patient monitoring in the operating room*. Unpublished MASc thesis, University of Toronto, Toronto.

Haken, H. (1988). *Information and self-organization: A macroscopic approach to complex systems*. Heidelberg: Springer-Verlag.

Halbach, W. R. (1994). Simulated breakdowns. In H. U. Gumbrecht & K. L. Pfeiffer (Eds.), *Materialities of communication* (pp. 335–343). Palo Alto, CA: Stanford University Press.

Hancock, P., Flach, J., Caird, J., & Vicente, K. J. (1995). *Local applications of the ecological approach to human-machine systems*. Hillsdale, NJ: Lawrence Erlbaum Associates.

Hanson, N. R. (1958). *Patterns of discovery*. Cambridge, England: Cambridge University Press.

Harrison, B. L., Ishii, H., Vicente, K. J., & Buxton, W. A. S. (1995). Transparent layered user interfaces: An evaluation of a display design to enhance focused and divided attention. In *Human Factors in Computing Systems: CHI '95 Conference Proceedings* (pp. 317–324). New York: ACM.

Hart, S. H., & Staveland, L. E. (1988). Development of NASA-TLX (Task Load Index): Results of empirical and theoretical research. In P. A. Hancock & N. Meshkati (Eds.), *Human mental workload* (pp. 139–183). Amsterdam: North-Holland.

Heath, C., & Luff, P. (1996). Convergent activities: Line control and passenger information on the London Underground. In Y. Engeström & D. Middleton (Eds.), *Cognition and communication at work* (pp. 96–129). Cambridge, England: Cambridge University Press.

Henderson, A., & Kyng, M. (1991). There's no place like home: Continuing design in use. In J. Greenbaum & M. Kyng (Eds.), *Design at work: Cooperative design of computer systems* (pp. 219–240). Hillsdale, NJ: Lawrence Erlbaum Associates.

Hirschhorn, L. (1984). *Beyond mechanization: Work and technology in a postindustrial age*. Cambridge, MA: MIT Press.

Hollnagel, E. (1993). *Human reliability analysis: Context and control*. London: Academic Press.

Hollnagel, E., & Woods, D. D. (1983). Cognitive systems engineering: New wine in new bottles. *International Journal of Man-Machine Systems, 18,* 583–600.

Holmqvist, B., & Andersen, P. B. (1991). Language, perspectives and design. In J. Greenbaum & M. Kyng (Eds.), *Design at work: Cooperative design of computer systems* (pp. 91–119). Hillsdale, NJ: Lawrence Erlbaum Associates.

Hovde, G. (1990). *Cognitive work analysis: Decision making in operation theatre planning* (Risø Working Report). Roskilde, Denmark: Risø National Laboratory.

Howie, D. E., & Vicente, K. J. (1997). Using self-explanation to exploit ecological interface design. In *Proceedings of the Human Factors and Ergonomics Society 41st Annual Meeting* (pp. 279–283). Santa Monica, CA: HFES.

Howie, D. E., & Vicente, K. J. (1998). Measures of operator performance in complex, dynamic microworlds: Advancing the state of the art. *Ergonomics, 41,* 385–400.

Hunter, C. N., Vicente, K. J., & Tanabe, F. (1996). Can "theoretical" training improve fault management performance? In *Proceedings of the 1996 American Nuclear Society International Topical Meeting on Nuclear Plant Instrumentation, Control and Human-Machine Interface Technologies* (pp. 683–690). La Grange Park, IL: American Nuclear Society.

Hurst, B., & Skilton, W. (1997). *Work domain analysis workbench: Requirements and a data model* (Report No. SCHIL-1998-04). Hawthorn, Australia: Swinburne University of Technology, School of Information Technology, Swinburne Computer-Human Interaction Laboratory.

Hutchins, E. (1990). The technology of team navigation. In J. Galegher, R. Kraut, & C. Egido (Eds.), *Intellectual teamwork: Social and technical bases of collaborative work* (pp. 191–220). Hillsdale, NJ: Lawrence Erlbaum Associates.

Hutchins, E. (1995a). *Cognition in the wild*. Cambridge, MA: MIT Press.

Hutchins, E. (1995b). How a cockpit remembers its speeds. *Cognitive Science, 19,* 265–288.

Hutchins, E. L., Hollan, J. D., & Norman, D. A. (1986). Direct manipulation interfaces. In D. A. Norman & S. W. Draper (Eds.), *User centered system design: New perspectives on human-computer interaction* (pp. 87–124). Hillsdale, NJ: Lawrence Erlbaum Associates.

Itoh, J., Sakuma, A., & Monta, K. (1995). An ecological interface for supervisory control of BWR nuclear power plants. *Control Engineering Practice, 3,* 231–239.

Itoh, J., Yoshimura, S., Ohtsuka, T., & Masuda, F. (1990). Cognitive task analysis of nuclear power plant operators for man-machine interface design. In *Proceedings of the ANS Topical Meeting on Advances in Human Factors Research on Man-Computer Interactions: Nuclear and Beyond* (pp. 96–102). La Grange Park, IL: American Nuclear Society.

Jamieson, G. A., & Vicente, K. J. (1998). Modeling techniques to support abnormal situation management in the petrochemical processing industry. In *Proceedings of the CSME Symposium on Industrial Engineering and Management* (pp. 249–256). Toronto: Ryerson Polytechnic University.

Janzen, M. E., & Vicente, K. J. (in press). Attention allocation within the abstraction hierarchy. *International Journal of Human-Computer Studies.*

Jensen, A., Rasmussen, J., & Timmerman, P. (1963). *Analysis of failure data for electronic equipment at Risø* (Risø Report No. 38). Roskilde, Denmark: Danish Atomic Energy Commission, Research Establishment Risø.

Johannson, R. (1993). *System modeling and identification.* Englewood Cliffs, NJ: Prentice-Hall.

Jones, P. M., & Mitchell, C. M. (1987). Operator modeling: Conceptual and methodological distinctions. In *Proceedings of the Human Factors Society 31st Annual Meeting* (pp. 31–35). Santa Monica, CA: Human Factors Society.

Jørgensen, S. S. (1993). *Fault diagnosis using generic multilevel flow modelling models.* Unpublished doctoral thesis, Technical University of Denmark, Lyngby.

Kanigel, R. (1997). *The one best way: Frederick Winslow Taylor and the enigma of efficiency.* New York: Viking.

Kaptelinin, V. (1996). Activity theory: Implications for human-computer interaction. In B. A. Nardi (Ed.), *Context and consciousness: Activity theory and human-computer interaction* (pp. 103–116). Cambridge, MA: MIT Press.

Karasek, R., & Theorell, T. (1990). *Healthy work: Stress, productivity, and the reconstruction of working life.* New York: Basic Books.

Kauffman, J. V., Lanik, G. F., Spence, R. A., & Trager, E. A. (1992). *Operating experience feedback report—Human performance in operating events* (Report No. NUREG-1275, Vol. 8). Washington, DC: U.S. Nuclear Regulatory Commission.

Kelley, M. R. (1994). Productivity and information technology: The elusive connection. *Management Science, 40,* 1406–1425.

Kelso, J. A. S. (1995). *Dynamic patterns: The self-organization of brain and behavior.* Cambridge, MA: MIT Press.

Kieras, D. E., & Meyer, D. E. (1998). *The role of cognitive task analysis in the application of predictive models of human performance* (Report No. TR-98/ONR-EPIC-11). Ann Arbor: University of Michigan.

Kim, I. S. (1994). Computerized systems for on-line management of failures: A state-of-the-art discussion of alarm systems and diagnostic systems applied in the nuclear industry. *Reliability Engineering and System Safety, 44,* 279–295.

Kirlik, A., Miller, R. A., & Jagacinski, R. J. (1993). Supervisory control in a dynamic and uncertain environment: A process model of skilled human-environment interaction. *IEEE Transactions on Systems, Man, & Cybernetics, SMC-23,* 929–952.

Kirwan, B. (1992). A task analysis programme for a large nuclear chemical plant. In B. Kirwan & L. K. Ainsworth (Eds.), *A guide to task analysis* (pp. 363–388). London: Taylor & Francis.

Kirwan, B., & Ainsworth, L. K. (1992). *A guide to task analysis.* London: Taylor & Francis.

Klein, G. A. (1989). Recognition-primed decisions. In W. B. Rouse (Ed.), *Advances in man-machine systems research* (Vol. 5, pp. 47–92). Greenwich, CT: JAI.

Klein, G. (1997). The recognition-primed decision (RPD) model: Looking back, looking forward. In C. E. Zsambok & G. Klein (Eds.), *Naturalistic decision making* (pp. 285–292). Mahwah, NJ: Lawrence Erlbaum Associates.

Klein, G. A., Orasanu, J., Calderwood, R., & Zsambok, C. (1993). *Decision making in action: Models and methods.* Norwood, NJ: Ablex.

Kolers, P. A., & Roediger, H. L., III (1984). Procedures of mind. *Journal of Verbal Learning and Verbal Behavior, 23,* 425–449.

Korf, R. E. (1987). Planning as search: A quantitative approach. *Artificial Intelligence, 33,* 65–88.

Krosner, S. P., Mitchell, C. M., & Govindaraj, T. (1989). Design of an FMS operator workstation using the Rasmussen abstraction hierarchy. In *Proceedings of the 1989 International Conference on Systems, Man and Cybernetics* (pp. 959–964). Piscataway, NJ: IEEE.

Kugler, P. N., Kelso, J. A. S., & Turvey, M. T. (1980). On the concept of coordinative structures as dissipative structures: 1. Theoretical lines of convergence. In G. E. Stelmach & J. Requin (Eds.), *Tutorials in motor behavior* (pp. 3–47). Amsterdam: North-Holland.

Kugler, P. N., Kelso, J. A. S., & Turvey, M. T. (1982). On the control and co-ordination of naturally developing systems. In J. A. S. Kelso & J. E. Clark (Eds.), *The development of movement control and co-ordination* (pp. 5–78). New York: Wiley.

Kugler, P. N., Shaw, R. E., Vicente, K. J., & Kinsella-Shaw, J. (1990). Inquiry into intentional systems: I. Issues in ecological physics. *Psychological Research, 52,* 98–121.

Kugler, P. N., & Turvey, M. T. (1987). *Information, natural law, and the self-assembly of rhythmic movement.* Hillsdale, NJ: Lawrence Erlbaum Associates.

Kuutti, K. (1996). Activity theory as a potential framework for human-computer interaction research. In B. A. Nardi (Ed.), *Context and consciousness: Activity theory and human-computer interaction* (pp. 17–44). Cambridge, MA: MIT Press.

Kyng, M. (1995). Making representations work. *Communications of the ACM, 38*(9), 46–55.

Landauer, T. K. (1995). *The trouble with computers: Usefulness, usability, and productivity.* Cambridge, MA: MIT Press.

Leape, L. L. (1994). Error in medicine. *Journal of the American Medical Association, 272,* 1851–1857.

Lee, D. N. (1976). A theory of visual control of braking based on information about time-to-collision. *Perception, 5,* 437–459.

Lees, F. P. (1983). Process computer alarm and disturbance analysis: Review of the state of the art. *Computers and Chemical Engineering, 7,* 669–694.

Leplat, J. (1989). Error analysis, instrument and object for task analysis. *Ergonomics, 32,* 813–822.

Leplat, J. (1990). Relations between task and activity: Elements for elaborating a framework for error analysis. *Ergonomics, 33,* 1389–1402.

Leveson, N. G. (1995). *Safeware: System safety and computers.* Reading, MA: Addison-Wesley.

Leveson, N. G. (in press). Intent specifications. *IEEE Transactions on Software Engineering.*

Luce, R. D. (1995). Four tensions concerning mathematical modeling in psychology. *Annual Review of Psychology, 46,* 1–26.

Mandel, M. (1989). *Making good time.* Riverside: CA: University of California, Calfornia Museum of Photography.

Marken, R. S. (1986). Perceptual organization of behavior: A hierarchical control model of coordinated action. *Journal of Experimental Psychology: Human Perception and Performance, 12,* 267–276.

Marsden, P. (1996). Procedures in the nuclear industry. In N. Stanton (Ed.), *Human factors in nuclear safety* (pp. 99–116). London: Taylor & Francis.

Matteson, M. T., & Ivancevich, J. M. (1987). *Controlling work stress: Effective human resource and management strategies.* San Francisco: Jossey-Bass.

Maynard, H. B. (1971). *Industrial engineering handbook* (3rd ed.). New York: McGraw-Hill.

McLeod, R. W., & Sherwood-Jones, B. M. (1992). Simulation to predict operator workload in a command system. In B. Kirwan & L. K. Ainsworth (Eds.), *A guide to task analysis* (pp. 301–310). London: Taylor & Francis.

Meister, D. (1985). *Behavioral analysis and measurement methods.* New York: Wiley.

Meister, D. (1989). *Conceptual aspects of human factors.* Baltimore: Johns Hopkins University Press.

Meister, D. (1995a). Cognitive behavior of nuclear reactor operators. *International Journal of Industrial Ergonomics, 16,* 109–122.

Meister, D. (1995b). *Divergent viewpoints: Essays on human factors questions.* Unpublished manuscript.

Mesarovic, M. D., Macko, D., & Takahara, Y. (1970). *Theory of hierarchical, multilevel, systems.* New York: Academic Press.

Meshkati, N., Buller, B. J., & Azadeh, M. A. (1994). *Integration of workstation, job, and team structure design in the control rooms of nuclear power plants.* Los Angeles: University of Southern California Press.

Miller, R. A. (1982). *Formal analytical structure for manned systems analysis.* Columbus: Ohio State University Press.

Miller, R. B. (1953). *A method for man-machine task analysis.* Wright–Patterson Air Force Base, OH: Wright Air Development Center, Air Research and Development Command.

Miller, T. E., & Woods, D. D. (1997). Key issues for naturalistic decision making researchers in system design. In C. E. Zsambok & G. Klein (Eds.), *Naturalistic decision making* (pp. 141–149). Mahwah, NJ: Lawrence Erlbaum Associates.

Mitchell, C. M. (1996). Models for the design of human interaction with complex dynamic systems. In *Proceedings of CSEPC 96: Cognitive Systems Engineering in Process Control* (pp. 230–237). Kyoto, Japan: Kyoto University.

Mitchell, C. M., & Saisi, D. L. (1987). Use of model-based qualitative icons and adaptive windows in workstations for supervisory control systems. *IEEE Transactions on Systems, Man, and Cybernetics, SMC-17,* 573–593.

Mizutani, E., & Dreyfus, S. E. (1998). Totally model-free reinforcement learning by actor-critic Elman networks in non-Markovian domains. In *Proceedings of the IEEE World Congress on Computational Intelligence.* Piscataway, NJ: IEEE.

Moray, N. (1988). Ex Risø semper aliquid antiquum: Sources of a new paradigm for engineering psychology. In L. P. Goodstein, H. B. Andersen, & S. E. Olsen (Eds.), *Tasks, errors, and mental models: A festschrift to celebrate the 60th birthday of Professor Jens Rasmussen* (pp. 12–17). London: Taylor & Francis.

Moray, N., & Huey, B. (1988). *Human factors research and nuclear safety.* Washington, DC: National Research Council, National Academy of Sciences.

Moray, N., Lootsteen, P., & Pajak, J. (1986). Acquisition of process control skills. *IEEE Transactions on Systems, Man, and Cybernetics, SMC-16,* 497–504.

Moray, N., Sanderson, P. M., & Vicente, K. J. (1992). Cognitive task analysis of a complex work domain: A case study. *Reliability Engineering and System Safety, 36,* 207–216.

Morone, J. G. (1993). *Winning in high-tech markets: The role of general management.* Boston: Harvard Business School Press.

Morris, C. D., Bransford, J. D., & Franks, J. J. (1977). Levels of processing versus transfer appropriate processing. *Journal of Verbal Learning and Verbal Behavior, 16,* 519–533.

Nardi, B. A. (1996a). Activity theory and human-computer interaction. In B. A. Nardi (Ed.), *Context and consciousness: Activity theory and human-computer interaction* (pp. 7–16). Cambridge, MA: MIT Press.

Nardi, B. A. (1996b). *Context and consciousness: Activity theory and human-computer interaction.* Cambridge, MA: MIT Press.

Nardi, B. A. (1996c). Studying context: A comparison of activity theory, situated action models, and distributed cognition. In B. A. Nardi (Ed.), *Context and consciousness: Activity theory and human-computer interaction* (pp. 69–102). Cambridge, MA: MIT Press.

Narendra, K. S. (1986). *Adaptive and learning systems: Theory and applications.* New York: Plenum.

Negroponte, N. (1995). *Being digital.* New York: Vintage Books.

Newell, A. (1973). You can't play 20 questions with nature and win: Projective comments on the papers of this symposium. In W. G. Chase (Ed.), *Visual information processing* (pp. 283–308). New York: Academic Press.

Newell, A., & Simon, H. A. (1972). *Human problem solving.* Englewood Cliffs, NJ: Prentice-Hall.

Newell, K. M. (1986). Constraints on the development of coordination. In M. G. Wade & H. T. A. Whiting (Eds.), *Motor development in children: Aspects of coordination and control* (pp. 341–360). Boston: Martinus Nijhoff.

Nickerson, R. S. (1992). *Looking ahead: Human factors challenges in a changing world.* Hillsdale, NJ: Lawrence Erlbaum Associates.

Nickerson, R. S. (1995). *Emerging needs and opportunities for human factors research.* Washington, DC: National Academy Press.

Norman, D. A. (1981). *Steps toward a cognitive engineering: System images, system friendliness, mental models* (Tech. Rep.). La Jolla: University of California Press.

Norman, D. A. (1986). Cognitive engineering. In D. A. Norman & S. W. Draper (Eds.), *User centered system design: New perspectives on human-computer interaction* (pp. 31–61). Hillsdale, NJ: Lawrence Erlbaum Associates.

Norman, D. A. (1993). *Things that make us smart: Defending human attributes in the age of the machine.* Reading, MA: Addison-Wesley.

Norman, D. A. (1998). *The invisible computer: Why good products fail, why the personal computer is so complex, and how to do it right.* Cambridge, MA: MIT Press.

Norris, G. (1995). Boeing's seventh wonder. *IEEE Spectrum, 32*(10), 20–23.

Norros, L. (1996). System disturbances as springboard for development of operators' expertise. In Y. Engeström & D. Middleton (Eds.), *Cognition and communication at work* (pp. 159–176). Cambridge, England: Cambridge University Press.

O'Donnell, R. D. & Eggemeier, F. T. (1986). Workload assessment methodology. In K. R. Boff, L. Kaufman, & J. P. Thomas (Eds.), *Handbook of perception and human performance* (Vol. 2, pp. 42–49). New York: Wiley.

OECD (1998). *Human capital investment: An international comparison.* Paris: Organization for Economic Co-operation and Development.

Olsen, S. E., & Rasmussen, J. (1989). The reflective expert and the prenovice: Notes on skill-, rule-, and knowledge-based performance in the setting of instruction and training. In L. Bainbridge & S. A. R. Quintanilla (Eds.), *Developing skills with information technology* (pp. 9–33). Chichester, England: Wiley.

Olsson, G., & Lee, P. L. (1994). Effective interfaces for process operators—A prototype. *The Journal of Process Control, 4,* 99–107.

Patrick, J. (1992). *Training: Research and practice.* London: Academic Press.

Pattee, H. H. (1972). The nature of hierarchical controls in living matter. In R. Rosen (Ed.), *Foundations of mathematical biology: Vol. 1. Subcellular systems* (pp. 1–21). New York: Academic Press.

Pawlak, W. S., & Vicente, K. J. (1996). Inducing effective operator control through ecological interface design. *International Journal of Human-Computer Studies, 44,* 653–688.

Payne, J. W., Bettman, J. R., & Johnson, E. J. (1993). *The adaptive decision maker.* Cambridge, England: Cambridge University Press.

Payne, S. J. (1991). Interface problems and interface resources. In J. M. Carroll (Ed.), *Designing interaction: Psychology at the human-computer interface* (pp. 128–153). Cambridge, England: Cambridge University Press.

Pejtersen, A. M. (1973). *Typology of fiction and its users* (Tech. Rep. in Danish). Copenhagen, Denmark: Royal School of Information and Library Science.

Pejtersen, A. M. (1979). Investigation of search strategies in fiction based on an analysis of 134 user-librarian conversations. In T. Henriksen (Ed.), *IRFIS 3 Conference Proceedings* (pp. 107–132). Oslo, Norway: Statens Biblioteks-och Informations Hoegskole.

Pejtersen, A. M. (1984). Design of a computer-aided user-system dialogue based on an analysis of users' search behaviour. *Social Science Information Studies, 4,* 167–183.

Pejtersen, A. M. (1988). Search strategies and database design for information retrieval from libraries. In L. P. Goodstein, H. B. Andersen, & S. E. Olsen (Eds.), *Tasks, errors, and mental models: A festschrift to celebrate the 60th birthday of Professor Jens Rasmussen* (pp. 171–190). London: Taylor & Francis.

Pejtersen, A. M. (1992). The BookHouse: An icon based database system for fiction retrieval in public libraries. In B. Cronin (Ed.), *The marketing of library and information services 2* (pp. 572–591). London: ASLIB.

Pejtersen, A. M., & Goodstein, L. P. (1990). Beyond the desk top metaphor: Information retrieval with an icon based interface. In M. J. Tauber & P. Gorny (Eds.), *Proceedings of the 7th Interdisciplinary Workshop on Informatics and Psychology on Visualization in Human Computer Interaction.* Berlin: Springer-Verlag.

Pejtersen, A. M., Sonnenwald, D. H., Buur, J., Govindaraj, T., & Vicente, K. J. (1997). The design explorer project: Using a cognitive framework to support knowledge exploration. *Journal of Engineering Design, 8,* 289–301.

Perrow, C. (1984). *Normal accidents: Living with high-risk technologies.* New York: Basic Books.

Petroski, H. (1995). The Boeing 777. *American Scientist, 83,* 519–522.

Polanyi, M. (1958). *Personal knowledge: Towards a post-critical philosophy.* Chicago: University of Chicago Press.

Pool, R. (1997). *Beyond engineering: How society shapes technology.* New York: Oxford University Press.

Port, R. F., & van Gelder, T. (1995). *Mind as motion: Explorations in the dynamics of cognition.* Cambridge, MA: MIT Press.

Potter, S. S., Roth, E. M., Woods, D. D., & Elm, W. C. (1998). *Toward the development of a computer-aided cognitive engineering tool to facilitate the development of advanced decision support systems for information warfare domains* (AFRL-HE-WP-TR-1998-0004). Wright-Patterson AFB, OH: United States Air Force Research Laboratory.

Powers, W. T. (1973a). *Behavior: The control of perception.* Chicago: Aldine.

Powers, W. T. (1973b). Feedback: Beyond behaviorism. *Science, 179,* 351–379.

Powers, W. T. (1978). Quantitative analysis of purposive systems: Some spadework at the foundations of scientific psychology. *Psychological Review, 85,* 417–435.

Raeithel, A., & Velichkovsky, B. M. (1996). Joint attention and co-construction: New ways to foster user-designer collaboration. In B. A. Nardi (Ed.), *Context and consciousness: Activity theory and human-computer interaction* (pp. 199–233). Cambridge, MA: MIT Press.

Raghupathi, W. (1997). Health care information systems. *Communications of the ACM, 40*(8), 81–82.

Raiffa, H. (1968). *Decision analysis: Introductory lectures on choices under uncertainty.* Reading, MA: Addison-Wesley.

Rasmussen, J. (1968a). *Characteristics of operator, automatic equipment and designer in plant automation* (Report No. Risø-M-808). Roskilde, Denmark: Danish Atomic Energy Commission, Research Establishment Risø.

Rasmussen, J. (1968b). *On the communication between operators and instrumentation in automatic process control plants* (Report No. Risø-M-686). Roskilde, Denmark: Danish Atomic Energy Commission, Research Establishment Risø.

Rasmussen, J. (1968c). *On the reliability of process plants and instrumentation systems* (Report No. Risø-M-706). Roskilde, Denmark: Danish Atomic Energy Commission, Research Establishment Risø.

Rasmussen, J. (1969). *Man-machine communication in the light of accident records* (Report No. S-1-69). Roskilde, Denmark: Danish Atomic Energy Commission, Research Establishment Risø.

Rasmussen, J. (1974). *The human data processor as a system component: Bits and pieces of a model* (Report No. Risø-M-1722). Roskilde, Denmark: Danish Atomic Energy Commission.

Rasmussen, J. (1976). Outlines of a hybrid model of the process plant operator. In T. B. Sheridan & G. Johannsen (Eds.), *Monitoring behavior and supervisory control* (pp. 371–383). New York: Plenum.

Rasmussen, J. (1978). *Operator/technician errors in calibration, setting, and testing nuclear power plant equipment* (Report No. N-17-78). Roskilde, Denmark: Risø National Laboratory, Electronics Department.

Rasmussen, J. (1979a). *Preliminary analysis of human error cases in U.S. licensee event reports* (Report No. N-8-79). Roskilde, Denmark: Risø National Laboratory, Electronics Department.

Rasmussen, J. (1979b). *On the structure of knowledge—A morphology of mental models in a man-machine system context* (Report No. Risø-M-2192). Roskilde, Denmark: Risø National Laboratory, Electronics Department.

Rasmussen, J. (1980). The human as a systems component. In H. T. Smith & T. R. G. Green (Eds.), *Human interaction with computers* (pp. 67–96). London: Academic Press.

Rasmussen, J. (1981). Models of mental strategies in process plant diagnosis. In J. Rasmussen & W. B. Rouse (Eds.), *Human detection and diagnosis of system failures* (pp. 241–258). New York: Plenum.

Rasmussen, J. (1983). Skills, rules, and knowledge; signals, signs, and symbols, and other distinctions in human performance models. *IEEE Transactions on Systems, Man, and Cybernetics, SMC-13*, 257–266.

Rasmussen, J. (1985). The role of hierarchical knowledge representation in decision making and system management. *IEEE Transactions on Systems, Man, and Cybernetics, SMC-15*, 234–243.

Rasmussen, J. (1986a). *A cognitive engineering approach to the modelling of decision making and its organization* (Report No. Risø-M-2589). Roskilde, Denmark: Risø National Laboratory.

Rasmussen, J. (1986b). A framework for cognitive task analysis in systems design. In E. Hollnagel, G. Mancini, & D. D. Woods (Eds.), *Intelligent decision support in process environments* (pp. 175–196). Berlin: Springer-Verlag.

Rasmussen, J. (1986c). *Information processing and human-machine interaction: An approach to cognitive engineering.* New York: North-Holland.

Rasmussen, J. (1988a). Cognitive engineering, a new profession? In L. P. Goodstein, H. B. Andersen, & S. E. Olsen (Eds.), *Tasks, errors, and mental models: A festschrift to celebrate the 60th birthday of Professor Jens Rasmussen* (pp. 325–334). London: Taylor & Francis.

Rasmussen, J. (1988b). Information technology: A challenge to the Human Factors Society? *Human Factors Society Bulletin, 31*(7), 1–3.

Rasmussen, J. (1990a). Mental models and the control of action in complex environments. In D. Ackermann & M. J. Tauber (Eds.), *Mental models and human-computer interaction 1* (pp. 41–69). Amsterdam: North-Holland.

Rasmussen, J. (1990b). A model for the design of computer integrated manufacturing systems: Identification of information requirements of decision makers. *International Journal of Industrial Ergonomics, 5*, 5–16.

Rasmussen, J. (1997a). Merging paradigms: Decision making, management, and cognitive control. In R. Flin, E. Salas, M. E. Strub, & L. Marting (Eds.), *Decision making under stress: Emerging paradigms and applications* (pp. 67–85). Aldershot, England: Ashgate.

Ramussen, J. (1997b). Risk management in a dynamic society: A modeling problem. *Safety Science, 27*, 183–213.

Rasmussen, J., & Batstone, R. (1991). *Towards improved safety control and risk management: Findings from the World Bank workshops.* New York: World Bank.

Rasmussen, J., & Goodstein, L. P. (1987). Decision support in supervisory control of high-risk industrial systems. *Automatica, 23*, 663–671.

Rasmussen, J., & Goodstein, L. P. (1988). Information technology and work. In M. Helander (Ed.), *Handbook of human-computer interaction* (pp. 175–201). Amsterdam: Elsevier.

Rasmussen, J., & Jensen, A. (1973). *A study of mental procedures in electronic trouble shooting* (Report No. Risø-M-1582). Roskilde, Denmark: Danish Atomic Energy Commission, Research Establishment Risø.

Rasmussen, J., & Jensen, A. (1974). Mental procedures in real-life tasks: A case study of electronic trouble shooting. *Ergonomics, 17,* 293–307.

Rasmussen, J., & Pejtersen, A. M. (1995). Virtual ecology of work. In J. Flach, P. Hancock, J. Caird, & K. J. Vicente (Eds.), *Global perspectives on the ecology of human-machine systems* (pp. 121–156). Hillsdale, NJ: Lawrence Erlbaum Associates.

Rasmussen, J., Pejtersen, A. M., & Goodstein, L. P. (1994). *Cognitive systems engineering.* New York: Wiley.

Rasmussen, J., & Timmerman, P. (1962). *Safety and reliability of reactor instrumentation with redundant instrument channels* (Risø Report No. 34). Roskilde, Denmark: Danish Atomic Energy Commission, Research Establishment Risø.

Rasmussen, J., & Vicente, K. J. (1989). Coping with human errors through system design: Implications for ecological interface design. *International Journal of Man-Machine Studies, 31,* 517–534.

Reason, J. (1990). *Human error.* Cambridge, England: Cambridge University Press.

Reed, E. S. (1996). *The necessity of experience.* New Haven, CT: Yale University Press.

Reising, D. V., & Sanderson, P. M. (1996). Work domain analysis of a pasteurization plant: Building an abstraction hierarchy representation. In *Proceedings of the Human Factors and Ergonomics Society 40th Annual Meeting* (pp. 293–297). Santa Monica, CA: Human Factors and Ergonomics Society.

Resnick, M. (1994). *Turtles, termites, and traffic jams: Explorations in massively parallel microworlds.* Cambridge, MA: MIT Press.

Rochlin, G. I., La Porte, T. R., & Roberts, K. H. (1987). The self-designing high-reliability organization: Aircraft carrier flight operations at sea. *Naval War College Review, 40*(4), 76–90.

Rosen, R. (1978). *Fundamentals of measurement and representation of natural systems.* New York: North-Holland.

Rosen, R. (1985a). *Anticipatory systems: Philosophical, mathematical, and methodological foundations.* Oxford, England: Pergamon.

Rosen, R. (1985b). Organisms as causal systems which are not mechanisms: An essay into the nature of complexity. In R. Rosen (Ed.), *Theoretical biology and complexity: Three essays on the natural philosophy of complex systems* (pp. 165–203). Orlando, FL: Academic Press.

Roth, E. M., Bennett, K. B., & Woods, D. D. (1987). Human interaction with an "intelligent" machine. *International Journal of Man-Machine Studies, 27,* 479–525.

Roth, E. M., & Mumaw, R. J. (1995). Using cognitive task analysis to define human interface requirements for first-of-a-kind systems. In *Proceedings of the Human Factors and Ergonomics Society 39th Annual Meeting* (pp. 520–524). Santa Monica, CA: HFES.

Roth, E. M., & Woods, D. D. (1988). Aiding human performance: I. Cognitive analysis. *Le Travail Humain, 51,* 39–64.

Rouse, W. B. (1980). *Systems engineering models of human-machine interaction.* New York: Elsevier.

Rouse, W. B. (1991). *Design for success: A human-centered approach to designing successful products and systems.* New York: Wiley.

Rouse, W. B., & Cody, W. J. (1988). On the design of man-machine systems: Principles, practices, and prospects. *Automatica, 24,* 227–238.

Rouse, W. B., Cody, W. J., Boff, K. R., & Frey, P. R. (1990). Information systems for supporting design of complex human-machine systems. In C. T. Leondes (Eds.), *Advances in control and dynamic systems* (pp. 1–60). Orlando, FL: Academic Press.

Rubinstein, E. (1979). The accident that shouldn't have happened. *IEEE Spectrum, 16,* 33–42.

Sachs, P. (1995). Tranforming work: Collaboration, learning, and design. *Communications of the ACM, 38*(9), 36–44.

Sanders, M. S., & McCormick, E. J. (1993). *Human factors in engineering and design* (7th ed.). New York: McGraw-Hill.

Sanderson, P. M., & Harwood, K. (1988). The skills, rules, and knowledge classification: A discussion of its emergence and nature. In L. P. Goodstein, H. B. Andersen, & S. E. Olsen (Eds.), *Tasks, errors, and mental models: A festschrift to celebrate the 60th birthday of Professor Jens Rasmussen* (pp. 21–34). London: Taylor & Francis.

Sanderson, P. M., Verhage, A. G., & Fuld, R. B. (1989). State-space and verbal protocol methods for studying the human operator in process control. *Ergonomics, 32,* 1343–1372.

Sarter, N. B., & Woods, D. D. (1994). Pilot interaction with cockpit automation: II. An experimental study of pilots' model and awareness of the Flight Management System. *International Journal of Aviation Psychology, 4,* 1–28.

Schmidt, K. (1991a). Computer support for cooperative work in advanced manufacturing. *International Journal of Human Factors in Manufacturing, 1,* 303–320.

Schmidt, K. (1991b). Cooperative work: A conceptual framework. In J. Rasmussen, B. Brehmer, & J. Leplat (Eds.), *Distributed decision making: Cognitive models for cooperative work* (pp. 75–110). Chichester, England: Wiley.

Schneider, W., & Shiffrin, R. M. (1977). Controlled and automatic human information processing: I. Detection, search, and attention. *Psychological Review, 84,* 1–66.

Schön, D. A. (1983). *The reflective practitioner: How professionals think in action.* New York: Basic Books.

Schraagen, J. M. C. (1998). *Report on the NATO-ONR workshop on cognitive task analysis* (TNO Tech. Rep. No. TM-98-B003). Soesterberg, Netherlands: TNO Human Factors Research Institute.

Schraagen, J. M. C., Chipman, S. E., Shute, V., Annett, J., Strub, M., Sheppard, C., Ruisseau, J.-Y., & Graff, N. (1997). *State-of-the-art review of cognitive task analysis techniques* (TNO Tech. Rep. TM-97-B1012). Soesterberg, Netherlands: TNO Human Factors Research Institute.

Selz, O. (1922). *Zur psychologie des produktiven denkens und des irrtums* [On the psychology of productive thinking and of error]. Bonn, Germany: Freidrich Cohen.

Seminara, J. L., Gonzalez, W. R., & Parsons, S. O. (1977). *Human factors review of nuclear power plant control room design* (Report No. EPRI NP-309). Palo Alto, CA: Electric Power Research Institute.

Senge, P. M. (1990). *The fifth discipline: The art and practice of the learning organization.* New York: Doubleday.

Sharp, T. D. (1996). *Progress towards a development methodology for decision support system for use in time-critical, highly uncertain, and complex environments.* Unpublished doctoral dissertation, University of Cincinnati, Cincinnati, OH.

Sharp, T. D., & Helmicki, A. J. (1996). Applying the abstraction hierarchy to intensive care medicine. In *Proceedings of the 3rd Annual Symposium on Human Interaction with Complex Systems* (p. 143). Los Alamitos, CA: IEEE Computer Society Press.

Shepherd, A. (1989). Analysis and training in information technology tasks. In D. Diaper (Ed.), *Task analysis for human-computer interaction* (pp. 15–55). Chichester, England: Ellis Horwood.

Shepherd, A. (1992). Maintenance training. In B. Kirwan & L. K. Ainsworth (Eds.), *A guide to task analysis* (pp. 327–339). London: Taylor & Francis.

Shepherd, A. (1993). An approach to information requirements specification for process control tasks. *Ergonomics, 36,* 1425–1437.

Sheridan, T. B. (1987). Supervisory control. In G. Salvendy (Ed.), *Handbook of human factors* (pp. 1243–1268). New York: Wiley.

Sheridan, T. B. (in press). Rumination on automation, 1998. In *Proceedings of the 7th IFAC Symposium on the Analysis, Design, and Evaluation of Man-Machine Systems.* Kyoto, Japan: IFAC.

Shiffrin, R. M., & Schneider, W. (1977). Controlled and automatic human information processing: II. Perceptual learning, automatic attending and a general theory. *Psychological Review, 84,* 127–190.

Shneiderman, B. (1983). Direct manipulation: A step beyond programming languages. *IEEE Computer, 16*(8), 57–69.

Siegel, D. (1997). The impact of computers on manufacturing productitivity growth: A multiple-indicators multiple-causes approach. *Review of Economics and Statistics, 79,* 68–78.

Simon, H. A. (1956). Rational choice and the structure of the environment. *Psychological Review, 63,* 129–138.

Simon, H. A. (1981). *The sciences of the artificial* (2nd ed.). Cambridge, MA: MIT Press.

Simonsen, J., & Kensing, F. (1997). Using ethnography in contextual design. *Communications of the ACM, 40*(7), 82–88.

Singleton, W. T. (1974). *Man-machine systems.* Middlesex, England: Penguin.

Sloan, R. P., Gruman, J. C., & Allegrante, J. P. (1987). *Investing in employee health.* San Francisco: Jossey-Bass.

Smith, C. L. (1992). A theory of situation assessment for decision support (ecological interface). *Dissertation Abstracts International, 53*(3-B).

Smith, C. L., & Sage, A. P. (1991). A theory of situation assessment for decision support. *Information and Decision Technologies, 17,* 91–124.

Smith, T. J., Henning, R. A., & Smith, K. U. (1994). Sources of performance variability. In G. Salvendy & W. Karwowski (Eds.), *Design of work and development of personnel in advanced manufacturing* (pp. 273–330). New York: Wiley.

Sperandio, J.-C. (1978). The regulation of working methods as a function of work-load among air traffic controllers. *Ergonomics, 21,* 195–202.

Suchman, L. A. (1987). *Plans and situated actions: The problem of human-machine communication.* Cambridge, England: Cambridge University Press.

Suchman, L. (1995). Representations of work (Introduction to Special Section). *Communications of the ACM, 38*(9), 33–35.

Taylor, F. W. (1911). *The principles of scientific management.* New York: Harper & Row.

Terkel, S. (1972). *Working: People talk about what they do all day and how they feel about what they do.* New York: Ballantine.

Thelen, E., & Smith, L. B. (1994). *A dynamic systems approach to the development of cognition and action.* Cambridge, MA: MIT Press.

Thompson, J. D. (1967). *Organizations in action: Social science bases of administrative theory.* New York: McGraw-Hill.

Thorndyke, P. W., & Goldin, S. E. (1983). Spatial learning and reasoning skill. In H. L. Pick, Jr. & L. P. Acredolo (Eds.), *Spatial orientation: Theory, research, and application* (pp. 195–217). New York: Plenum.

Torenvliet, G. L., Jamieson, G. A., & Vicente, K. J. (1998). Making the most of ecological interface design: The role of cognitive style. In *Proceedings of the Fourth Annual Symposium on Human Interaction with Complex Systems* (pp. 214–225). Los Alamitos, CA: IEEE Computer Society Press.

Turvey, M. T., Shaw, R. E., & Mace, W. (1978). Issues in the theory of action: Degrees of freedom, coordinative structures and coalitions. In J. Requin (Ed.), *Attention and performance VII* (pp. 557–595). Hillsdale, NJ: Lawrence Erlbaum Associates.

Ujita, H., Kawano, R., & Yoshimura, S. (1995). An approach for evaluating expert performance in emergency situations. *Reliability Engineering and System Safety, 47,* 163–173.

U.S. Department of Health and Human Services (1988). *National health spending report.* Washington, DC: U.S. Government Printing Office.

USNRC. (1981). *Guidelines for CR design reviews* (Report No. NUREG-0700). Washington, DC: U.S. Nuclear Regulatory Commission.

USNRC. (1996). *Human-system interface design review guideline* (Report No. NUREG-0700, Rev. 1). Washington, DC: U.S. Nuclear Regulatory Commission.

van Passen, R. (1995). New visualization techniques for industrial process control. In *Proceedings of the 6th IFAC/IFIP/IFORS/IEA Symposium on Analysis, Design, and Evaluation of Man-Machine Systems* (pp. 457–462). Cambridge, MA: MIT Press.

Venda, V. F., & Venda, Y. V. (1995). *Dynamics in ergonomics, psychology, and decisions: Introduction to ergodynamics.* Norwood, NJ: Ablex.

Vicente, K. J. (1988). Adapting the memory recall paradigm to evaluate interfaces. *Acta Psychologica, 69,* 249–278.

Vicente, K. J. (1990a). Coherence- and correspondence-driven work domains: Implications for systems design. *Behaviour and Information Technology, 9,* 493–502.

Vicente, K. J. (1990b). Ecological interface design as an analytical evaluation tool. In *Proceedings of the American Nuclear Society Topical Meeting on Advances in Human Factors Research: Nuclear and Beyond* (pp. 259–265). La Grange Park, IL: American Nuclear Society.

Vicente, K. J. (1991). *Supporting knowledge-based behaviour through ecological interface design.* Unpublished doctoral dissertation, University of Illinois, Urbana.

Vicente, K. J. (1992a). Memory recall in a process control system: A measure of expertise and display effectiveness. *Memory & Cognition, 20,* 356–373.

Vicente, K. J. (1992b). Multilevel interfaces for power plant control rooms: I. An integrative review. *Nuclear Safety, 33,* 381–397.

Vicente, K. J. (1995a). Cognitive work analysis: Implications for microworld research on human-machine interaction. In *Proceedings of the 1995 IEEE International Conference on Systems, Man, and Cybernetics* (pp. 3432–3436). Piscataway, NJ: IEEE.

Vicente, K. J. (1995b). A few implications of an ecological approach to human factors. In J. Flach, P. Hancock, J. Caird, & K. J. Vicente (Eds.), *Global perspectives on the ecology of human-machine systems* (pp. 54–67). Hillsdale, NJ: Lawrence Erlbaum Associates.

Vicente, K. J. (1995c). Task analysis, cognitive task analysis, cognitive work analysis: What's the difference? In *Proceedings of the Human Factors and Ergonomics Society 39th Annual Meeting* (pp. 534–537). Santa Monica, CA: HFES.

Vicente, K. J. (1996). Improving dynamic decision making in complex systems through ecological interface design: A research overview. *System Dynamics Review, 12,* 251–279.

Vicente, K. J. (1997). Heeding the legacy of Meister, Brunswik, and Gibson: Toward a broader view of human factors research. *Human Factors, 39,* 323–328.

Vicente, K. J., Burns, C. M., Mumaw, R. J., & Roth, E. M. (1996). How do operators monitor a nuclear power plant? A field study. In *Proceedings of the 1996 American Nuclear Society International Topical Meeting on Nuclear Plant Instrumentation, Control and Human-Machine Interface Technologies* (pp. 1127–1134). LaGrange Park, IL: ANS.

Vicente, K. J., Christoffersen, K., & Hunter, C. N. (1996). Response to Maddox critique. *Human Factors, 38,* 546–549.

Vicente, K. J., Christoffersen, K., & Pereklita, A. (1995). Supporting operator problem solving through ecological interface design. *IEEE Transactions on Systems, Man, and Cybernetics, SMC-25,* 529–545.

Vicente, K. J., & Rasmussen, J. (1990). The ecology of human-machine systems: II. Mediating "direct perception" in complex work domains. *Ecological Psychology, 2,* 207–250.

Vicente, K. J., & Rasmussen, J. (1992). Ecological interface design: Theoretical foundations. *IEEE Transactions on Systems, Man, and Cybernetics, SMC-22,* 589–606.

Vicente, K. J., & Tanabe, F. (1993). Event-independent assessment of operator information requirements: Providing support for unanticipated events. In *Proceedings of the American Nuclear Society Topical Meeting on Nuclear Plant Instrumentation, Control and Man-Machine Interface Technologies* (pp. 389–393). LaGrange Park, IL: ANS.

Vicente, K. J., & Wang, J. H. (1998). An ecological theory of expertise effects in memory recall. *Psychological Review, 105,* 33–57.

Vicente, K. J., & Williges, R. C. (1988). Accommodating individual differences in searching a hierarchical file system. *International Journal of Man-Machine Studies, 29,* 647–668.

Waldrop, M. M. (1987). Computers amplify Black Monday. *Science, 238,* 602–604.

Wegner, P. (1997). Why interaction is more powerful than algorithms. *Communications of the ACM, 40*(5), 81–91.

Weinberg, G. M. (1982). *Rethinking systems analysis and design.* Boston: Little, Brown.

Weiner, J. (1994). *The beak of the finch: A story of evolution in our time.* New York: Vintage Books.

Wickens, C. D. (1984). *Engineering psychology and human performance.* Columbus, OH: Merrill.

Wickens, C. D. (1992). *Engineering psychology and human performance* (2nd ed.). New York: Harper-Collins.

Wickens, C. D., & Carswell, M. (1995). The proximity compatibility principle: Its psychological foundation and relevance to display design. *Human Factors, 37,* 473–494.

Wiener, E. L. (1988). Cockpit automation. In E. L. Wiener & D. C. Nagel (Eds.), *Human factors in aviation* (pp. 433–461). San Diego, CA: Academic Press.

Willcocks, L., & Lester, S. (1996). Beyond the IT productivity paradox. *European Management Journal, 14,* 279–290.

Woods, D. D. (1988). Coping with complexity: The psychology of human behaviour in complex systems. In L. P. Goodstein, H. B. Andersen, & S. E. Olsen (Eds.), *Tasks, errors, and mental models: A festschrift to celebrate the 60th birthday of Professor Jens Rasmussen* (pp. 128–148). London: Taylor & Francis.

Woods, D. D. (1991). The cognitive engineering of problem representations. In G. S. Weir & J. L. Alty (Eds.), *Human-computer interaction and complex systems* (pp. 169–188). San Diego, CA: Academic Press.

Woods, D. D. (1998). *Visualizing function: The theory and practice of representation design in the computer medium.* Columbus: Ohio State University, Cognitive Systems Engineering Laboratory, Manuscript in preparation.

Woods, D. D., & Hollnagel, E. (1987). Mapping cognitive demands in complex problem-solving worlds. *International Journal of Man-Machine Studies, 26,* 257–275.

Woods, D. D., & Roth, E. M. (1988a). Aiding human performance: II. From cognitive analysis to support systems. *Le Travail Humain, 51,* 139–172.

Woods, D. D., & Roth, E. M. (1988b). Cognitive engineering: Human problem solving with tools. *Human Factors, 30,* 415–430.

Xiao, Y. (1994). *Interacting with complex work environments: A field study and a planning model.* Unpublished doctoral dissertation, University of Toronto, Toronto.

Xiao, Y., Milgram, P., & Doyle, D. J. (1997). Planning behavior and its functional roles in the interaction with complex systems. *IEEE Transactions on Systems, Man, and Cybernetics, Part A: Systems and Humans, 27*, 313–324.

Xu, W. (1996). *Externalizing a work domain structure on a hypertext interface using an abstraction hierarchy: Supporting complex search tasks and problem solving activities.* Unpublished doctoral dissertation, Miami University, Oxford, OH.

Xu, W., Dainoff, M., & Mark, L. (1996). Externalizing a work domain structure using an abstraction hierarchy: An alternative approach for hypertext interface designs. In *Proceedings of the Human Factors and Ergonomics Society 40th Annual Meeting* (p. 1263). Santa Monica, CA: Human Factors and Ergonomics Society.

Yamaguchi, Y., Furukawa, H., & Tanabe, F. (1997). Simulator study on supporting operators' intellectual activities in NPP. In *Proceedings of the International Symposium on Artificial Intelligence, Robotics and Intellectual Human Activity Support for Nuclear Applications* (pp. 67–79). Wako-shi, Japan: Institute of Physical and Chemical Research.

Yu, X., Chow, R., Jamieson, G. A., Khayat, R., Lau, E., Torenvliet, G. L., Vicente, K. J., & Carter, M. W. (1997). *Research on the characteristics of long-term adaptation* (JAERI Final Report No. CEL 97–04). Toronto: University of Toronto, Cognitive Engineering Laboratory.

Zhang, J., & Norman, D. A. (1994). Representations in distributed cognitive tasks. *Cognitive Science, 18,* 87–122.

Zsambok, C. E., & Klein, G. (1997). *Naturalistic decision making.* Mahwah, NJ: Lawrence Erlbaum Associates.

Zuboff, S. (1988). *In the age of the smart machine: The future of work and power.* New York: Basic Books.

AUTHOR INDEX

SUBJECT INDEX

All entries that are in **bold** are defined in the Glossary (pp. 3–10).